ATLAS OF THE BRYOPHYTES
OF BRITAIN AND IRELAND

3. MOSSES
(Diplolepideae)

ATLAS OF THE
BRYOPHYTES
OF
BRITAIN AND IRELAND

VOLUME 3. MOSSES
(DIPLOLEPIDEAE)

EDITED BY

M. O. HILL
Institute of Terrestrial Ecology

C. D. PRESTON
Institute of Terrestrial Ecology

AND

A. J. E. SMITH
University of Wales Bangor

HARLEY
BOOKS

1994

Published by Harley Books
(B. H. & A. Harley Ltd.)
Martins, Great Horkesley,
Colchester, Essex, CO6 4AH, England
for
the British Bryological Society
c/o Department of Botany,
National Museum of Wales,
Cardiff, CF1 3NF

with the support of
The Natural Environment Research Council
and
The Nature Conservancy Council

Designed by Geoff Green
Text set in Ehrhardt by Saxon Graphics Ltd, Derby
Printed in Great Britain by St Edmundsbury Press Ltd,
Bury St Edmunds, Suffolk

British Library Cataloguing-in-Publication Data

I. Hill, M. O. (Mark O.) *1945–*   II. Preston, C. D. (Christopher David) *1955–*
III. Smith, A. J. E. (Anthony John Edwin) *1935–*
Atlas of the Bryophytes of Britain and Ireland
Volume 3. *Mosses* Part 2: *Diplolepideae*
588.0941

ISBN *0 946589 31 3*

CONTENTS

PREFACE

This is the third and last volume of dot-distribution maps showing the occurrence of bryophytes in Britain and Ireland. It completes the British Bryological Society's Mapping Scheme, begun a third of a century ago in 1960. The first two volumes (Hill *et al.*, 1991, 1992) dealt with liverworts and with mosses other than the Diplolepideae. The present volume maps the diplolepidous mosses, which form about half the total number of mosses in northwest Europe. Most Diplolepideae possess a double peristome. Many of them are pleurocarpous, having a prostrate growth-form, with gametangia produced on dwarf lateral branches. The pleurocarps, even more than other bryophytes, tend to have wide world ranges, typically including much of the Northern Hemisphere and often the temperate Southern Hemisphere as well.

The database from which the maps have been plotted is held by the Biological Records Centre at Monks Wood. We would encourage research workers to exploit the opportunities which it offers for analysis. The introductory chapter in this volume has been written by M. O. Hill, one of the editors, and Felipe Domínguez Lozano of Madrid, and illustrates some of the possible uses of the database in analytical work. Collaboration on this chapter was made possible by a British Council fellowship awarded to F. D. L. under the auspices of the Anglo-Spanish Cultural Foundation. In 1991 he spent the autumn at the Institute of Terrestrial Ecology's research station at Monks Wood, returning subsequently to his work in the botany department of the Universidad Autónoma de Madrid. We are very grateful to him for allowing his work to be published here.

Although publication of the *Atlas* has now been completed, recording the distribution of British bryophytes continues. It is clear from the maps that there are several areas in Great Britain that are seriously under-recorded, and that there remains a great deal of scope for fieldwork in Ireland. It is also clear from the species accounts that many bryophytes are mobile, and whereas the range of some species has contracted in recent years others have spread into new areas or have re-occupied areas from which they had previously been eliminated by high levels of air pollution. This not only means that it is worth continuing to record bryophytes in areas that are thought to be well-worked, but emphasizes the need to ensure that new records are precisely dated. The logging of a list of additions to the maps simply dated as 1950 onwards would result in the loss of much valuable information.

Whilst we have pleasure in bringing the BBS Mapping Scheme to a conclusion, yet this pleasure is not unmixed, for we have enjoyed the work. It has been a privilege to serve as coordinator (A. J. E. S.), data manager (C. D. P.) and principal editor (M. O. H.) for this large venture. The commitment of so many bryologists to the project over so many years has been inspiring. As a result, the bryophyte flora of our islands is now clearly on record, before it is inevitably modified by rapid environmental change in the next century.

M. O. HILL

C. D. PRESTON

A. J. E. SMITH

Monks Wood and Bangor

October 1993

ACKNOWLEDGEMENTS

Many people have been involved in the BBS Mapping Scheme from its launch in 1960 to the publication of the final volume of this *Atlas*. Almost all the fieldwork for the project has been carried out by members of the BBS. The names of the main contributors to this remarkable campaign of systematic recording are given in the list of recorders (p. 414). It is, however, appropriate to mention here the major contribution of A. B. G. Averis, H. J. B. Birks, T. L. Blockeel, M. F. V. Corley, U. K. Duncan, J. W. Fitzgerald, R. D. Fitzgerald, J. C. Gardiner, M. O. Hill, E. M. Lobley, D. G. Long, J. A. Paton and A. R. Perry. This small group of bryologists has submitted over 250,000 records, many of them from remote areas where resident bryologists are lacking.

Records have been computerized and mapped at the Biological Records Centre (BRC), data processing costs being met jointly by BRC's parent body, the Natural Environment Research Council (NERC), and by the Nature Conservancy Council (NCC) and its successor body the Joint Nature Conservation Committee (JNCC). J. Hellawell, Mrs M. A. Palmer, Dr D. A. Ratcliffe and M. J. Wigginton of NCC/ JNCC have dealt with the project at various times. Thanks to their commitment, the project has steered serenely through the bureaucratic difficulties resulting from the reorganisation of the nature conservation agencies.

At Monks Wood, Paul Harding, head of BRC, has not only negotiated the contracts with NCC/JNCC but has been a constant source of wise counsel. Dorothy Greene, Claire Appleby and Julian Dring, BRC data managers, have planned the data processing. Mrs Greene's determination to press on with the job and her capacity for hard work provided an essential impetus at BRC. At several stages of the project we were also assisted by Henry Arnold. The computerization of the records, their subsequent validation and the production of distribution maps was carried out by Mrs J. M. Croft, Dr R. A. Finch, Mrs W. Forrest, Miss N. M. Gomes, Dr C. M. Hine, N. G. Hodgetts, Mrs L. Ling, Miss A. E. Newton, Miss S. E. Yates and M. J. M. Yeo. We are greatly indebted to them for carrying out these unglamorous but vital tasks.

After he left BRC, Marcus Yeo undertook to trace records of species from vice-counties from which they were recorded in the *Census Catalogue* (Corley & Hill, 1981) but for which there was no record in the database – a far from easy job, involving much burrowing in half-forgotten literature, which he performed with remarkable diligence and accuracy. Another valuable source of records was the data collected by Nick Stewart during the preparation of a Red Data Book for bryophytes. Many of these records have been incorporated into the database, although the maps had to be finalized before data collection for the Red Data Book had been completed. Altitude data often presented difficulties, as they can rarely be obtained from the record cards. We are grateful to Ben Averis, David Long, Peter Pitkin, Gordon Rothero and, especially, Martin Corley for supplying these.

We are indebted to Dr D. A. Ratcliffe for his foreword to Volume 1, to A. C. Crundwell for the introductory chapter to Volume 2 and to F. Domínguez Lozano (Madrid) for collaborating on the introduction to Volume 3. The British Council supported his visit to Monks Wood. T. L. Sparks and M. G. Le Duc gave valuable advice on statistics and computing. The preparation of the introductory chapter to Volume 1 was greatly facilitated by the fact that we had access to the library of the Department of Plant Sciences of the University of Cambridge.

Twenty-nine authors contributed short articles to accompany species maps. The numbers of the volumes to which they contributed are indicated in brackets: B. Averis (1), J. W. Bates (3), H. J. B. Birks (1,2,3), T. H. Blackstock (1,2), T. L. Blockeel (1,2,3), J. S. Burley (2), D. F. Chamberlain (2), M. F. V. Corley (1,2), A. C. Crundwell (1,2,3), J. G. Duckett (1), R. A. Finch (2,3), M. O. Hill (1,2,3), N. G. Hodgetts (1,3), D. G. Long (1,2,3), R. E. Longton (3), B. M. Murray (2), M. E. Newton (1,2,3), P. H. Pitkin (2), R. D. Porley (2), C. D. Preston (1,2,3), M. C. F. Proctor (2,3), D. A. Ratcliffe (1), G. P. Rothero (3), F. J. Rumsey (2,3), A. J. E. Smith (1,2,3), N. F. Stewart (3), H. L. K. Whitehouse (2,3), M. J. Wigginton (3), M. M. Yeo (1). Almost all these contributors read and commented on some of the draft captions. We wish particularly to thank David Long and Jean Paton for their help with the liverworts, and Tom Blockeel, Martin Corley and Alan Crundwell for helping with the mosses.

The maps of environmental factors included in Volume 1 were plotted by N. J. Brown, Miss N. Shadbolt and Mrs J. Ullyett. The Warren Spring Laboratory kindly supplied the data on atmospheric sulphur dioxide. The cover photographs are two-dimensional prints taken from stereoscopic photographs; we thank Dr H. L. K. Whitehouse for allowing us to select them from his collection.

Publication of the *Atlas* was planned by a small committee chaired by Paul Harding; we are grateful to Dr R. E. Longton and Dr M. E. Newton, successive secretaries of the BBS, for serving on this committee. Grants towards the publication costs were obtained from the Linnean Society of London and the Royal Society.

Finally, it is a pleasure to thank our publishers, Basil and Annette Harley, for the time and trouble they have devoted to the production of these volumes.

A NUMERICAL ANALYSIS OF THE DISTRIBUTION OF LIVERWORTS IN GREAT BRITAIN

M. O. HILL AND F. DOMINGUEZ LOZANO

INTRODUCTION

Numerical analysis of plant distributions has been recognized for some time as a potentially powerful adjunct to traditional methods of biogeography (Birks, 1987). Numerical analysis is most profitable where it allows patterns of species distributions to be related to environment. For this to be possible, data must be available both on the location of species and on the environment of those locations. In the past, detailed environmental data have rarely been available, so that biogeographers have contented themselves with ascribing species distributions to broad categories, which were only subsequently related to environment.

The distribution of Atlantic liverworts in Britain and Ireland has already been studied in some depth by Ratcliffe (1968), and the vice-county distributions of liverworts have been subjected to a numerical analysis by Proctor (1967). However, neither of these two authors had the results of the British Bryological Society's Mapping Scheme available to them; nor did they have access to detailed environmental information at the scale of the 10-km grid square.

The possibility of analysing species and environmental data jointly at the 10-km square scale has already been explored by one of us (Hill, 1991), who made a pilot analysis of the distribution of birds and vascular plants. Results were sufficiently encouraging for us to attempt a larger analysis, this time of liverwort distributions. In addition to analysing patterns of distribution, we wished to predict not only where species were likely to be found but also how many liverworts would be expected in each 10-km square.

DATA AND METHODS

Species data for analysis were the presence and absence of liverwort taxa in 10-km squares, ignoring date classes. Old as well as new records were included, using the data published in the first volume of this *Atlas* (Hill *et al.*, 1991), together with a small number of subsequent additions. In order to restrict the problem to a manageable size, a one-in-eight systematic selection of species was made, with the additional restriction that each selected species should occur in at least 50 of the 10-km squares in Britain. This selection criterion resulted in a sample of 37 species for study (Table 1).

Irish data were omitted, partly because of relatively poor recording in Ireland, but mainly because good environmental data were not available for Ireland.

Environmental data for 10-km squares were taken from a national database compiled by Ball *et al.* (1983). The subset of variables selected included most of those

Table 1. Species of liverwort selected for analysis; these are a one-in-eight sample of those species that occur in at least fifty 10-km squares in Britain.

Adelanthus decipiens	*Fossombronia incurva*	*Mylia anomala*
Anastrophyllum donnianum	*Frullania teneriffae*	*Nardia scalaris*
Anthelia julacea	*Gymnocolea inflata*	*Odontoschisma sphagni*
Apometzgeria pubescens	*Gymnomitrion concinnatum*	*Phaeoceros laevis*
Barbilophozia hatcheri	*Herbertus aduncus*	*Plagiochila asplenioides*
Blasia pusilla	*Jungermannia sphaerocarpa*	*Plagiochila exigua*
Calypogeia neesiana	*Lejeunea cavifolia*	*Radula aquilegia*
Cephalozia connivens	*Lepidozia reptans*	*Riccia cavernosa*
Cephalozia divaricata	*Leptoscyphus cuneifolius*	*Riccia glauca*
Chiloscyphus polyanthus	*Lophozia bicrenata*	*Scapania scandica*
Colura calyptrifolia	*Lophozia sudetica*	*Scapania uliginosa*
Diplophyllum obtusifolium	*Marsupella sprucei*	
Fossombronia foveolata	*Metzgeria furcata*	

Table 2. Environmental variables used in canonical correspondence analysis of liverwort distributions, together with canonical coefficients used to derive ordination axes from standardized variables. Maximum and minimum mean temperatures in squares do not refer to daily maxima and minima but to estimated spatial maximum and minimum temperatures in the square. For many squares these values were equal, because values were estimated to the nearest 1°C.

Environmental variable	Canonical coefficients	
	Axis 1	Axis 2
Climate		
Annual mean temperature (T)	−34	−32
Minimum mean T in square, July–Sept.	233	186
Maximum mean T in square, July–Sept.	150	−72
Minimum mean T in square, Jan.–Mar.	−34	−75
Maximum mean T in square, Jan.–Mar.	−57	−87
Annual mean daily sunshine hours	55	−16
Annual mean precipitation	−77	−169
Physiography		
Occurrence of coast in square	−34	−45
Minimum altitude in square	31	25
Maximum altitude in square	−102	204
Presence of steep (>11°) slope	−5	−35
Geology		
Presence of chalk bedrock	37	−10
Presence of soft clayey bedrock	−37	−11
Presence of basic igneous rock	−31	−37
Quartzose metamorphic rock	−15	24
Precambrian to Devonian age rock	35	32
Carboniferous to Jurassic rock	32	28

thought to be relevant to liverwort distributions (Table 2). Climate data were derived from monthly means for the three-year period from April 1978 to March 1981, estimated on a 40-km grid. The annual number of 'wet days' (i.e. days with at least 1 mm rain), deemed by Ratcliffe (1968) to be an especially strong determinant of liverwort distributions, was not available in the database and was not used.

The first stage of analysis was to calculate 'canonical axes' of variation, using the technique of Canonical Correspondence Analysis (Ter Braak, 1988; the abbreviation CCA is used below). The essence of CCA is to construct a new variable, which is a linear combination of the given environmental variables, and which on average predicts the species distributions as well as possible. The method closely resembles multiple regression analysis, but, in multiple regression a particular dependent variable is predicted, whereas in CCA the aim is to derive a predictor variable with good average predictive power for the whole set of species under study (Hill, 1991).

The 'detrending' option of CCA was used with detrending by segments. The purpose here was to ensure that the canonical axes were as nearly independent of one another as possible. A small technical difficulty was that some 10-km squares had not been visited by bryologists, so that liverwort data were lacking. These squares were omitted from the analysis, but the canonical axes, being linear combinations of environmental variables, were calculated for all squares in Britain.

Using the top two derived canonical axes, two further analyses were made. First, the probability of occurrence of an individual species, *Barbilophozia hatcheri*, was estimated using the GENSTAT computer package to fit quadratic logistic regression (Jongman et al., 1987). The fitted model was

$$\log\left(\frac{p}{1-p}\right) = a + a_1 x_1 + a_2 x_2 + a_{11} x_1^2 + a_{22} x_2^2$$

where p is the probability of the species being found in a 10-km square, x_1 and x_2 are the values of the grid square on canonical axes 1 and 2, and a, a_1, a_2, a_{11} and a_{22} are coefficients to be estimated by generalized regression.

Finally, the number of liverwort species expected in a 10-km square was estimated by regression, also using the GENSTAT computer package to fit the model

$$y = b + b_1 x_1 + b_2 x_2 + b_{11} x_1^2 + b_{22} x_2^2$$

where y is the number of species, x_1 and x_2 are defined as above, and b_1 etc. are a different set of coefficients to be estimated by regression.

CANONICAL CORRESPONDENCE ANALYSIS

Only the first two canonical axes of CCA proved to be interpretable. The others were very weak and did not correspond to a general pattern of variation; their eigenvalues, which measure their strength, were 0.025 and 0.020, compared with 0.261 and 0.108 for axes 1 and 2. Axis values have been plotted for each 10-km square on a grey scale such that low values appear black and high values appear white (Fig. 1).

The first axis distinguishes squares that are warm, sunny and dry from those that are cool, cloudy and wet (which also tend to have steep land and old, crystalline rocks). The second axis distinguishes coastal areas with high winter temperatures from land with high minimum altitude and low winter temperatures (Table 2).

The most striking thing about the first axis is how closely the distribution of axis values (Fig. 1a) resembles that of wet days (Volume 1, p. 333). Thus, the omission of the annual number of wet days as an explanatory variable has scarcely affected the outcome of the analysis, because a similar pattern could be obtained from a linear combination of the variables supplied. This apparent paradox reinforces the point that a correlation is not an explanation. Liverwort distributions are strongly correlated with wet days, but other explanations of their pattern of occurrence are possible.

The second axis (Fig. 1b) lacks such a simple climatic explation. Winter warmth obviously plays a part, but comparison with January mean temperature (Volume 1, p. 330) shows that the large black area in western Scotland is comparable in winter temperature to much of Wales, which is not picked out in black. The extreme west of Scotland, however, does have a distinctive Atlantic liverwort flora, corresponding to the dark area in Fig. 1b. Thus, the second axis has succeeded in picking out an environmental pattern that is strongly related to liverwort distributions.

The environmental tendencies of species can be inferred from the species ordination (Fig. 2). In this diagram, the plotted position of each species is determined by the species' mean position on the canonical axes. For example, *Riccia cavernosa* has a high value on axis 1, indicating that it occurs mainly in parts of the country that themselves have high values on axis 1, i.e. the lowlands (Fig. 1a). Its value on axis 2 is near the average, because it has no marked tendency to occur near the coast or where winters are especially warm or cold.

Barbilophozia hatcheri, which occurs mainly in upland areas, has a moderately low

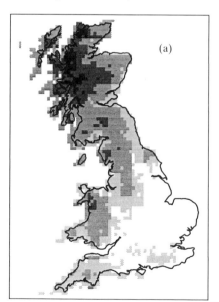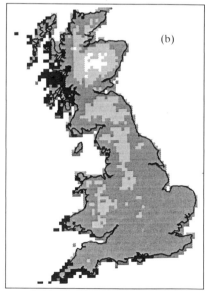

Fig. 1. Distribution of canonical axis values resulting from canonical correspondence analysis, showing (a) axis 1 and (b) axis 2. Values are plotted as differing grey scales ranging from black to white, with bands as follows. Axis 1: <150 (black), 150–200, 250–280, >280 (white). Axis 2: <200 (black), 200–250, 250–300, >300 (white).

value on axis 1 and a high value on axis 2, signifying that it grows mainly in the winter-cold regions of eastern Scotland. In this respect it resembles several other boreal plants such as *Trientalis europaea*, *Lophozia longidens* and *Dicranum spurium*.

At the other extreme, with the lowest values on the second axis, are Atlantic species such as *Frullania teneriffae* and *Leptoscyphus cuneifolius*. These have distributions concentrated on the dark area in Fig. 1b. *F. teneriffae*, with a wider distribution in south-west Britain, has a higher score on axis 1 than *L. cuneifolius*.

Species that are not extreme on either axis appear near the centre of Fig. 2. They include *Cephalozia connivens*, *Diplophyllum obtusifolium*, *Lejeunea cavifolia*, *Lepidozia reptans* and *Nardia scalaris*. These liverworts occur widely in Britain, with no marked climatic limitation, although all except *L. cavifolia* are restricted to acid substrata.

Predicting species occurrence and species number

It is instructive to compare the actual and predicted distributions of *B. hatcheri* (Fig. 3). Here, the predicted distribution has been inferred solely from the first two canonical axes using the logistic equation set out above. Because the predicted distribution depends only on these two axes, it is of necessity generalized and there are discrepancies of detail. The predicted distribution suggests that *B. hatcheri* should be more frequent in the Southern Uplands of Scotland than it actually is, and conversely the prediction under-estimates the frequency of this species in the eastern lowlands.

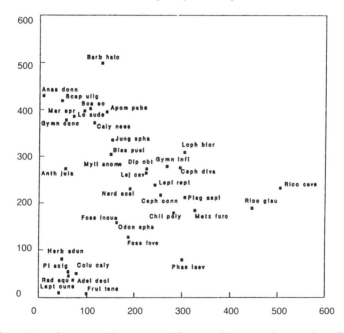

Fig. 2. Disposition of species in relation to axes of canonical correspondence analysis. Canonical axis 1 appears as the *x*-axis and canonical axis 2 as the *y*-axis. The species are those listed in Table 1. The interpretation of species' positions in this diagram is given in the text.

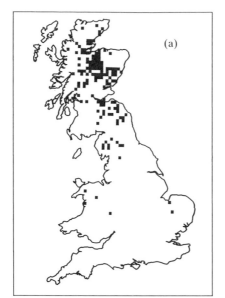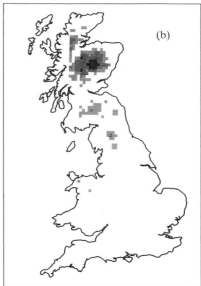

Fig. 3. Actual (a) and predicted (b) occurrence of *Barbilophozia hatcheri* in 10-km squares of Britain. Predicted values give the probability of finding the species in a 10-km square. This is indicated on a grey scale as follows: <0.2 white, 0.2–0.4 pale grey, 0.4–0.6 mid-grey, 0.6–0.8 dark grey

When the number of species predicted for a square is compared with the actual number found (Fig. 4), the predicted number is seen to vary more smoothly and not to reach such high values.

The map of the ratio between observed and predicted numbers (Fig. 5) can be used to pick out under-recorded areas, taking account of local environment. This map gives a better impression of recording intensity than the map of raw species numbers (Fig. 4a). The improvement is greatest in the lowlands, where liverwort numbers are expected to be low. Many of the under-recorded areas mentioned in Volume 1 (pp. 30, 31) are displayed clearly, namely the English counties of Nottinghamshire, Shropshire, Staffordshire and Suffolk, and the former Scottish counties of Aberdeenshire, Ayrshire and Fife. The Flow Country of northern Scotland, whose bogs are one of the great wildlife habitats of Britain (Ratcliffe & Oswald, 1988), was never recorded systematically during the Mapping Scheme and also appears as under-recorded.

Norfolk and Galloway, on the other hand, which in Volume 1 were said to be poorly recorded, do not appear to be so in Fig. 5. Norfolk, indeed, appears unexpectedly rich in liverworts. This is an error due to the technique of linear regression, which permitted the estimated number of species for some Norfolk 10-km squares to fall to near zero, and even allowed negative values to be estimated for two squares. Galloway, on the other hand, is in reality only moderately under-recorded. Indeed, it is much better recorded than the tract of land to the north of it.

Fig. 4. (Facing Page) Actual (a) and predicted (b) numbers of liverwort species in 10-km squares of Britain. The prediction is based solely on the values of the canonical axes. Values are indicated as follows: <25 white, 25–49 open circles, 50–99 half open circles, ≥100 solid circles.

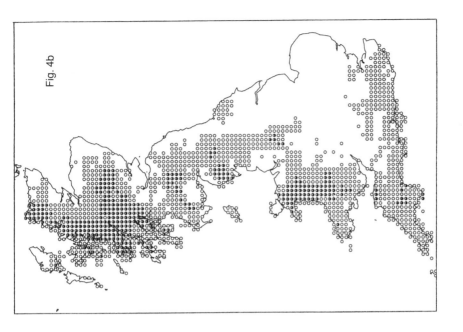

Fig. 4b

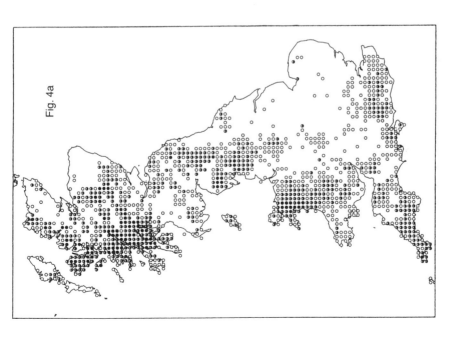

Fig. 4a

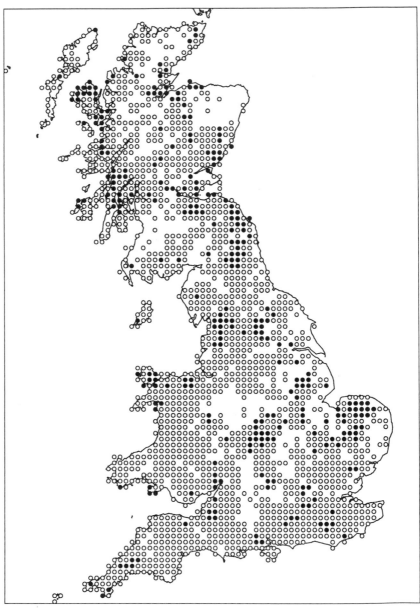

Fig. 5. Ratio of actual to predicted numbers of liverwort species in 10-km squares. Blank squares, with less than half the predicted number, are likely to be under-recorded. Squares with open circles have an actual value within a factor of 2 of the predicted value. Squares with closed circles have more than twice as many species in them as predicted.

DISCUSSION AND CONCLUSIONS

This analysis has confirmed a phenomenon that has long been known to phytogeo-graphers, namely that large-scale pattern is determined mainly by climate; soil is a major determinant only at smaller scales (Cain, 1944). In future analyses at the scale of Britain, it would probably be better simply to ignore geological and edaphic variables. They contributed little to the results of the liverwort CCA, and, likewise they contributed little to an earlier analysis of birds and vascular plants (Hill, 1991).

The physiographic variables of altitude and coast contributed substantially to both analyses. In the liverwort analysis, altitude played only a small part in determining the first axis, but was strongly influential on the second. The fact that coast was also a significant variable for this axis suggests either that many liverworts have a requirement for coastal habitats or that the coastal climate, with warmer winters and cooler summers, plays a part in determining liverwort occurrence at the local scale.

No British liverwort is halophytic; the few coastal species such as *Fossombronia angulosa* all occur inland in other parts of their range. Thus local climate is almost certainly the main coastal influence on liverworts. This influence could not be discerned in the rather crude climate data at our disposal. The short period, April 1978 to March 1981, to which the means refer was probably not important, but the coarse spatial scale of the 40-km grid is clearly visible in Fig. 4.

Another result, which also emerged from the analysis of birds and vascular plants (Hill, 1991), is the finding that well-known major climatic factors may be recon-structed, even if they were not entered as data. In the liverwort analysis, the wet-day variable was reconstituted; in the bird-and-plant analysis, winter temperature, not supplied as an input variable, was reconstituted. This can happen because climate variables are strongly correlated with one another, so that approximate values of a more appropriate variable can be derived as a linear combination of others that are supplied.

Because of this redundancy, we infer that although numerical analyses can provide useful predictions of where species are likely to be found, there is room for further simplification. For example, the climatic requirements of species may be better represented by the three main bioclimatic variables – winter cold, summer warmth and annual moisture availability – recommended by Prentice *et al.* (1992). Simplicity is highly desirable; the more readily interpretable an analysis, the more likely it is to allow predictions of where species will occur when the climate changes in future.

At present, such extreme simplification is not possible, because each 10-km square contains a diversity of possible climates, corresponding to the range of altitudes and aspects in the square. Unless this range can be specified explicitly, CCA, with altitude as an input variable, remains a good approach to interpretation.

One of the main virtues of an analysis such as that given above is that it allows the separation of the special features of species distributions from standard general patterns. In the case of *Barbilophozia hatcheri*, the analysis drew attention to its apparently anomalous occurrence in the eastern lowlands. In fact, this pattern is shown by other boreal bryophytes, most notably by the moss *Brachythecium reflexum*, which is a strictly montane species in Britain but occurs also in the Netherlands. There must be a reason why some – but by no means all – boreal bryophytes become montane towards the western edge of their range.

The prediction of expected species richness has already served to highlight areas of under-recording. However, the analysis could profitably be taken further. In the first place, the regression could be redone using a statistical model that did not permit negative values. A suitable model would be

$$\log y = b + b_1 x_1 + b_2 x_2 + b_{11} x_{11}^2 + b_{22} x_{22}^2$$

which could be fitted by generalized linear regression.

Another improvement would be for the expected number of species to be re-estimated, after the more extremely under-recorded squares had been removed. Such an analysis would not be difficult and could be a useful guide for recorders in future mapping schemes; but it is beyond the scope of the present study.

In conclusion, there are many possibilities for using numerical methods to interpret bryophyte distributions. The foregoing analysis presents some results and suggests further options. These will no doubt be explored further as improved computing facilities and better environmental data become available. Better bryophyte data cannot, however, be expected; after nearly 35 years of the BBS Mapping Scheme, the liverwort distribution data are as good as can be achieved in our generation.

MAPS OF MOSS DISTRIBUTIONS

Explanation of maps and accompanying notes

Records are mapped in the 10 × 10-km squares of the Ordnance Survey National Grid in Great Britain and in the Ordnance Survey/Suirbheireacht Ordonais National Grid in Ireland. Records from the Channel Islands are mapped in the 10 × 10-km squares of the Universal Transverse Mercator Grid. The symbols used are

○   Record made before 1950, or undated
●   Record made in or after 1950.

Where very few symbols appear on a map, the symbols have been encircled to make them more conspicuous. A few outlying symbols that might otherwise be overlooked are marked by arrows.

The numbering and nomenclature of species is basically that of *The Moss Flora of Britain and Ireland* (Smith, 1978). We have retained Smith's names as far as possible, even if later research has suggested that they are incorrect. We have, however, cited in synonymy all the names used in recent checklists for North America (Anderson *et al.*, 1990) and Europe (Corley *et al.*, 1981, Corley & Crundwell, 1991). In the large pleurocarpous families Amblystegiaceae, Brachytheciaceae and Hypnaceae many of the traditional genera are rather ill defined. Recently, some of these genera have been revised, and it has been shown that several species should not only be removed from their traditional genus but even assigned to a different family. We have cited most proposed new names in the synonymy, and have often referred in the text to a recent revision. Species that have been discovered in Britain and Ireland since the publication of Smith's *Moss Flora* are interpolated at an appropriate place.

Numerous mosses mapped in these volumes were not known to British bryologists in 1950. Many of these are small acrocarpous species of *Bryum* or *Pohlia* which reproduce asexually by bulbils in the leaf-axils or by rhizoidal tubers. Although we have normally mapped undated records as open circles, we have departed from this rule for twelve mosses which were discovered since 1950 and have since proved to be widespread (*Bryum bornholmense, B. gemmiferum, B. klinggraeffii, B. microerythrocarpum, B. rubens, B. ruderale, B. sauteri, B. subelegans, B. tenuisetum, B. violaceum, Dicranella staphylina* and *Fissidens celticus*). For these species, and for two introduced mosses which have spread rapidly since 1950 (*Campylopus introflexus* and *Orthodontium lineare*) we have assumed that undated records were made after 1950 and mapped them with solid symbols.

Each map is accompanied by notes on the taxon mapped. The first paragraph describes the habitat. The altitudinal range is given (in metres). 'Lowland' denotes altitudes below 300 m. The paragraph ends with a formula indicating the number of 10-km grid squares in which the taxon is mapped. GB 18+11*, IR 6+8* indicates that the plant has been recorded in or after 1950 in 18 grid squares in Great Britain and 6 in Ireland. There are pre-1950 records from an additional 11 squares in Great Britain and 8 in Ireland. Channel Island records are not included in these totals; squares in the Isle of Man are counted as British.

The second paragraph describes the sexuality of the taxon, the frequency of sporophytes and indicates whether it has specialized means of vegetative spread, e.g. by gemmae (usually few-celled, more or less undifferentiated structures), bulbils, branchlets, tubers or fragile leaves. The following terms may be used to describe sexuality:

Sterile: antheridia (male sex organs) and archegonia (female sex organs) not produced;

Dioecious: antheridia and archegonia borne on separate plants which are genetically distinct and arise from separate spores;

Pseudodioecious: antheridia and archegonia borne on separate plants which are genetically identical and arise from the same protonema; and

Monoecious: antheridia and archegonia borne on the same plant. Some authors have distinguished separate categories of monoeciousness:

Autoecious: antheridia and archegonia borne in separate inflorescences;

Paroecious: antheridia naked in the axils of the leaves immediately below the female inflorescence; and

Synoecious: antheridia and archegonia borne in the same inflorescence.

The third paragraph describes the distribution of the taxon outside the British Isles. For many species, information on European distribution is derived from the valuable compilations of Duell (1985, 1992). Information on the extra-European distribution of British and Irish species is readily available for North America, but for other continents is sometimes of doubtful accuracy. Species reported from other continents may not be the same as the European plant given that name, or the European plant may be known by another name elsewhere. Readers should bear in mind this element of uncertainty.

Additional comments may be given in a fourth paragraph.

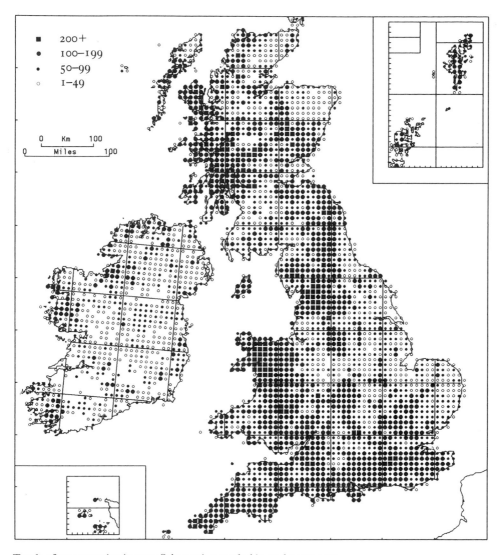

Totals of moss species (except *Sphagnum*) recorded in 10-km squares

The map shows the total number of moss species, other than *Sphagnum*, recorded in each 10-km square. Records have been counted without regard to date, including old records as well as new ones. Infraspecific taxa have not been counted separately. A map showing the total number of *Sphagnum* species in each 10-km square was published in Volume 2, p. 19.

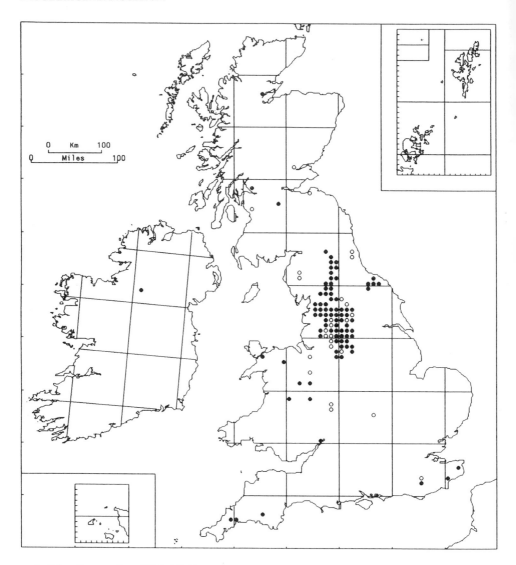

70/1. **Discelium nudum** (Dicks.) Brid.

A colonist of newly bared reservoir mud and acid clayey banks by streams and rivers, sometimes forming extensive mats. It has also been found on clay banks by roads, on china-clay waste, on fine-textured mud in a slate quarry, on clay in brickworks, on bare earth in a woodland ride, and on a muddy farm track. The colonies, with scattered bud-like gametophores emerging from a persistent protonema, are often pure, but can be associated with a wide range of other bryophytes, including some that regenerate from spores and others regenerating from tubers. A particularly frequent associate is *Dicranella rufescens*. 0–520 m (Harthope Moor). GB 85 + 25*, IR 1.

Pseudodioecious; capsules often abundant, ripe in spring. Tubers have been found on rhizoid-like filaments at the base of the persistent protonema (Side & Whitehouse, 1987), and are probably common.

Circumboreal. Scattered and occasional in N. Europe from the Arctic south to N. France, N. Germany and Bohemia (Czech Republic). Very rare in Asia, known in Siberia from Tobolsk and the Yenisey region, and in the Far East from Sakhalin and Japan; rare and scattered in N. America south to California and Pennsylvania.

For a longer list of associates and a more detailed account of its habitat in N. England, refer to Duckett (1973).

M. O. HILL

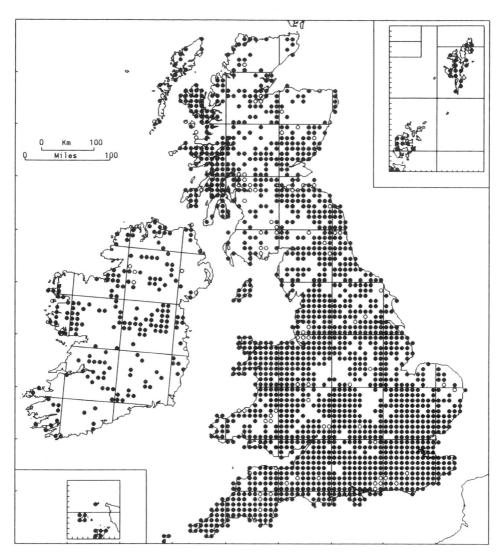

71/1. Funaria hygrometrica Hedw.

In yellowish green patches or as scattered plants on nutrient-rich soil, especially characteristic of the sites of fires, where it may be abundant and the only species present. It is also common on waste ground, on old walls and buildings, in quarries, on disturbed roadsides, in flowerpots and in cultivated fields. It is commoner on basic than on acid soils. Mainly lowland, ascending to 715 m (Ben Lawers). GB 1698+85*, IR 195+8*.

Autoecious; fruit common, ripening all the year round. Gemmae are produced on the protonema in culture (Bopp *et al.*, 1991).

Cosmopolitan.

The spores can survive for up to two years in the soil under field conditions (During, 1986)

A. C. CRUNDWELL

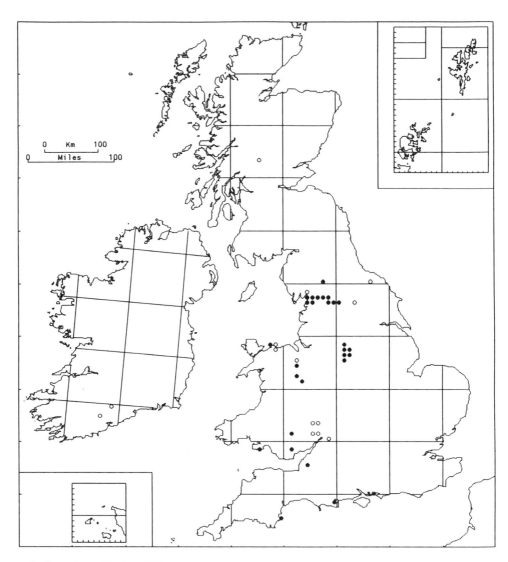

71/2. Funaria muhlenbergii Turn.

In yellowish green patches or as scattered shoots; calcicole. It occurs on bare soil among rocks, in turf, on ant-hills, or on thinly earth-covered rocks. 0–380 m (Llangattock). GB 27+13*, IR 2*.

Autoecious; capsules common, ripe in May.

Europe east to Czechoslovakia and Hungary, north to southern Scandinavia. Macaronesia, N. Africa, S.W. Asia, Caucasus, western N. America.

In the past this species has been confused with others, especially *F. pulchella*. The distribution shown in the map is likely to be substantially correct; the overseas distribution is taken from Nyholm (1989).

A. C. CRUNDWELL

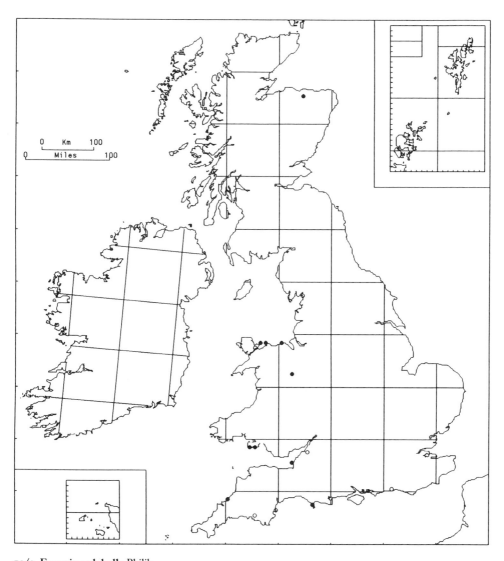

71/3. **Funaria pulchella** Philib.

A calcicole, similar to *F. muhlenbergii* both in appearance in the field and in ecology. Lowland. GB 10+6*.
　Autoecious; capsules common, ripe in May.
　All round the Mediterranean and north to the Rhineland, Switzerland, Austria and Hungary. Also known from Macaronesia, C. Asia (Samarkand) and western N. America (Arizona).
　Not distinguished from *F. muhlenbergii* in Britain until some time after the Mapping Scheme had started (Crundwell & Nyholm, 1974). Its British and foreign distribution are certainly incomplete.

A. C. CRUNDWELL

27

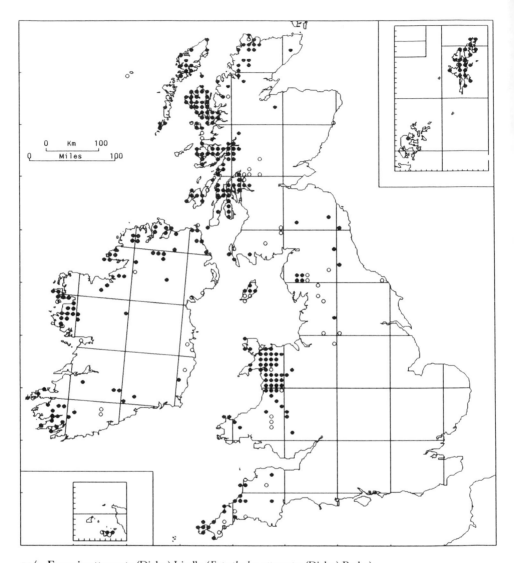

71/4. **Funaria attenuata** (Dicks.) Lindb. (*Entosthodon attenuatus* (Dicks.) Bryhn)

As green patches or scattered stems on damp acid but not waterlogged soils on banks by streams and ditches, in woodland and moorland, and on earth-covered rocks and rock-ledges. 0–700 m (Carneddau). GB 232+46*, IR 70+12*.

Autoecious; capsules common, ripe spring to autumn.

A Mediterranean-Atlantic species extending through western Europe south from Iceland and the Faeroes to both sides of the Mediterranean and east to Israel and Turkey. Also known from Macaronesia and from western N. America (Oregon, California, Arizona).

A. C. CRUNDWELL

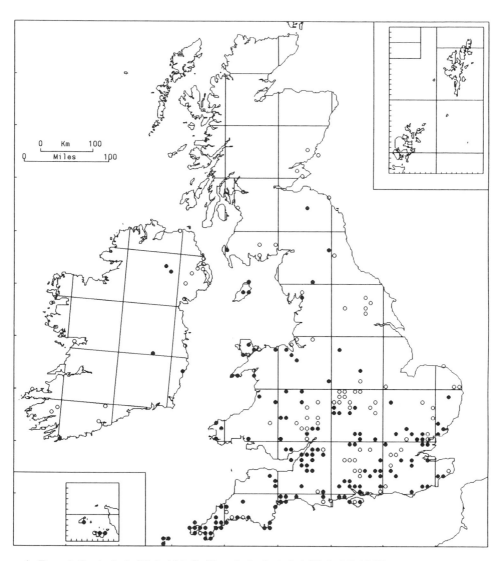

71/5. **Funaria fascicularis** (Hedw.) Lindb. (*Entosthodon fascicularis* (Hedw.) C. Müll.)

As green patches or as scattered plants on non-calcareous soil in stubble-fields and grass leys, on disturbed ground in young forestry plantations, on banks and waste ground, on tracks and paths and in meadows. Lowland. GB 144+80*, IR 5+16*.

Autoecious; capsules common, ripe spring.

Europe north to southern Scandinavia. Morocco, Algeria, Turkey, Caucasus, western N. America.

The resemblance of this species to *Physcomitrium pyriforme* is now too well known to cause many errors.

A. C. CRUNDWELL

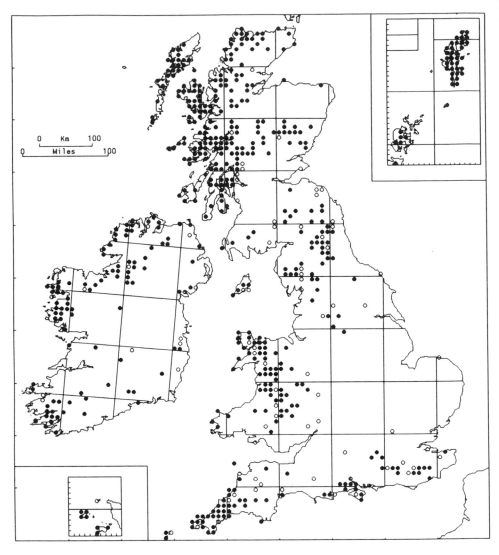

71/6. Funaria obtusa (Hedw.) Lindb. (*Entosthodon obtusus* (Hedw.) Lindb.)

As green patches or scattered stems on damp acid or neutral peaty or gravelly soils, on heaths and moorlands, stream- and ditch-banks, occasionally on rock-ledges. Ascends to 780 m near Dalwhinnie. GB 419+67*, IR 94+13*.

Autoecious; capsules ripening from winter to summer.

A suboceanic-submediterranean species. From Iceland, the Faeroes and southern Scandinavia south to Macaronesia, Algeria, Tunisia and Turkey.

A. C. CRUNDWELL

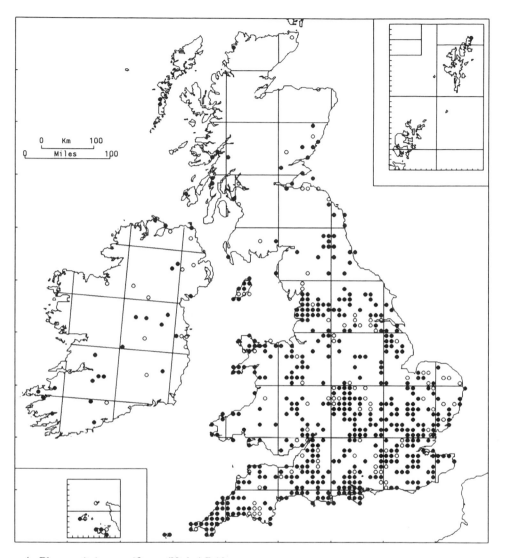

72/1. **Physcomitrium pyriforme** (Hedw.) Brid.

As green patches or scattered shoots on damp soil or mud in marshy meadows, in moist cultivated fields, on the banks of ditches, streams and rivers and frequently also on mud dredged up from them. Lowland. GB 499+100*, IR 23+13*.

Autoecious; capsules abundant, ripe in spring and summer.

From southern Scandinavia south to Macaronesia, N. Africa and Turkey. Caucasus, east Siberia, temperate N. America south to Mexico, Australia (where probably introduced).

A. C. CRUNDWELL

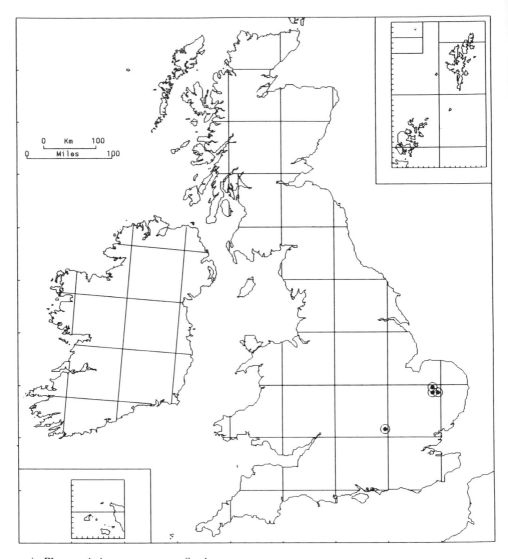

72/2. **Physcomitrium eurystomum** Sendtn.

As green patches or scattered shoots on mud exposed on the margins of Breckland meres with a fluctuating water-table and beside a drying-out reservoir near Tring. In the Breckland its most frequent associates are *Riccia cavernosa*, *Leptobryum pyriforme* and *Physcomitrium pyriforme*. Lowland. GB 4.

Autoecious; capsules abundant, ripe autumn and winter.

From Denmark south to France and C. Europe. Also known from tropical Africa, western Turkey, C. and E. Asia, India and New Guinea.

This species was first seen in Britain in 1961 (Ducker & Warburg, 1961). It is of sporadic occurrence and appears only after a period of prolonged drought.

A. C. CRUNDWELL

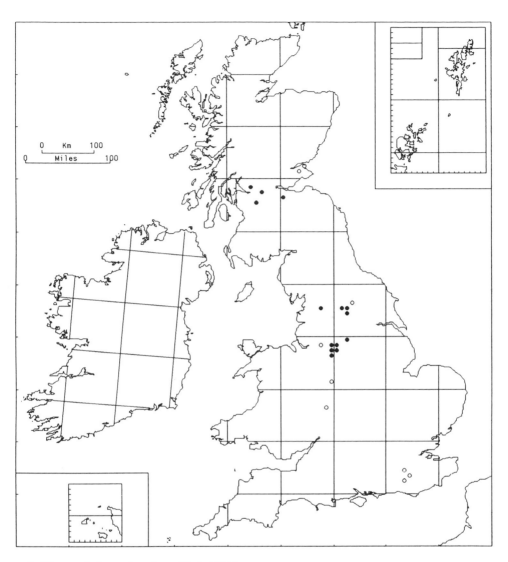

72/3. **Physcomitrium sphaericum** (Hedw.) Brid.

As green patches or scattered shoots on mud exposed on the margins of lakes and reservoirs and on the beds of dried-up ponds. Lowland. GB 14+9*.

Autoecious; capsules abundant, ripe autumn and winter.

In Europe from southern Finland and Denmark south to Spain, France and C. Europe. India, E. Asia, Japan.

Like *P. eurystomum*, this can develop and mature only after a long period of drought or when the water has been artificially drained, the plants surviving meanwhile in the form of spores in the submerged mud (Furness & Hall, 1981). Many of the records are old, but this does not necessarily imply extinction, for in most years the plants do not develop. The continued existence of the species is, however, threatened in a number of its sites by pollution of the water and by artificial maintenance of a high water-level for yachting and other purposes.

A. C. CRUNDWELL

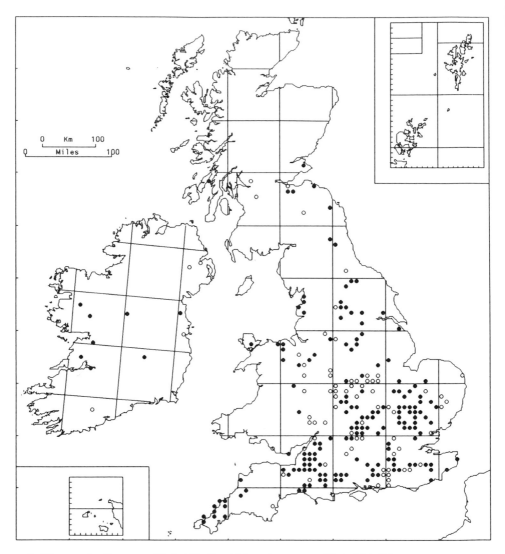

73/1. **Physcomitrella patens** (Hedw.) Br. Eur. (*Aphanorhegma patens* (Hedw.) Lindb.)

As green patches or scattered shoots on mud exposed at the margins of lakes and reservoirs or on the bottoms of dried-up ponds; also on the sides of ditches and streams, on wet farm tracks and woodland rides, and in wet meadows. Lowland. GB 168+71*, IR 8+4*.

Paroecious or synoecious; capsules abundant, ripe summer and autumn.

In Europe north of the Mediterranean region to southern Scandinavia. W. Siberia, western N. America (British Columbia), eastern N. America.

A. C. CRUNDWELL

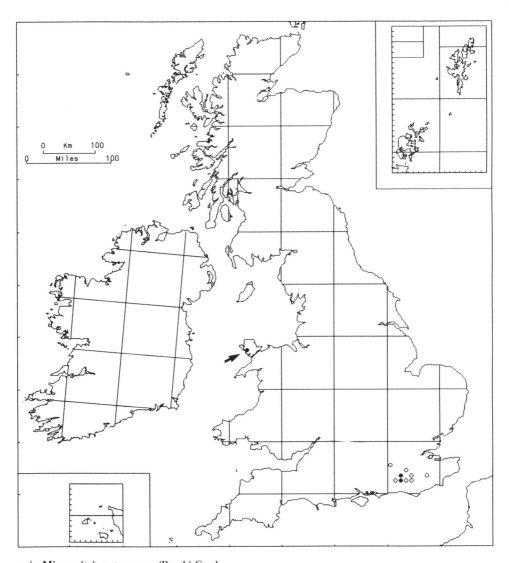

74/1. Micromitrium tenerum (Bruch) Crosby

A sporadic colonist, usually plentiful when present, of drying non-calcareous mud around ponds and reservoirs. Lowland. GB 3+7*.

Autoecious; capsules abundant, ripe in autumn.

Known in Europe from S. Sweden, Belgium, France, Germany and Czechoslovakia, nowhere common. Assam, China, Korea, Japan, western N. America (British Columbia).

An endangered species, declining because of destruction of its habitats.

A. C. CRUNDWELL

35

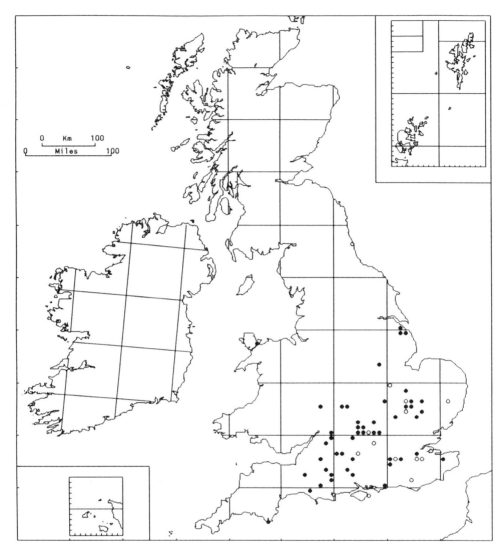

75/1. **Ephemerum recurvifolium** (Dicks.) Boul.

An ephemeral species of calcareous clay and, less often, bare, moist soils on chalk and limestone. It occurs on woodland rides, on soil in beech-woods, in arable fields, on ant-hills and in bare places in grassland. Lowland. GB 48+13*.

   Pseudodioecious; occasionally sterile, but capsules usually abundant, ripening in autumn and winter.

   In Europe from southern Scandinavia southward. N. Africa, Israel, Turkey.

   Like all species of *Ephemerum*, under-recorded both because of its small size and because it is not normally visible during the summer months.

A. C. CRUNDWELL

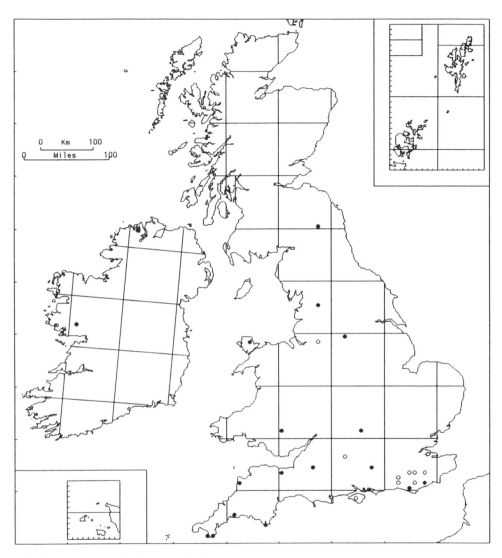

75/2. Ephemerum sessile (Br. Eur.) C. Müll.

An ephemeral species of neutral and acid soils, often associated with *E. serratum*. Found mainly on woodland rides and on mud at the edges of reservoirs, but also recorded from damp heathland, a sea-cliff and a rut in a field. Rarely if ever in arable fields. Lowland. GB 16+9*, IR 2.

Pseudodioecious; capsules abundant, ripe in autumn and winter.

In Europe from southern Scandinavia southward. Morocco, Israel, Turkey.

A. C. Crundwell

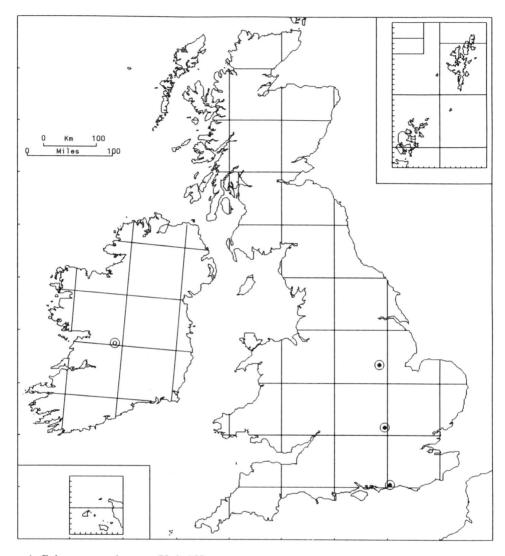

75/3. **Ephemerum cohaerens** (Hedw.) Hampe

A pioneer species of non-calcareous soils on moist banks and on mud at the edges of reservoirs. Lowland. GB 3, IR 1*.

Pseudodioecious; capsules abundant, mature in autumn.

In Europe from Germany and Poland south to Sardinia and Yugoslavia. Turkey, eastern N. America from Ontario to Florida.

A. C. CRUNDWELL

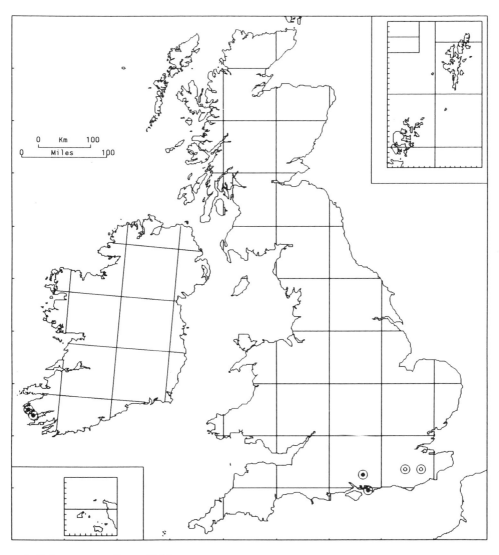

75/4. **Ephemerum stellatum** Philib.

An ephemeral of acid and neutral soils. Habitats include woodland rides, a cliff path and the earth-covered parapet of a bridge. Lowland. GB 2+2*, IR 2.

Pseudodioecious; capsules abundant, ripe in autumn.

In Europe known only from Portugal, France and Germany (very rare). Turkey. Reported from N. Africa. A very rare moss, known perhaps from less than ten localities in the world.

A. C. CRUNDWELL

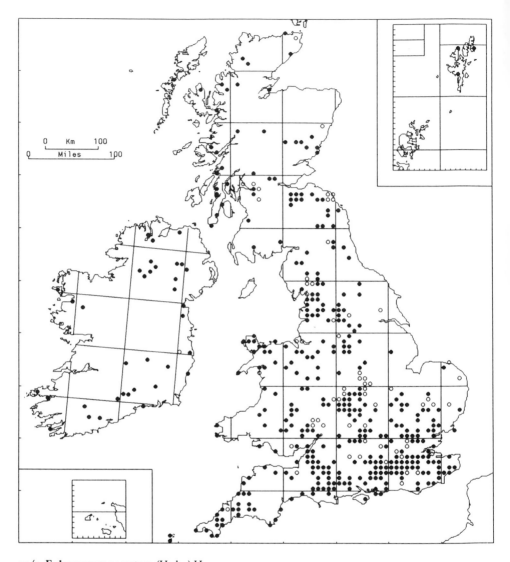

75/5. **Ephemerum serratum** (Hedw.) Hampe

An ephemeral of damp non-calcareous soils, usually of clay or fine sand. Most records are from arable fields, woodland rides and the drying margins of lakes and reservoirs, where it may sometimes form extensive pure sheets. It also occurs on molehills and in bare patches in grassland, in churchyards, on stream-banks and damp lane-banks, and on cliffs and other places where the soil is occasionally disturbed. 0–350 m (Loch Ericht). GB 372+62*, IR 35+2*.

Pseudodioecious; capsules abundant, usually ripe in autumn, but occasionally at other times of year.

In Europe from southern Scandinavia southward. N. Africa, Israel, Turkey; widespread in N. America.

Probably just over half the records are based on var. *minutissimum* (Lindb.) Grout, but determinations of this are rash without ripe spores and not always certain with them. Var. *minutissimum* is commoner in the south of the country than var. *serratum*, is much the more frequent variety in arable fields, and apparently never occurs around lakes and reservoirs. No mixed gatherings of the two varieties have been reported. Var. *praecox* Walth. & Mol. has been found in several localities in Sussex but not seen recently.

A. C. CRUNDWELL

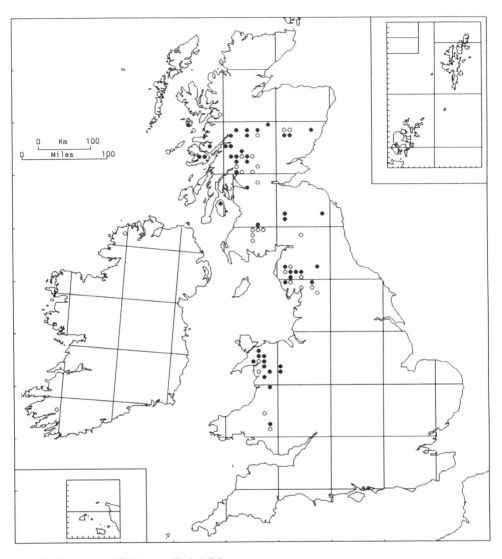

76/1. Oedipodium griffithianum (Dicks.) Schwaegr.

In loose, pale green tufts or as scattered plants on moist humus-rich soil in shady rock crevices and in block-screes in mountain regions. Mainly at moderate and high altitudes, to 1200 m on Ben Nevis. GB 48+30*, IR 2*.

Autoecious or synoecious. Capsules frequent, ripening in summer. There is vegetative reproduction by multicellular discoid gemmae.

In Scandinavia, where mainly in the mountains, the Russian Far East, Japan, Canada (British Columbia, Yukon, Newfoundland), Greenland, Falkland Islands.

A. C. Crundwell

41

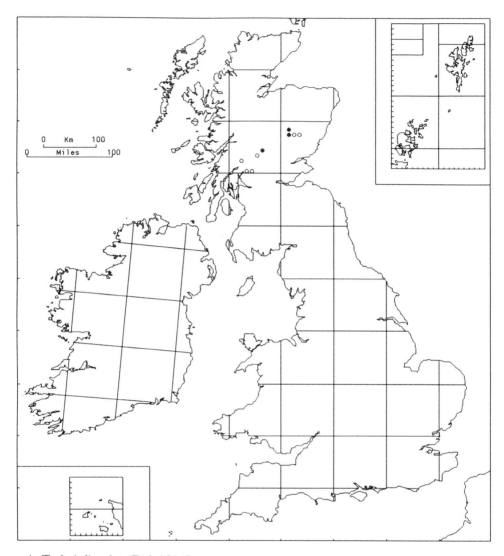

77/1. **Tayloria lingulata** (Dicks.) Lindb.

As dense, pale green tufts in basic montane flushes, usually with dominant or abundant *Saxifraga aizoides*. It reaches 850 m in Clova. GB 3+6*.

Autoecious or synoecious; capsules frequent, ripe in summer.

From Svalbard, Iceland and Scandinavia south to C. Europe. Turkey, Siberia, China (Yunnan), Korea, Japan, N. America, Greenland.

A. C. CRUNDWELL

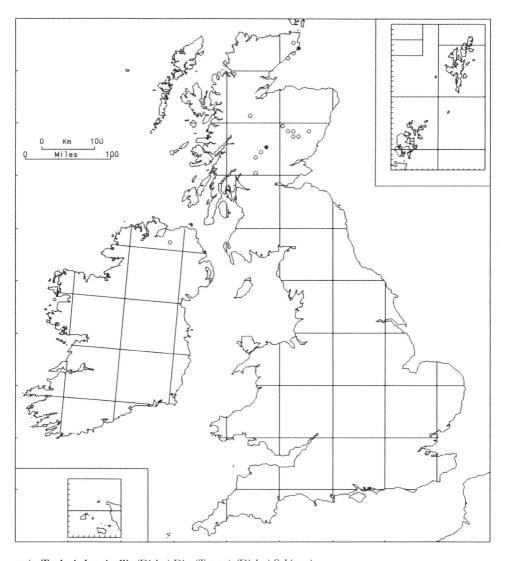

77/2. **Tayloria longicollis** (Dicks.) Dix. (*T. tenuis* (Dicks.) Schimp.)

As loose, green tufts on damp decaying vegetable matter in montane habitats. GB 2+14*, IR 1*.

Autoecious or synoecious; capsules frequent, ripe in summer.

From Scandinavia south to France and northern Italy. Siberia, northern N. America, Greenland.

This species is often confused with the non-British *T. serrata* (Hedw.) B. & S. so the details of its world distribution remain uncertain. It seems to have disappeared from almost all its British localities. The reason for this is unknown.

A. C. CRUNDWELL

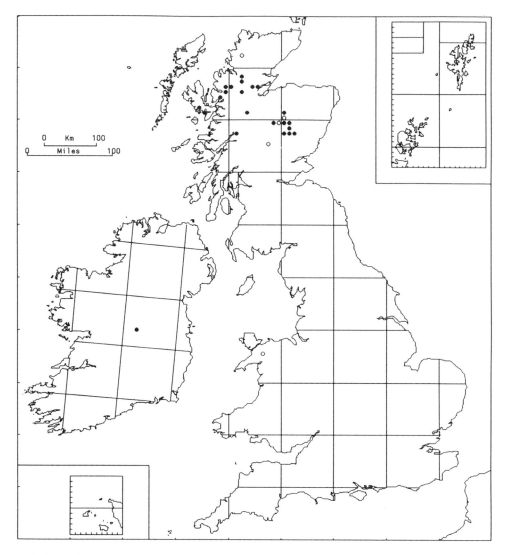

78/1. **Tetraplodon angustatus** (Hedw.) Br. Eur.

As compact, pale green tufts on decaying bones of deer and sheep, less frequently on dung, in wet places, especially along tracks, in the mountains, reaching 910 m in Glen Feshie. GB 21+5*, IR 1.

Autoecious; capsules common, ripe in summer.

From Scandinavia south to France and northern Italy. Siberia, China, Japan, northern N. America, Greenland.

Marino (1991b), working in Alberta, Canada, where *T. angustatus* often colonizes dung, could find no difference between the habitat requirements of *T. angustatus* and *T. mnioides*. There was, however, a seasonal difference; the capsules of *T. angustatus* matured in May, a month earlier than those of *T. mnioides*.

A. C. CRUNDWELL

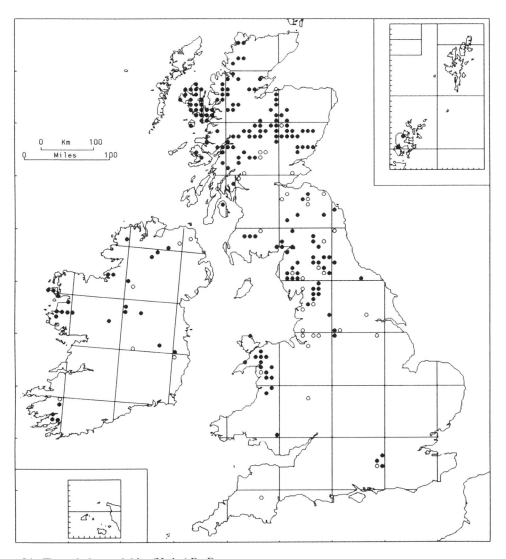

78/2. **Tetraplodon mnioides** (Hedw.) Br. Eur.

In pale green tufts on the decaying bones of sheep and deer, less often on dung, on moorland, bogs and mountains, in wet places especially by paths and tracks. Ascends to 950 m (Cairngorm Mountains). GB 171+37*, IR 26+7*.

Autoecious; capsules common, ripe in spring and summer.

From Svalbard south to Spain, Italy and C. Europe. Through Asia to Russian Far East (Sakhalin), Japan, Borneo and New Guinea; N. America, Greenland, Colombia, Ecuador, southern S. America, Tanzania.

Of the three widely distributed British mosses characteristic of animal remains, *Tetraplodon mnioides* is most often on bones but is occasionally on dung. *Splachnum sphaericum*, which probably has a greater moisture requirement, is most often on dung but occasionally on bones. *S. ampullaceum*, which appears to need most moisture of the three, is always on dung, never on bones. Marino (1991a), working on *T. angustatus*, *T. mnioides*, *S. ampullaceum* and *S. luteum* Hedw. colonizing dung in Alberta, found that *Tetraplodon* spp. had no preference for carnivore (wolf) dung over herbivore (moose) dung, but that in dry habitats in the field *Tetraplodon* spp. produced more gametophytes than *Splachnum* spp. and often eliminated them; the reverse was true in wet habitats.

A. C. CRUNDWELL

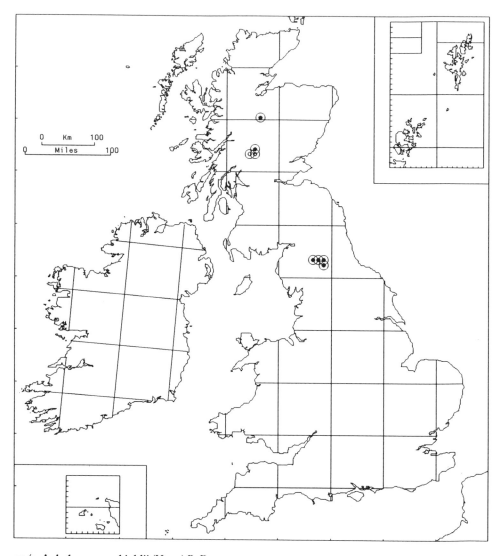

79/1. **Aplodon wormskjoldii** (Horn.) R. Br.

As pale green tufts or patches on dung in wet peaty places at high altitudes. At 580 m on Knock Fell. GB 6+2*.
Autoecious; capsules frequent, ripe in summer.
Svalbard and northern Scandinavia. Arctic Siberia, northern N. America, Greenland.
The occurrence of this Arctic species in Britain is rather surprising.

A. C. CRUNDWELL

46

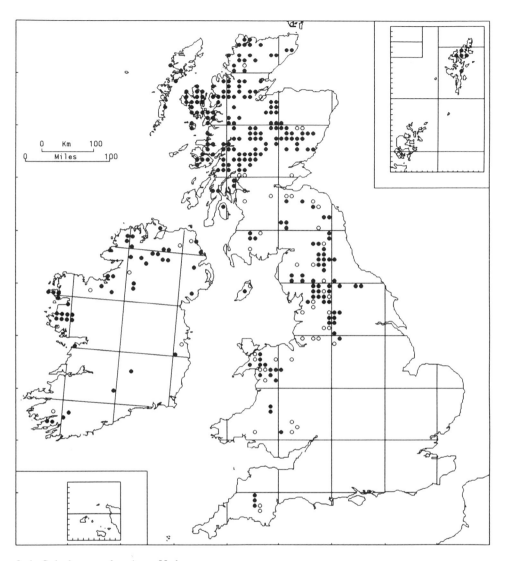

80/1. **Splachnum sphaericum** Hedw.

As pale green tufts on the dung of cattle, sheep and deer on wet heaths, moors and bogs, mainly in the mountains. Sea-level (Skye, Shetland) to 920 m (Ben Lawers). GB 233 + 48*, IR 46 + 6*.

Dioecious; fruit common, ripe from spring to autumn. Gemmae are produced on the protonema in culture (Whitehouse, 1987).

From Iceland, the Faeroes and Scandinavia south to northern Italy. Siberia, N. America, Greenland.

The spores of *Splachnum* species are held in clumps by a sticky material and dispersed by insects. This helps to ensure that both sexes of the dioecious species are dispersed simultaneously (Koponen, 1990).

A. C. CRUNDWELL

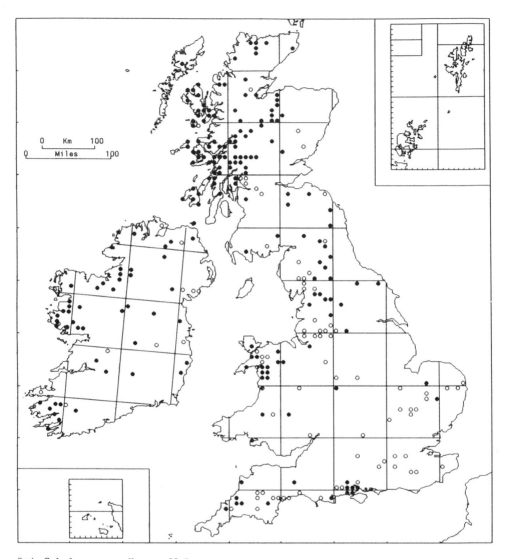

80/2. Splachnum ampullaceum Hedw.

It forms pale green tufts on the dung of cattle, sheep and deer on wet heaths, moorlands and bogs. A predominantly lowland species, rarely found at the higher elevations favoured by *S. sphaericum*, but ascending to 740 m (Burnhope Head). GB 151+76*, IR 48+11*.

Autoecious; capsules common, ripe in summer.

From Iceland, the Faeroes and Scandinavia south to Spain and Italy. Caucasus, Siberia, E. Asia, Japan, Celebes, N. America.

The distribution of *Splachnum ampullaceum* illustrates the fate of an ecologically specialized species in a changing environment. Its virtual disappearance from S.E. England can be attributed to the destruction of its habitat by drainage and agricultural improvement and to the drying out of the remaining areas of wet heath because of falling water-tables.

A. C. CRUNDWELL

48

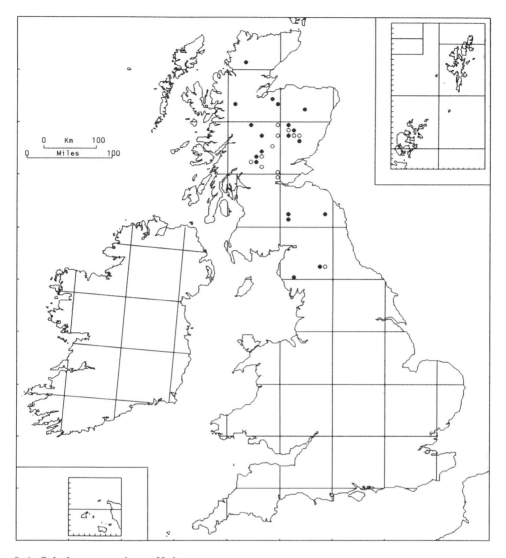

80/3. **Splachnum vasculosum** Hedw.

In pale green loose patches on dung in springs and flushes at high altitudes. According to McVean & Ratcliffe (1962), *Saxifraga stellaris*, *Scapania uliginosa*, *Philonotis fontana* and *Pohlia wahlenbergii* are typical associates. 760 m (Sgurr na Lapaich) to 870 m (A' Bhuidheanach). GB 19+12*.

Dioecious; capsules frequent, ripe in summer.

From Svalbard, Iceland and northern Scandinavia south to Germany and Estonia. Arctic and northern Siberia, northern N. America, Greenland.

A. C. CRUNDWELL

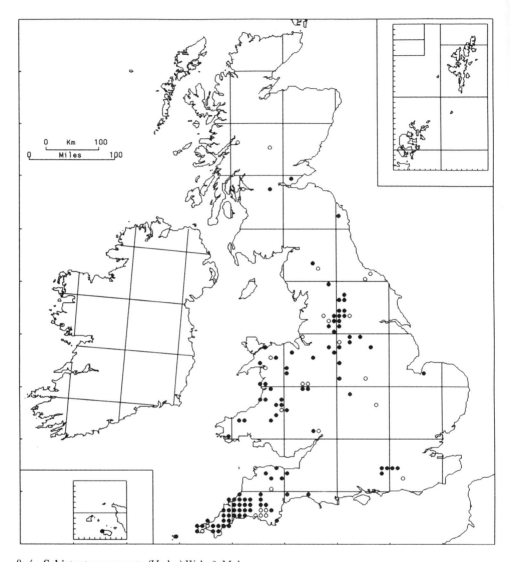

81/1. **Schistostega pennata** (Hedw.) Web. & Mohr

A calcifuge species occurring in deep shade on soft, often crumbling acid soil in dark recesses of shaded lanes and hedgebanks, particularly in S.W. England, under overhanging banks in woodlands and quarries, entrances to caves and mine-shafts, disused rabbit burrows, and deep crevices between granite blocks and in granite tors and sandstone cliffs. Associates are rare, and include *Lepidozia reptans* and *Isopterygium elegans*. Lowland. GB 105+29*.

Pseudodioecious, with male and female plants arising separately from the same protonema; capsules rare or occasional, mature in spring or summer. Gemmae are produced abundantly on the protonema; these are sticky and probably dispersed by mites and flies (Edwards, 1978).

Suboceanic-montane, occuring in W., C. and E. Europe and extending north of the Arctic Circle in Torne Lappmark (Sweden). N.E. Asia, Japan, N. America.

A curiously local species rarely present in any abundance. It appears luminescent when growing in very dark places such as crevices in cliffs, because of the light-reflective properties of the convex cells of its protonema.

H. J. B. Birks

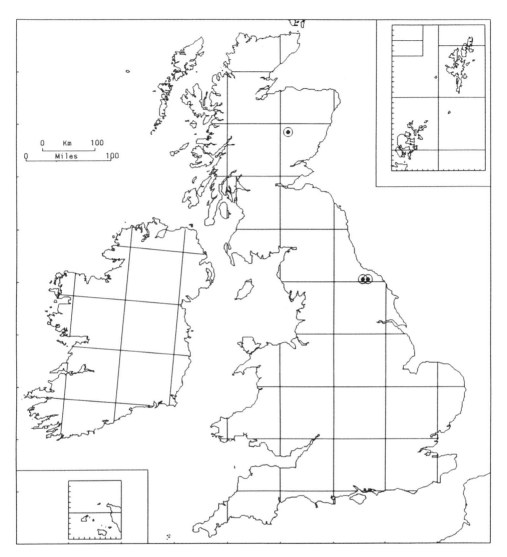

82/1. **Mielichhoferia elongata** (Hoppe & Hornsch. ex Hook.) Hornsch. (*M. mielichhoferiana* (Funck) Loeske var. *elongata* (Hoppe & Hornsch. ex Hook.) Wijk & Marg.)

A plant of deeply shaded, N.- to N.E.-facing, strongly acid (pH 2.7–4.2), permanently moist rocks rich in heavy-metal sulphides. In the Scottish site, it is found on calcopyritic metamorphosed rock, associated with *Gymnocolea inflata* and *Grimmia atrata*. In Yorkshire it occurs on vertical faces of shale in a series of narrow, steep gullies with constantly trickling streams. 230–300 m (Yorkshire), and 690–820 m (Scotland). GB 3.

Dioecious; fruit last found 1930, capsules undehisced in July; gametangia and fruit not found in Yorkshire. Vegetative propagation occurs through detached tufts and fragmentation of the fragile stems.

Spain, France, Italy, Switzerland, Austria, Norway, Sweden, Svalbard; very local and rare throughout its range. Scattered in N. America from the Arctic south to the western U.S.A.

Discovered in Scotland in 1830. Early gatherings may all originate from the Corrie Kander area (Coker, 1971), where it is now very scarce. It may have been introduced to its English locality as a result of medieval alum-shale extraction (Coker, 1968a).

F. J. Rumsey

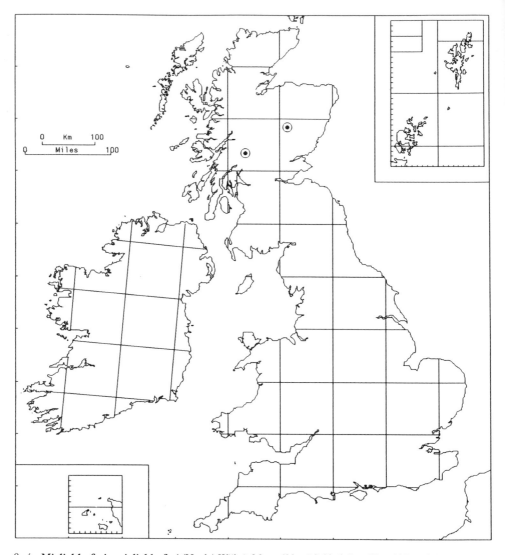

82/2. **Mielichhoferia mielichhoferi** (Hook.) Wijk & Marg. (*M. mielichhoferiana* (Funck) Loeske)

A plant of damp, shaded, metal-rich, acidic lithosols facing north-east, possibly less copper-tolerant than *M. elongata* (Coker, 1968b). At its site in eastern Scotland it is associated with *Gymnocolea inflata*, *Grimmia atrata*, *M. elongata* and several small hepatics. 700 m (Corrie Kander) and 830 m (Beinn Dorain). GB 2.

Dioecious; one shoot from Corrie Kander is male; female plants and capsules not definitely known from Britain. Vegetative propagation can occur by stem fragmentation; axillary gemmae have been reported by Shaw & Crum (1984) from the Yukon (Canada).

Rare over its whole range, but distribution uncertain owing to confusion with *M. elongata*. Pyrenees, Alps, Norway, Sweden, Caucasus, Siberia, N. America. The closely related *M. japonica* Besch., treated by some as a variety of *M. mielichhoferi*, occurs in Japan.

Less distinctive than *M. elongata*, it may sometimes be overlooked as a sterile *Pohlia*. First definitely recorded as British after its discovery by Coker (1968b) in Corrie Kander.

F. J. Rumsey

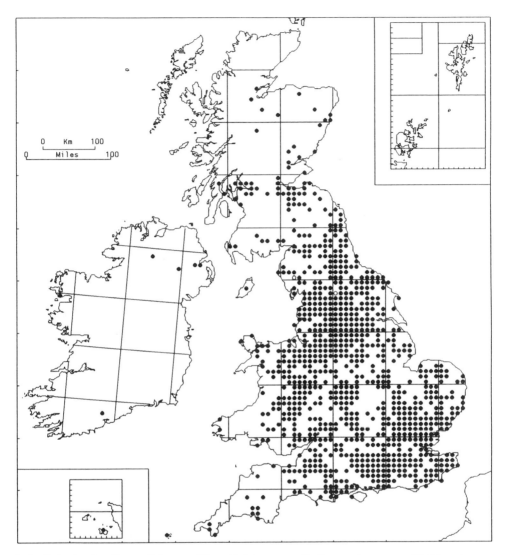

83/1. Orthodontium lineare Schwaegr. (*O. gracile* Schwaegr. ex Br. Eur. var. *heterocarpum* W. Watson)

An invasive species characteristic of shaded base-poor habitats. It occurs on sandy and peaty banks, on tree-boles, old logs and stumps, and on siliceous rocks and crags, particularly on vertical surfaces. It is tolerant of atmospheric pollution, and in some regions, for example on Millstone Grit in the southern Pennines, it occurs prolifically in rocky woods, associated with *Lophozia ventricosa, Isopterygium elegans, Tetraphis pellucida* and many other species. Tree-bases and stumps are the commonest habitat in southern and eastern districts. 0–500 m (Pennines). GB 917+2*, IR 6+1*.

Autoecious; capsules abundant, maturing in spring and early summer. In Cambridgeshire, protonemal gemmae are produced abundantly (Whitehouse, 1964); they are probably abundant elsewhere.

Introduced in Europe, now widespread in W. and C. Europe, north to S. Scandinavia (Hedenäs *et al.*, 1989). Australia, New Zealand, S. Africa.

It was collected in Cheshire in 1910 (Burrell, 1940) but was not recognized as distinct from *O. gracile* till about 1920. Margadant & Meijer (1950) established its identity with *O. lineare* from the Southern Hemisphere, from where it is thought to have been introduced to Britain, perhaps with imported timber.

T. L. BLOCKEEL

53

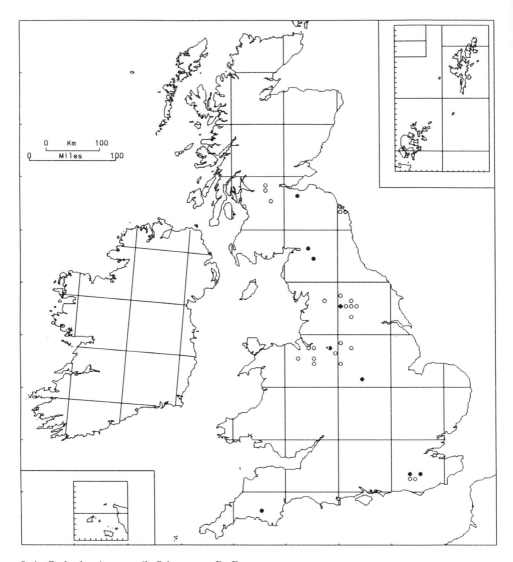

83/2. **Orthodontium gracile** Schwaegr. ex Br. Eur.

This species has been recorded from shaded sandstone and gritstone rocks in lowland woods in a number of scattered localities. It usually occurs in sites where it is subject to little competition, for example on the vertical surfaces of rocks and crags. Lowland. GB 9+25*.

Paroecious; capsules frequent, but less so than in *O. lineare* and less reliably maturing.

A rare species, recorded in continental Europe only from N.W. France. Outside Europe known from western N. America (California) and from many parts of the tropics and sub-tropics.

Apparently always rare in Britain, it has decreased markedly during the present century as its habitat has been invaded by the more aggressive *O. lineare*. This process was observed in the field at a site in Wharfedale (Burrell, 1940), and the survival of the species is now a matter of concern.

T. L. BLOCKEEL

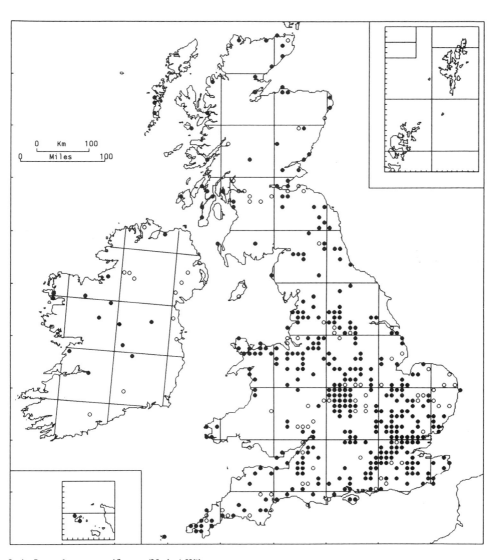

84/1. **Leptobryum pyriforme** (Hedw.) Wils.

Often an abundant weed on soil in flowerpots in glasshouses and gardens, this species is occasionally found in damp natural habitats. It colonizes open to lightly shaded, bare or disturbed ground by streams and ponds and in quarries, arable fields and meadows, and can be abundant on cinders and burnt sites. It grows on banks and ledges of rocks and cliffs, especially sandstone, less often limestone; rarely on tufa, in salt-marsh, and, in humid sites, on wood. It occurs on stone or brick walls; and in inner cities where aerial pollution eliminates most species, can be prominent on old walls. It can tolerate a mean aerial SO_2 content exceeding 85 µg S m^{-3} (Gilbert, 1970). This mainly lowland species ascends to 610 m (Uisgnaval Mor). GB 375+83*, IR 13+12*.

Synoecious; capsules common, ripe spring and summer in the open, all year in glasshouses. Rhizoidal gemmae abundant; detached protonemal cells also act as a means of dispersal.

Cosmopolitan, reaching beyond 75°N and 60°S.

Probably native; a Flandrian subfossil associated with mesolithic flints in the Isle of Wight (Clifford, 1937) suggests it has been in Britain for at least 3500 years.

R. A. Finch

55

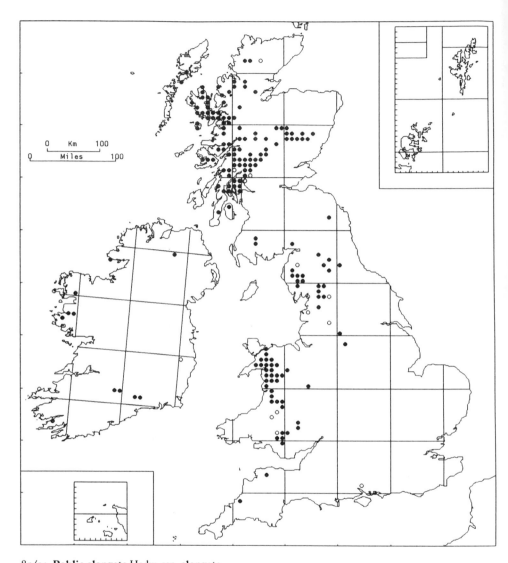

85/1a. Pohlia elongata Hedw. ssp. **elongata**

It occurs in shaded locations on thin soil in crevices of rock-walls, on ledges, and on bare soil on banks. It prefers an acidic substratum, and is normally found on humus-rich, skeletal or peaty soils, often in fairly dry sites. Associated species may include *Diplophyllum albicans*, *Bartramia pomiformis* and *Isopterygium elegans*. It occurs more rarely on mildly basic soils, and associates in such locations include *Saxifraga hypnoides*, *Amphidium mougeotii* and *Anomobryum filiforme*. 0–920 m (Ben Lawers). GB 161+13*, IR 11+4*.

Monoecious; sporophytes frequent, spring to autumn.

In northern and montane areas through much of the world from northern Norway south through Europe and C. Africa to the Cape, from Siberia south through S.E. Asia to New Guinea, from northern N. America and Greenland south to Mexico, and on the Galapagos and Kerguelen Island.

Var. *elongata*, the common form, is paroecious. Var. *acuminata*, which is not mapped separately, differs only in its inflorescence, which is autoecious. It is very similar in its ecological requirements.

M. J. WIGGINTON

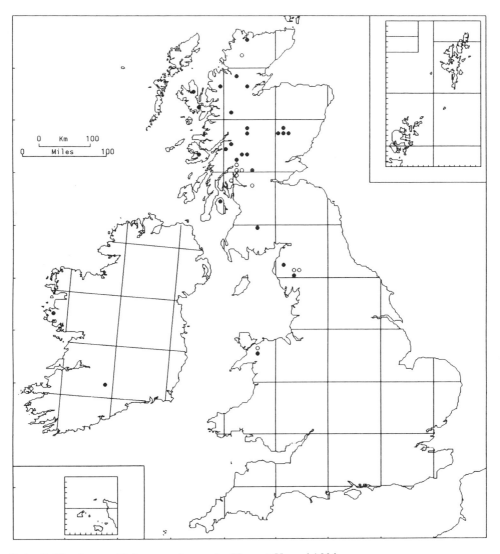

85/1c. **Pohlia elongata** Hedw. ssp. **polymorpha** (Hoppe & Hornsch.) Nyh.

This subspecies occurs in similar habitats to ssp. *elongata*, though it typically occurs at higher altitudes. It grows on skeletal peaty or gritty soils on rock-ledges, in crevices, and occasionally on rocky stream-banks. It also grows on soil amongst boulders on mountain summits, where it may be associated with such species as *Kiaeria falcata*. 400 m (The Storr) to 1050 m (Carnedd Llewelyn). GB 25+11*, IR 2+1*.

Paroecious; sporophytes frequent, late summer, autumn.

It has a subarctic-alpine distribution in Europe, north to Iceland and Svalbard. Turkey, Caucasus, Himalaya, China, N. America, Greenland.

M. J. WIGGINTON

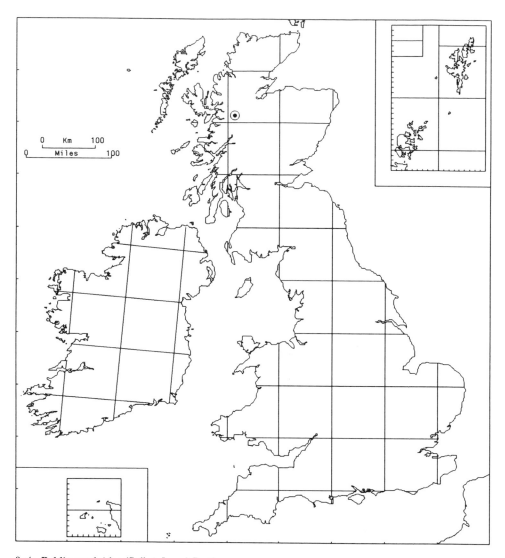

85/2. **Pohlia crudoides** (Sull. & Lesq.) Broth.

This mainly Arctic species is known in Britain in only a single locality, where it was found growing in crevices of montane rocks at about 900 m. GB 1.

Dioecious; only male plants known in Britain.

Scandinavia, N. Russia, Svalbard. N., C. and E. Asia, Arctic N. America, Greenland.

This species was found in Scotland in 1968 (Wallace, 1972) and has apparently not been seen since then. In Scandinavia, it is said to prefer a siliceous soil (Nyholm, 1958).

M. J. WIGGINTON

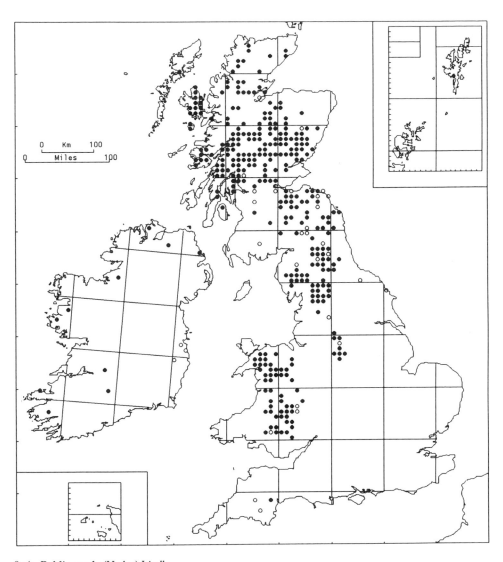

85/3. **Pohlia cruda** (Hedw.) Lindb.

A plant of dry or moist, mildly acidic to basic epilithic habitats, usually in the montane region. It occurs typically in rather dry, shaded crevices and recesses of montane rock cliffs. Where the substratum is basic, associates may include *Tritomaria quinquedentata*, *Anoectangium aestivum* and *Mnium stellare*; where it is more acid, they may be *Bartramia pomiformis*, *Isopterygium elegans* and *Mnium hornum*. It also occurs on damp montane ledges among tall herbs, in association with mosses such as *Dicranum scoparium*, *Hylocomium splendens*, *Racomitrium ericoides* and *Rhytidiadelphus loreus*. Other habitats include basic coastal rocks, walls, stream ravines, and blocky talus slopes. 0–1180 m (Ben Lawers). GB 317+30*, IR 10+4*.

Dioecious; sporophytes occasional, summer.

This species has a worldwide distribution in cool regions and on mountains, and is particularly common in the Arctic. Azores and Europe north to Iceland and Svalbard, N. Africa, southern Africa (Lesotho, Natal), N., C. and E. Asia, Hawaii, N. and S. America, Australasia, Kerguelen Island, Antarctica.

M. J. WIGGINTON

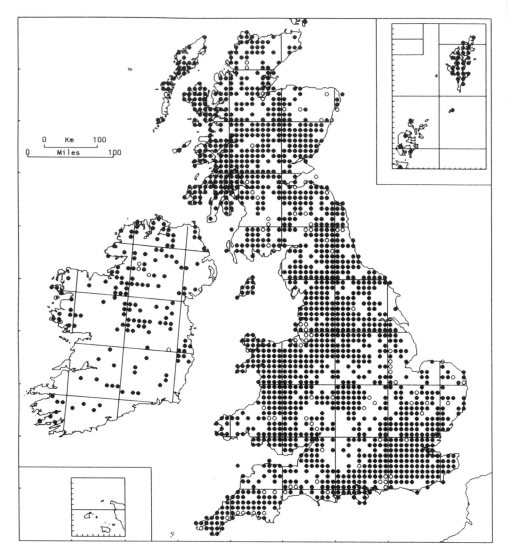

85/4. Pohlia nutans (Hedw.) Lindb.

An exclusively calcifuge species which occurs, often in abundance, in a very wide range of wet and dry habitats. It occurs in lowland and montane heaths, in dunes, in ombrotrophic bogs and flushes, on peat hags, on sandy banks in exposed and shaded sites, on peat and humus on boulders and rock outcrops, in rock crevices, on masonry and wall-tops, on rotting logs and stumps, in acidic grassland, on bark, and in a variety of anthropogenic habitats including sand-, gravel- and coal-pits, ashed ground and derelict industrial sites. On sandy soils on heaths and dunes, common associated species include *Barbilophozia floerkei*, *Cephaloziella* spp., *Campylopus introflexus*, *Ceratodon purpureus*, *Dicranum scoparium*, *Hypnum jutlandicum* and *Cladonia* spp. It is also common in mires, often growing in *Sphagnum* carpets and hummocks with such species as *Calypogeia fissa*, *Lophozia ventricosa*, *Mylia anomala* and *Odontoschisma sphagni*. 0–1340 m (Ben Nevis). GB 1591+91*, IR 151+6*.

Paroecious; sporophytes common, summer.

A species with a worldwide distribution in cool and cold regions. Europe, including Svalbard and Iceland, Asia, N. America, southern S. America, S. Africa, Australasia, Kerguelen Island, Antarctica.

M. J. WIGGINTON

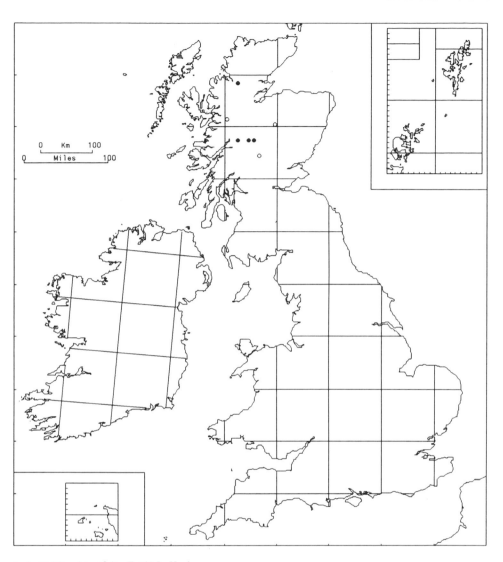

85/5. **Pohlia obtusifolia** (Brid.) L. Koch

A plant of wet habitats, including alpine bogs and flushes, soil flushed with meltwater from snow-patches, and periodically wet rocks. At one site it has been found in a large area of flush on a summit plateau with *Scapania uliginosa*, *Calliergon sarmentosum* and *Drepanocladus exannulatus*. It has also been found on a periodically irrigated calcareous schist crag. It occurs in the upper montane region, ascending to 1100 m (Ben Alder Forest). GB 4+3*.

Paroecious; sporophytes frequent, late summer.

An arctic-alpine, occuring in northern and alpine Europe, including Svalbard. N. and C. Asia, Japan, N. America, Greenland.

M. J. WIGGINTON

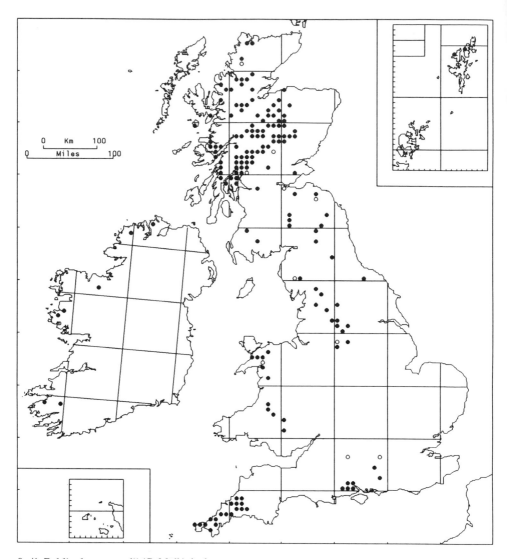

85/6. **Pohlia drummondii** (C. Müll.) Andrews

Like other bulbiliferous *Pohlia* species, it is a relatively poor competitor and normally occupies habitats that are kept open by disturbance, erosion or inundation. It grows on acidic sandy or loamy soils in moist or periodically dry habitats. It is found on sandy or gravelly alluvium in rock crevices and on banks by streams and rivers, on shingle-banks, roadsides and moorland tracks, in old gravel-pits, and in a range of montane and alpine habitats. On sandy tracks and banks, it may be associated with *Cephalozia bicuspidata*, *Nardia scalaris*, *Dicranella heteromalla*, *Polytrichum piliferum* and other propaguliferous *Pohlia* species. On summit debris, in snow-patches and in alpine moss-heaths, recorded associates include *Gymnomitrion concinnatum*, *Conostomum tetragonum*, *Kiaeria starkei*, *Oligotrichum hercynicum*, *Pohlia ludwigii* and *Racomitrium lanuginosum*. 0–1220 m (Ben Macdui). GB 158+10*, IR 7.

Dioecious; sporophytes occasional in Scotland, rare elsewhere, summer. Propagation mainly by axillary bulbils.

Europe north to Svalbard, montane in the south. Asia, N. America, southern S. America.

M. J. WIGGINTON

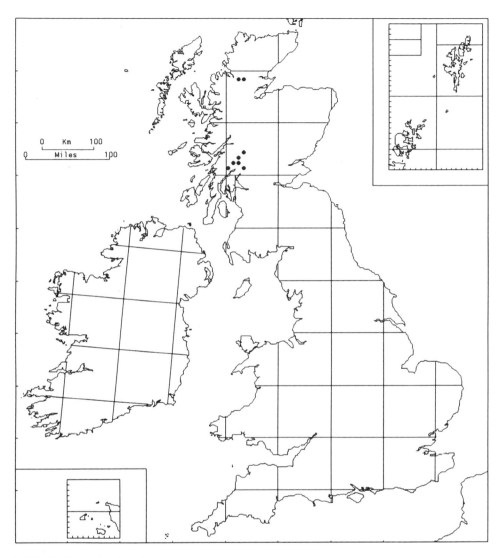

85/6A. **Pohlia scotica** Crundw.

On silt among rocks and on damp sandy and stony ground. All localities are close to water, beside lochs and streams. 95 m (R. Orchy below Beinn Udlaidh) to 650 m (Ben Vorlich). GB 8.

Sterile; gametangia and sporophytes unknown. It lacks any specialized means of dispersal.

Apparently endemic, but further searches may reveal its presence elsewhere in the boreal or Arctic regions of Europe.

First collected in 1964, it was for a time thought to be a non-bulbiliferous form of *P. drummondii* but was later described as a new species (Crundwell, 1982).

M. J. Wigginton

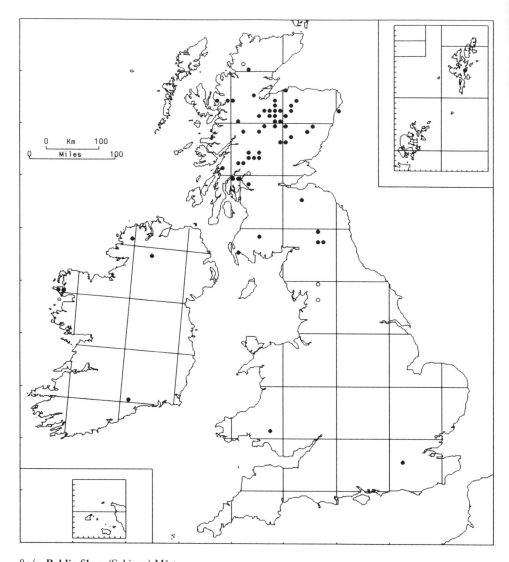

85/7. **Pohlia filum** (Schimp.) Mårt.

A calcifuge species almost exclusively confined to soils of very low organic content. It grows mostly on damp sandy or gravelly soils, by streams and rivers, on rocks, on sandy tracks and roadsides, on dunes and in sand-pits. It has also been recorded growing on damp cindery soil. Mainly at low altitudes, to 330 m (Kielder Forest). GB 53+4*, IR 5.

Dioecious; sporophytes rare, late spring. Propagation mainly vegetative, by caducous bulbils.

Arctic and northern Europe including Iceland and Svalbard, and in mountains further south except Iberia. Caucasus, Asia, N. America, Greenland.

It has been found only once in southern England, where its occurrence may have been casual.

M. J. WIGGINTON

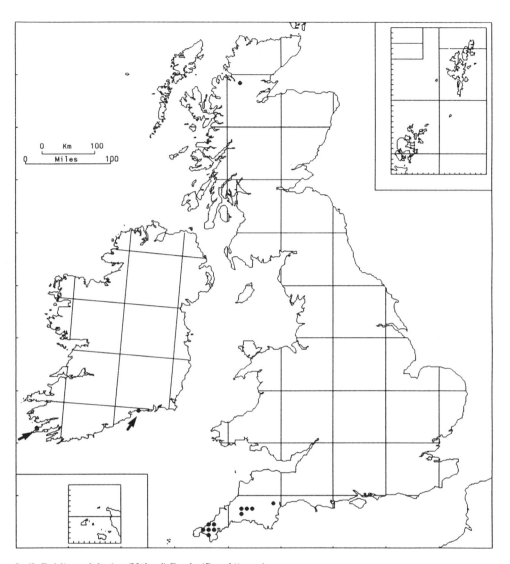

85/8. **Pohlia andalusica** (Höhnel) Broth. (*P. rothii* auct.)

In Ireland and S.W. England, it grows on damp sandy acidic soils, on heathy gravelly tracks, and on china-clay and metalliferous mine-waste. In western Scotland it was found on bare soil at the foot of a cliff at 600 m (Seana Bhraigh). GB 12, IR 2.

Dioecious; sporophytes unknown in Britain. Propagation vegetative, by caducous bulbils.

The species has a sub-oceanic boreal-montane distribution in Europe, from Iceland and the Arctic southwards, and from Spain east to the Baltic and C. Europe. Azores, N. Asia, N. America.

Although distinguished clearly (under the incorrect name *P. rothii*) by Smith (1978), it was misunderstood by most British bryologists until its distribution was revised for a recent *Census Catalogue* (Corley & Hill, 1981). It is rare in western Britain, but must surely be commoner than the map suggests. For a full description and discussion of nomenclature, refer to Shaw (1981a).

M. J. WIGGINTON

65

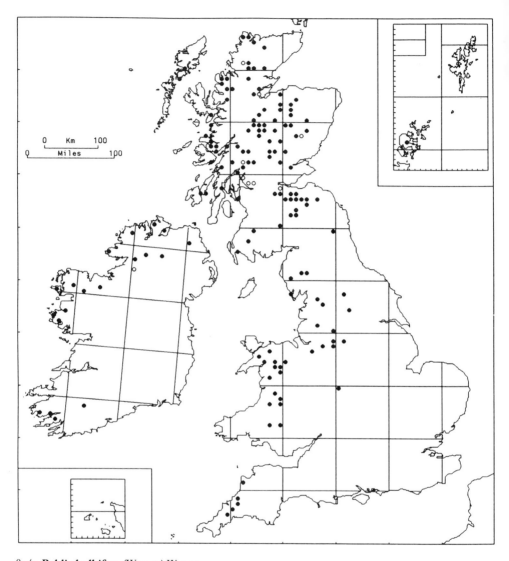

85/9. Pohlia bulbifera (Warnst.) Warnst.

Like most bulbiliferous species of *Pohlia* in Britain, it typically grows on open damp non-calcareous sandy and loamy soils. It occurs on alluvium by streams, on the margins of lakes and reservoirs, on sandy river-banks, on tracks and roadsides, in ditches and disused quarries, and on cliff-ledges. On open, flat sandy ground, e.g. reservoir margins, accompanying species may include *Blasia pusilla*, *Nardia scalaris*, *Archidium alternifolium*, *Pohlia annotina* and tuberous species of *Bryum*. On steep sandy banks it may occur with such species as *Dicranella rufescens*, *Ditrichum cylindricum*, *Pohlia camptotrachela* and *Pseudephemerum nitidum*. 0–350 m (R. Quoich, near Braemar). GB 129+8*, IR 19+1*.

Dioecious; sporophytes very rare. Propagation mainly by caducous bulbils.

Widespread in the Northern Hemisphere, especially in the boreal zone. Azores, Iceland, Europe, Turkey, Caucasus, N. Asia, Japan, N. America, Greenland.

M. J. WIGGINTON

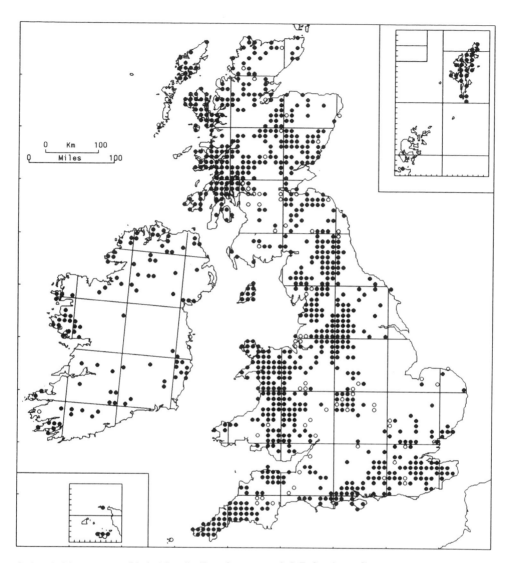

85/10. **Pohlia annotina** (Hedw.) Lindb. (*P. proligera* sensu A. J. E. Smith, 1978).

A species of open habitats, growing on acidic to mildly basic moist (sometimes periodically wet) sandy, loamy or gravelly soils in a variety of habitats. These include banks of streams and rivers, reservoir margins, ditch-banks, tracks, roadsides, disused quarries, gravel-pits, waste ground, woodland rides, and wet shale and sandstone rock-scars. More unusual habitats include canal towpaths and colliery slag-heaps. Typical associates on sandy banks and tracks include *Jungermannia gracillima*, *Nardia scalaris*, *Scapania irrigua*, *Ditrichum cylindricum*, *Pohlia camptotrachela* and *P. lutescens*. 0–700 m (near Loch Kander). GB 850+71*, IR 99+4*.

Dioecious; sporophytes very rare, early summer. Propagation mainly vegetative, by caducous bulbils.

Widespread in N. and C. Europe. Azores, Turkey, Caucasus, N. Asia, Japan, N. America, Greenland.

There has been frequent taxonomic and nomenclatural confusion with *P. proligera*. In Smith's (1978) flora, *P. proligera* and *P. annotina* are treated as synonyms. Shaw (1981a) revised the propaguliferous species of *Pohlia* in N. America and confirmed that both *P. annotina* and *P. proligera* occur in Britain. They are mapped separately in this *Atlas*.

M. J. WIGGINTON

67

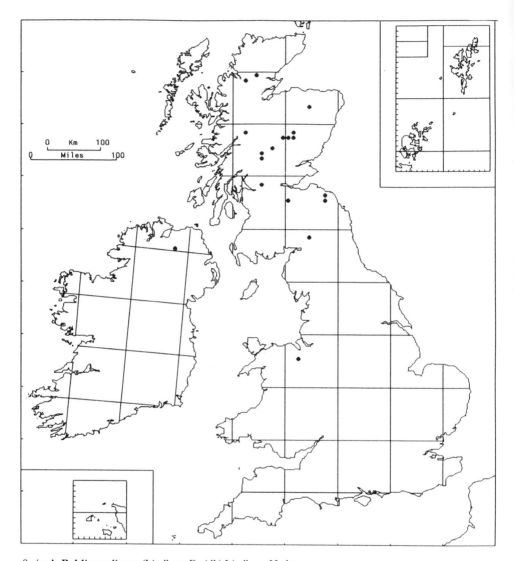

85/10A. **Pohlia proligera** (Lindb. ex Breidl.) Lindb. ex H. Arn.

As green patches or groups of stems among other mosses in disturbed or open habitats, apparently, as in western N. America (Shaw, 1981b), more often on soils of relatively high pH (above 6.0) than related species. Ascends to 800 m (Creag an Duine). GB 17, IR 1.

Dioecious; fruit not known in Britain. Axillary filiform bulbils usually abundant.

In Europe from Scandinavia and N. Russia (Kola Peninsula) south to Italy. N. America, Greenland.

A species that has been much misunderstood. The map is based upon specimens, many of them seen by E. F. Warburg, A. J. Shaw or both of them, in BBSUK, E and the private herbaria of A. C. Crundwell and J. A. Paton. Old accounts of the species, such as that in Dixon's *Handbook* (1924), are based on forms of *P. annotina* (q.v. for further information). Shaw (1981a) showed that it is a good species distinguished from *P. annotina* by its slender filiform twisted bulbils with never more than two leaf primordia and by some other less obvious characters. Warburg had earlier realized that the name *proligera* applied to this plant and not to any form of *P. annotina*, but he published nothing about it.

A. C. CRUNDWELL

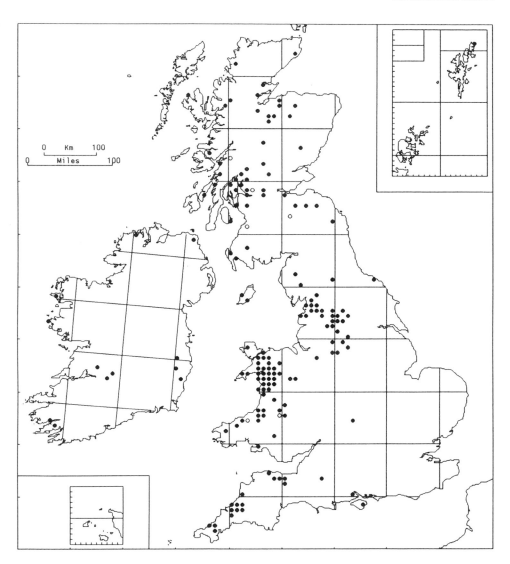

85/11. **Pohlia camptotrachela** (Ren. & Card.) Broth.

This species grows preferentially on damp neutral to acidic sandy and loamy soils, though it also occurs on those that are more peaty. It is found on banks by ponds, streams and rivers, at the edges of lakes and reservoirs, on paths and tracks, and in damp fields. In such habitats it is often found with *P. annotina*. Other frequent associates include *Bryum rubens*, *B. sauteri*, *Ceratodon purpureus*, *Dicranella rufescens*, *Ditrichum cylindricum*, *Ephemerum serratum*, *Pleuridium acuminatum* and *Pseudephemerum nitidum*. Mainly at low altitudes, it presumably ascends above 300 m although this is not recorded. GB 148+7*, IR 12.

Dioecious; sporophytes unknown in Britain. Vegetative propagation by caducous axillary bulbils.

Europe including the Faeroes and Scandinavia. Western N. America.

Not generally distinguished by British bryologists until the revision by Lewis & Smith (1978). It must therefore be under-recorded.

M. J. WIGGINTON

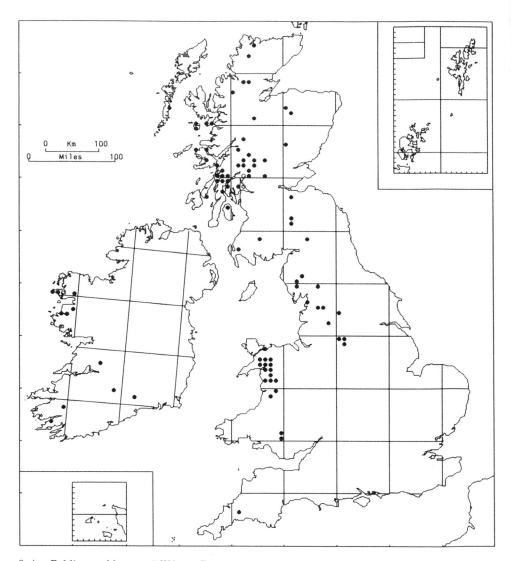

85/12. **Pohlia muyldermansii** Wilcz. & Dem.

This species grows on moist sandy or gritty soils in rocky places by streams and waterfalls in montane regions, often in very sheltered locations which are noted for the occurrence of hygrophilous liverworts. It also occurs on steep sandy river-banks with such species as *Diplophyllum albicans, Scapania scandica, Dichodontium pellucidum* and *Pohlia annotina*. It is of less frequent occurrence in rock crevices, and is rarely found on sandy roadsides and in lowland fields. Ascends to 800 m (Ben More). GB 78+2*, IR 11.

Dioecious; sporophytes very rare. Propagation by axillary bulbils.

Apparently a European endemic, recorded from Ireland, Britain, Belgium, the Netherlands, Switzerland and Austria.

Doubtless under-recorded; it was not distinguished by most British bryologists until the notice of its occurrence by Lewis & Smith (1978). British, Irish and alpine plants show small but consistent differences from those in the Low Countries , and have been distinguished as var. *pseudomuyldermansii* Arts, Nordhorn-Richter & A. J. E. Smith (1987).

M. J. Wigginton

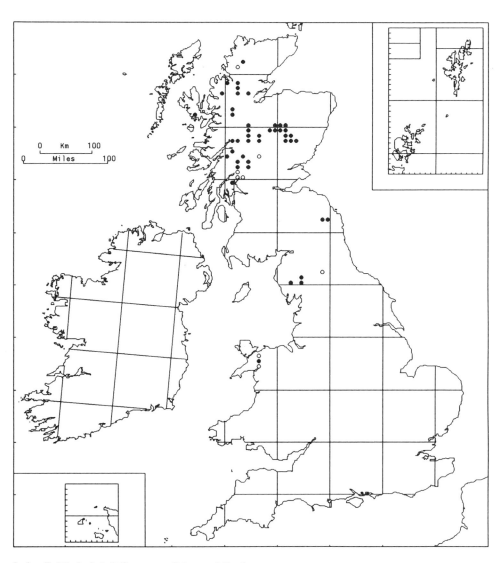

85/13. **Pohlia ludwigii** (Spreng. ex Schwaegr.) Broth.

It grows on moist or wet, sandy and gritty soils on banks and by streams, on soil overlying rock, on rock-ledges and in scree. It is particularly characteristic of late snow-bed bryophyte-dominated vegetation, and accompanies species such as *Anthelia juratzkana*, *Pleurocladula albescens*, *Oligotrichum hercynicum*, *Pohlia drummondii* and *Polytrichum sexangulare*. It is also locally frequent in montane bryophyte springs, in which associated species may include *Anthelia julacea*, *Scapania uliginosa*, *Calliergon stramineum*, *Philonotis fontana* and *Pohlia wahlenbergii* var. *glacialis*. Mainly at higher altitudes, ascending to 1340 m (Ben Nevis). GB 44 + 8*.

Dioecious; sporophytes very rare, summer.

Northern and montane Europe, including Iceland, Faeroes and Svalbard. Turkey, Caucasus, Asia, western N. America, Greenland.

M. J. WIGGINTON

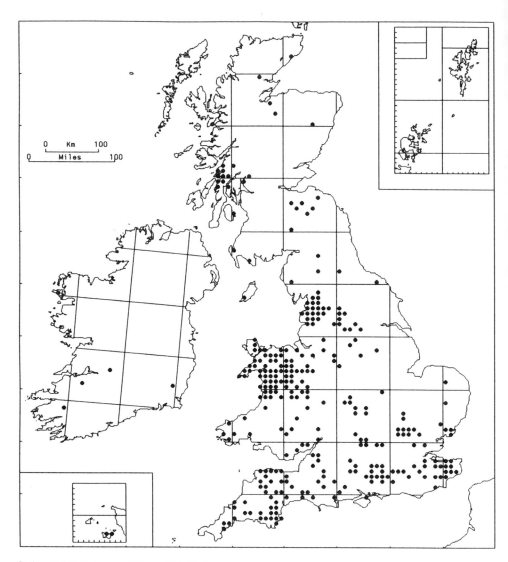

85/14. Pohlia lutescens (Limpr.) Lindb. f.

This species is characteristic of friable, neutral to mildly acid clay-loam and sandy-loam soils, particularly in rather open habitats by paths, ditches and streams, at the entrance of rabbit burrows, and in hedgebanks. A common associate on the more freely draining acidic soils is *Dicranella heteromalla*, often accompanied by *Ceratodon purpureus, Isopterygium elegans* and *Mnium hornum*. Sometimes just a few individual stems of *P. lutescens* occur in a mat of *D. heteromalla*. On loamy banks, the species is a member of a characteristic assemblage, which may include *Atrichum undulatum, Bryum rubens, Dicranella schreberana, D. staphylina, Ditrichum cylindricum* and *Pseudephemerum nitidum*. It has very rarely been recorded growing directly on sandstone rock. Mainly lowland, ascending to 450 m (Berwyn Mountains). GB 299, IR 5.

Dioecious; female plants predominate in Britain, male plants are apparently rare, and sporophytes are unknown in Britain or Ireland. Vegetative propagation by rhizoid tubers.

C. Europe, north to Ireland, Britain and the Baltic region.

Probably under-recorded. It was recorded from Britain and Ireland by Watson (1968), but identification presented difficulties until Whitehouse (1973) described the tubers.

M. J. WIGGINTON

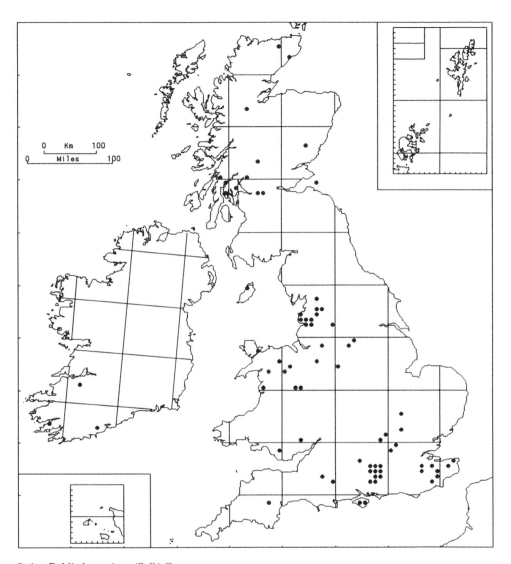

85/15. Pohlia lescuriana (Sull.) Grout

This species prefers moisture-retentive soils, and is therefore most commonly found on clays, though not those that are strongly basic. It occurs on banks of streams and ditches, at the margins of ponds and reservoirs, on bare soil in old fields, in ruts, on moist compacted soil by paths, and on open clay soil in woodland rides. Associated species in these habitats may include *Lophocolea bidentata*, *Pellia endiviifolia*, tuberous *Bryum* spp., *Dicranella schreberana*, *Ditrichum cylindricum*, *Eurhynchium swartzii* and *Pleuridium acuminatum*. Mainly lowland, ascending to 400 m (near Llangollen). GB 70, IR 3.

Dioecious; sporophytes occasional to rare, spring to autumn. Vegetative propagation by rhizoid tubers.

W., C. and N. Europe, Caucasus, N. Asia, Japan, N. America.

Probably under-recorded. Although it was recognized as British by Warburg (1965), identification presented difficulties until Whitehouse (1973) described the tubers.

M. J. WIGGINTON

73

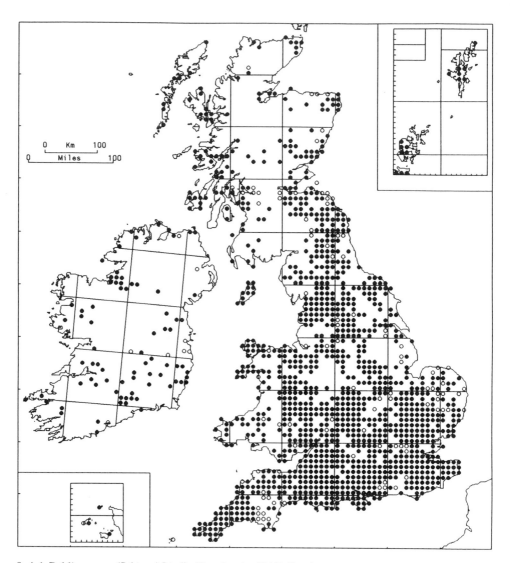

85/16. **Pohlia carnea** (Schimp.) Lindb. (*P. melanodon* (Brid.) Shaw)

A plant of moist or wet, usually clay soils which are circum-neutral to basic, sometimes in dense shade. It is characteristic of steep, shaded clay banks of ditches and streams, particularly in the flood zone or where the bank is kept wet by seepage, often forming pure stands in such locations. It is also a common constituent of the typical clay-bank community which includes such species as *Lunularia cruciata*, *Pellia endiviifolia*, *Barbula tophacea*, *Bryum gemmiferum*, tuberous *Bryum* spp., *Fissidens bryoides* and *Pseudephemerum nitidum*. This species also grows on moist ground by paths, in fields, clay-pits, marshes and woodland rides, on sea-cliffs, and at the entrance to caves and mines. Lowland. GB 1200+97*, IR 77+9*.

Dioecious; sporophytes occasional, winter to spring. Rhizoid tubers frequent, but very difficult to find unless the material is stained by iodine solution (Arts, 1986).

Europe. Macaronesia, N. Africa, Caucasus, Turkey, Asia, Japan, N. America.

M. J. WIGGINTON

74

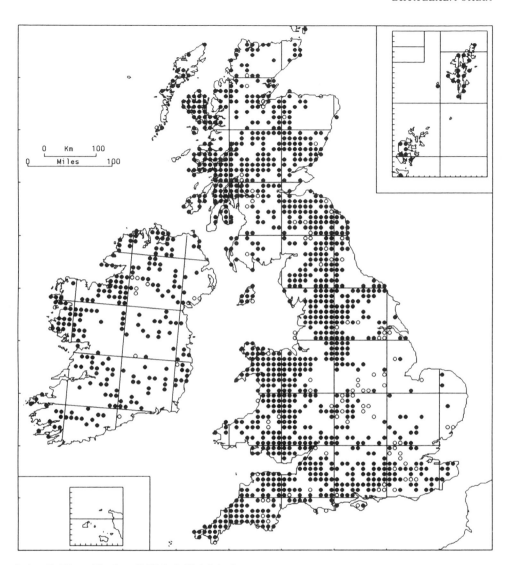

85/17. **Pohlia wahlenbergii** (Web. & Mohr) Andrews

This species occurs in a wide variety of moist or wet habitats. In the mountains it grows in wet places, usually in or near oligotrophic bryophyte springs and flushes, with such species as *Saxifraga stellaris*, *Scapania undulata*, *Calliergon stramineum*, *Philonotis fontana*, *Pohlia ludwigii* and *Sphagnum* spp. At lower altitudes it frequently grows on moist sandy or gritty soil on roadsides, tracks and waste ground. It is also a characteristic species of seeping or dripping rock-scars in ravines; other habitats include flushes and marshes, floors of sand-pits, earthy ditch-banks, damp woodland rides, and, uncommonly, cultivated fields. 0–1200 m (Cairngorm Mountains). GB 1101+101*, IR 225+12*.

Dioecious; sporophytes rare, summer.

Europe including Iceland and Svalbard. Algeria, Turkey, Caucasus, Asia, N., C. and S. America, Greenland, Australasia, Kerguelen Island, Antarctica.

Large montane forms are often distinguished as var. *glacialis* (Schleich. ex Brid.) Warb., but this taxon has not been recorded with sufficient consistency to be mapped separately.

M. J. WIGGINTON

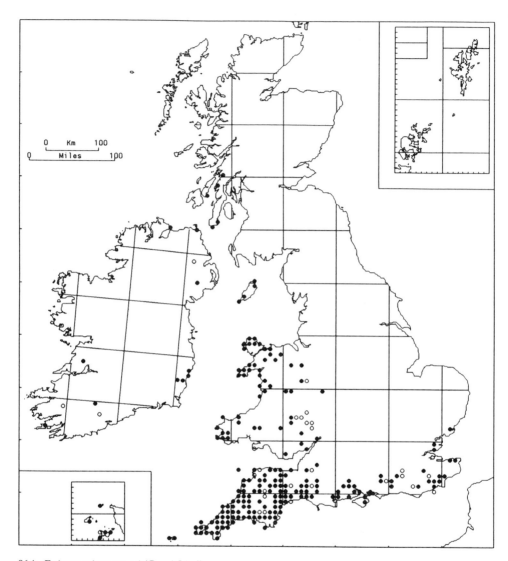

86/1. **Epipterygium tozeri** (Grev.) Lindb.

A species of warm, sheltered, often humid, disturbed, loamy non-calcareous soil. It is found primarily on stream- and river-banks, but also on lanesides, in crevices of Cornish 'hedges' and on coastal undercliffs. Frequent associates include *Calypogeia arguta*, *Fossombronia pusilla*, *Phaeoceros laevis*, *Bryum sauteri*, *Fissidens bryoides*, *F. viridulus*, *Pleuridium acuminatum*, *Pohlia lutescens* and *Weissia controversa*. Lowland. GB 201+24*, IR 10+3*.

Dioecious; capsules rare, spring to early summer. Rhizoidal tubers have been found in a specimen from Essex and may be frequent, especially in shaded conditions (Arts & Nordhorn-Richter, 1986). Bulbiform stem-bases with large attached axillary bulbils probably act as organs of perennation during summer drought.

Mediterranean and Atlantic parts of Europe north to Ireland, Scotland and northern France. Macaronesia, Morocco, Algeria, Lebanon, Turkey, Caucasus, S.W. Himalaya, Taiwan, S.W. Japan, western N. America.

Its world distribution, which is similar to that of the xerophilous Mediterranean vegetation type, has been mapped by Arts & Nordhorn-Richter (1986). They suggest that it is a Tertiary relict, from a southern coastal Laurasian origin. The species fruits much more frequently in S. Europe.

F. J. Rumsey

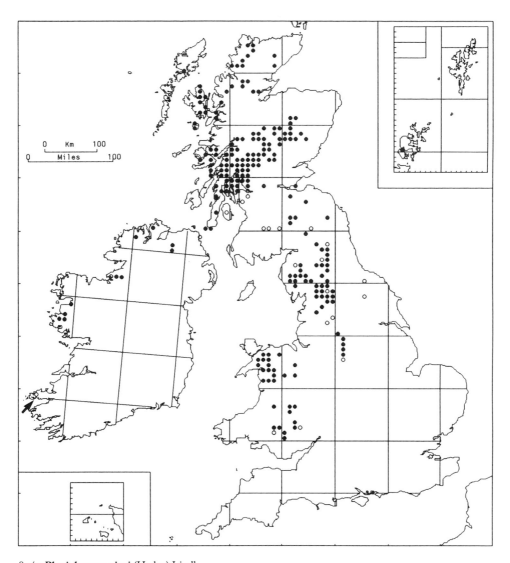

87/1. **Plagiobryum zieri** (Hedw.) Lindb.

Locally frequent as tufts in damp, shaded crevices of N.- or E.-facing basic montane cliffs (limestone, mica-schist, volcanic tuff, gabbro, basalt, calcareous sandstone), often growing with *Anoectangium aestivum*, *Ctenidium molluscum*, *Pohlia cruda* and *Tortella tortuosa*, and on bare soil and fine gravel in gullies on shaded cliffs, growing with *Poa alpina* and *Saxifraga oppositifolia*. More rarely it occurs in crevices of basic rock-walls of ravines and gullies, often associated with *Bryoerythrophyllum ferruginascens*, and on bare soil amongst rocks influenced by base-rich seepage. Mainly above 300 m, ascending to 1205 m on Ben Lawers, rarely descending to sea-level (Danna). GB 200+19*, IR 10+3*.

Dioecious; capsules frequent, May to September.

A subarctic-subalpine plant, occurring in W., C. and N. Europe, including Iceland and Faeroes. N. Africa, Turkey, Caucasus, N. and C. Asia, N. America, Greenland.

H. J. B. BIRKS

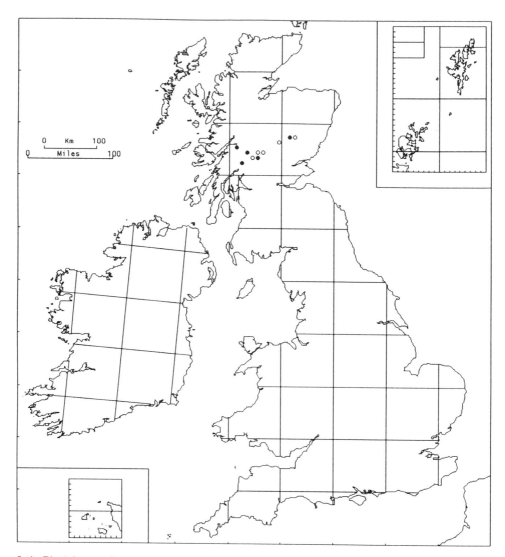

87/2. Plagiobryum demissum (Hook.) Lindb.

Occurs as scattered tufts on damp or periodically irrigated outcrops and in shaded crevices of soft calcareous mica-schist cliffs. It favours, but is not restricted to, N. and E. aspects. Associates include *Saxifraga oppositifolia*, *Bryoerythrophyllum ferruginascens*, *Encalypta alpina*, *Hypnum bambergeri*, *Mnium spinosum*, *Myurella julacea* and *Pohlia cruda*. Restricted to high altitudes, from 760 m (Caenlochan Glen) to 1150 m (Ben Lawers). GB 5+5*.
  Dioecious; capsules frequent, summer.
  An arctic-alpine species of montane areas in N. and C. Europe, including Iceland. Caucasus, C. Asia, China, N. America, Greenland.
  Possibly commoner in the C. Highlands than the map indicates.

H. J. B. Birks

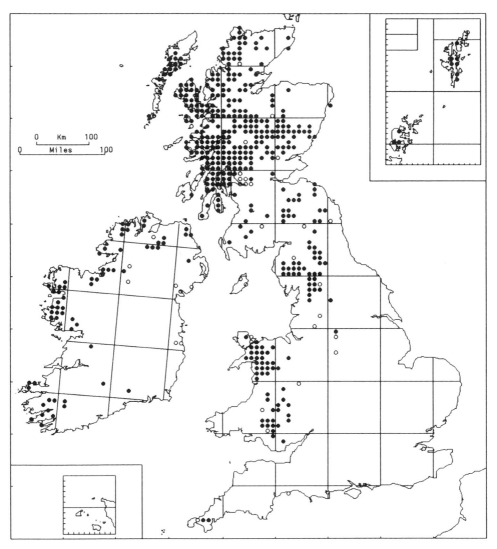

88/1a. Anomobryum filiforme (Dicks.) Solms-Laub. var. **filiforme** (*A. julaceum* (Gaertn., Meyer & Scherb.) Schimp.)

Frequent but usually present in small quantity as tufts or scattered stems in moist sand, silt or gravel by lakes and streams, and in crevices of mildly basic rocks by streams, in gullies and by waterfalls, often growing with *Jungermannia* spp. and *Dicranella palustris*. It also occurs on irrigated mildly basic montane rock-faces associated with *Blindia acuta*, in open basic gravel flushes, on damp mine-waste, and on moist sand or gravel in disused quarries and gravel-pits, growing with *Blasia pusilla*, *Haplomitrium hookeri* and *Riccardia incurvata*. 0–950 m (Snowdon). GB 397+33*, IR 63+10*.

Dioecious; capsules rare, ripe from late spring to autumn.

Most of Europe, including Iceland and Faeroes. Macaronesia, W. Africa (Cameroon), C. and E. Africa, Arabia, Caucasus, C., N. and E. Asia, N. and C. America, Greenland.

H. J. B. BIRKS

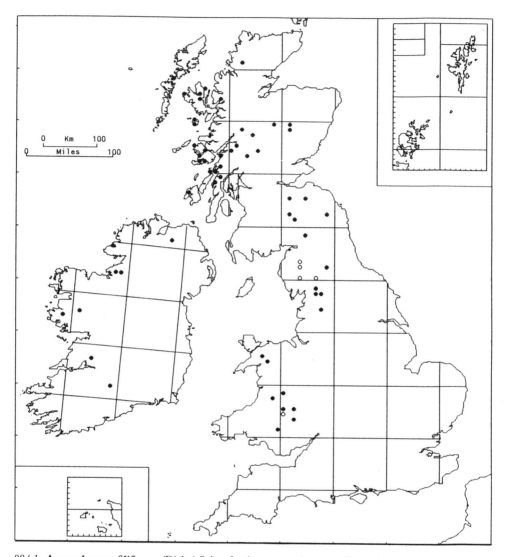

88/1b. Anomobryum filiforme (Dicks.) Solms-Laub. var. **concinnatum** (Spruce) Loeske

A rare taxon of dry earthy, often crumbling ledges of sun-exposed basic mica-schist, metamorphosed limestone, or basalt cliffs, growing with *Barbula icmadophila, Encalypta rhaptocarpa, Schistidium apocarpum, S. strictum* and *Tortula subulata*. It also occurs in damp crevices of basic slate rocks and on irrigated mildly basic montane rock-faces, with *Eremonotus myriocarpus, Jungermannia* spp. and *Scapania* spp. 380 m (Dugwm Rock) to 1100 m (Breadalbane). GB 49+6*, IR 8.

Dioecious; capsules unknown.

Subarctic-subalpine, occurring in Iceland, Pyrenees, C. and N. Europe. Macaronesia, Turkey, Caucasus, N., C. and E. Asia, N. America, Greenland.

A poorly defined taxon, probably under-recorded because of its similarity to var. *filiforme*.

H. J. B. BIRKS

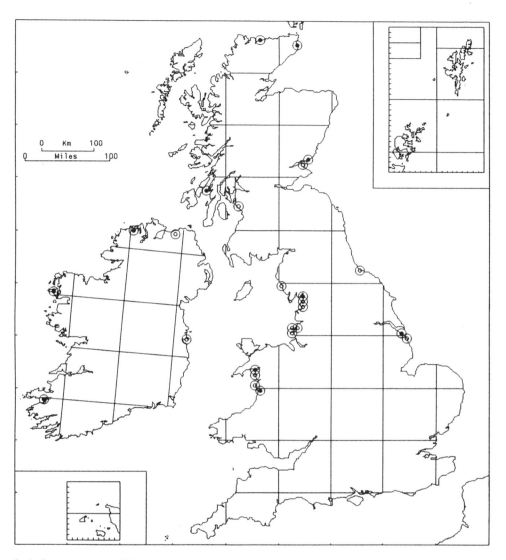

89/1. Bryum marratii Wils.

A plant of damp, usually calcareous, sandy or muddy ground by the sea, and of dune-slacks. It is evidently tolerant of slight salinity, but avoids saltmarshes. *B. calophyllum* and *B. warneum* are frequent associates. Lowland. GB 8+12*, IR 3+2*.

Autoecious; fruit occasional, ripe in late summer and autumn.

Scattered here and there on coasts of W. Europe from Normandy to Arctic Norway and on shores all round the Baltic. It is also found in N. America on the coast of Newfoundland and at inland stations in Alberta and N. Dakota.

A distinctive species not likely to be overlooked, though easily confused with *B. calophyllum*. It has disappeared, through human activity, from several of its former British sites. The record for the Faeroes is based on misidentified *B. calophyllum* (Boesen *et al.*, 1975).

A. C. CRUNDWELL

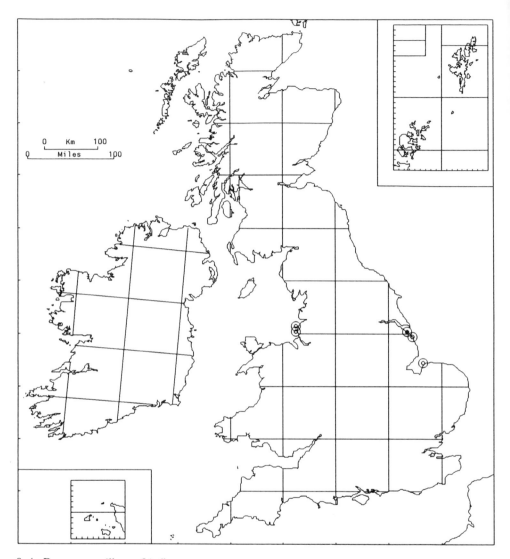

89/2. Bryum mamillatum Lindb.

A plant of dune-slacks and damp sandy soil near the sea. In its Norfolk locality it grew among *Juncus gerardii*. Lowland. GB 1+4*.

Autoecious; capsules ripe in early summer.

Known from around the shores of the Baltic and from Svalbard; reported also from Greenland. Atypical plants from Switzerland and Austria have also been given this name.

Apparently extinct in at least two, possibly all three of its British localities. It is not nameable without ripe capsules; one of the diagnostic features is the very large spores. There are sometimes, however, equally large spores mixed with smaller and abortive ones in abnormal capsules belonging to other species; these can lead to misidentifications.

A. C. CRUNDWELL

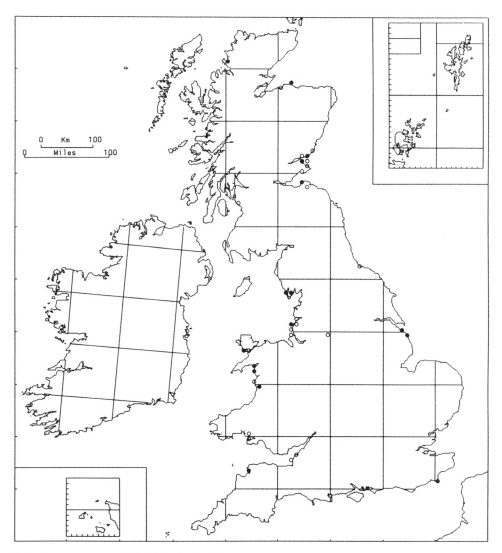

89/3. Bryum warneum (Röhl.) Bland. ex Brid.

A plant of dune-slacks and damp sand by dunes, where it sometimes grows among *Juncus maritimus*, and, more rarely, in gravel-pits. It is often associated with other species of *Bryum*, especially *B. algovicum*, *B. calophyllum*, *B. inclinatum*, *B. intermedium*, *B. knowltonii* and *B. marratii*. Lowland. GB 16+20*, IR 2*.

Autoecious; capsules mature in late summer and autumn.

Known from the western and Baltic coasts of Europe, from N. France to Arctic Norway; also, much more rarely, by lakes and rivers in inland localities in C. and E. Europe. Himalaya, Altai Mts, Quebec (Canada).

Not nameable without mature capsules.

A. C. CRUNDWELL

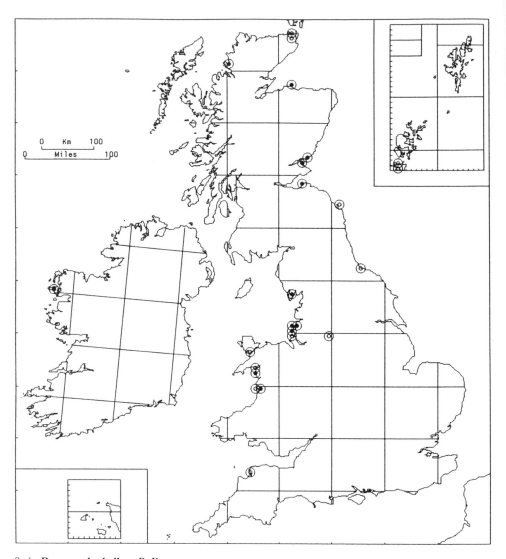

89/4. Bryum calophyllum R. Br.

A plant of calcareous dune-slacks, occasionally on damp sand in estuaries or in gravel-pits by the sea; often associated with other species of *Bryum*, especially *B. marratii* and *B. warneum*. Lowland. GB 10+11*, IR 1.

Autoecious; capsules frequent, ripe from late summer to early winter.

Circumpolar. Coasts of N. Europe from the Netherlands north to Iceland, Faeroes, Arctic Norway and Svalbard; also known from Sardinia and Romania. Arctic and N. Siberia, C. Asia, Tibet, N. America, Greenland.

A distinctive species, though occasionally confused with *B. marratii*. Like other mosses with similar ecology it has disappeared from many of its former British sites. In W. and N. Europe it is almost exclusively coastal but in most of the rest of its range it is more often away from the sea, often growing along rivers or by lakes.

A. C. CRUNDWELL

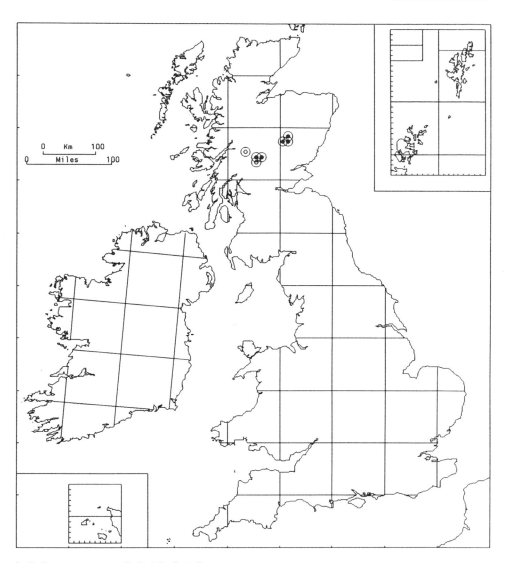

89/6. **Bryum arcticum** (R. Br.) B., S. & G.

Forming reddish tufts on basic soil among rocks on mountains. 700–1000 m. GB 5+2*.
  Synoecious; capsules ripe in late summer.
  Circumboreal, from the High Arctic south to the Alps, Urals, Tibet and northern U.S.A.

A. C. CRUNDWELL

85

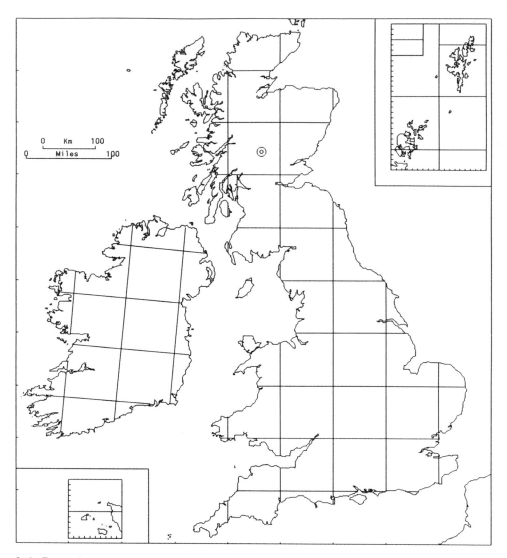

89/7. **Bryum lawersianum** Philib.

Found as dull green patches on damp, bare, micaceous soil at about 1070 m on Ben Lawers. GB 1*.
Synoecious; capsules ripe in summer. No gemmae known.
Endemic.
Seen twice in the year of its discovery, 1899, and again in 1912 and 1924, but not found recently. Relationships to *B. arcticum* and *B. pallens* have been suggested but nearly all who have seen material have thought it a good species.

A. C. CRUNDWELL

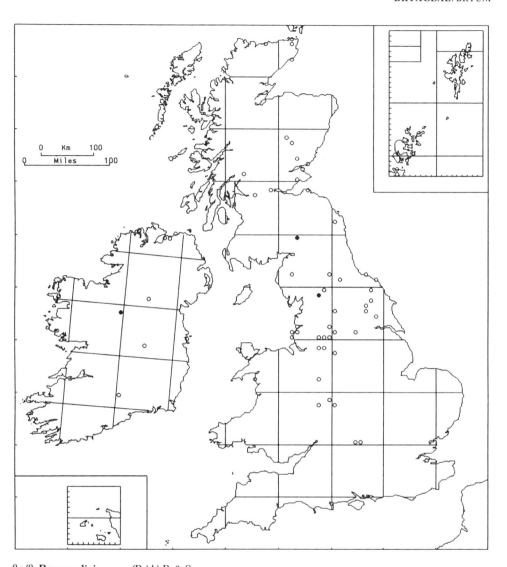

89/8. **Bryum uliginosum** (Brid.) B. & S.

A mildly calcicolous species forming greenish patches on damp soil by streams and in dune-slacks. Lowland. GB 2+47*, IR 1+8*.

Autoecious; capsules maturing in late summer and autumn. Gemmae not known.

N. Italy and Yugoslavia north to Iceland and S. Scandinavia. N. Asia east to Japan and the Russian Far East (Sakhalin and Kamchatka), N. America, Greenland, Chile, Argentina (Tierra del Fuego), S. Georgia, New Zealand.

It has probably always been rare in Britain, but there are many old records from places where it cannot now be found. While a few of these may result from misidentifications – it is not nameable without ripe capsules and has been confused in the past with *B. inclinatum*, *B. intermedium*, *B. pallens* and *B. pseudotriquetrum* – it has clearly decreased very greatly in the last hundred years, much more than would be expected from normal human activities.

A. C. CRUNDWELL

87

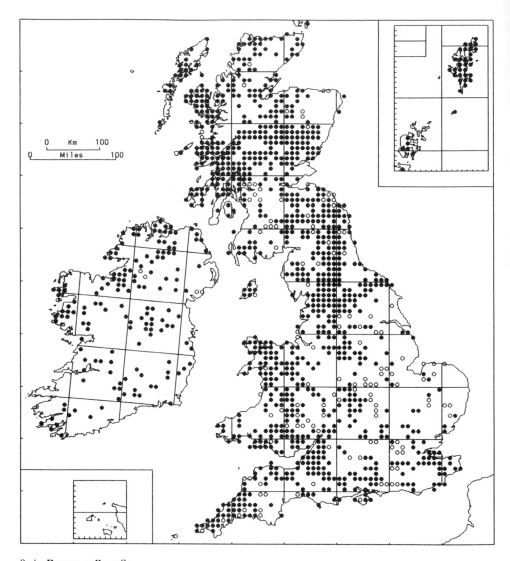

89/9. Bryum pallens Sw.

In pink, green or brownish patches on damp slightly acid to strongly basic soil on banks of streams and rivers, in old brick pits, on woodland rides, in flushes and springs, cultivated fields and meadows, chalk grassland and waste ground; also on wet rocks and on damp walls. 0–1180 m (Ben Lawers). GB 1009+135*, IR 154+6*.

Dioecious; fruit rare, especially in the south, ripening in summer. Filamentous gemmae occasionally present in the leaf axils; gemmae are produced on the protonema in culture (Whitehouse, 1987).

Circumboreal with extensions to the tropics and Southern Hemisphere. From Iceland and Svalbard south to N. Italy, Yugoslavia and Bulgaria. In Asia south to the Caucasus, Iran, Nepal and Japan; N. America, Greenland. Also in the mountains of Rwanda, Ecuador and Peru, and in Argentina (Tierra del Fuego).

An easily recognized species, but variable and wide-ranging ecologically. The plant of the southern English chalk may prove to be taxonomically distinct from northern plants.

A. C. CRUNDWELL

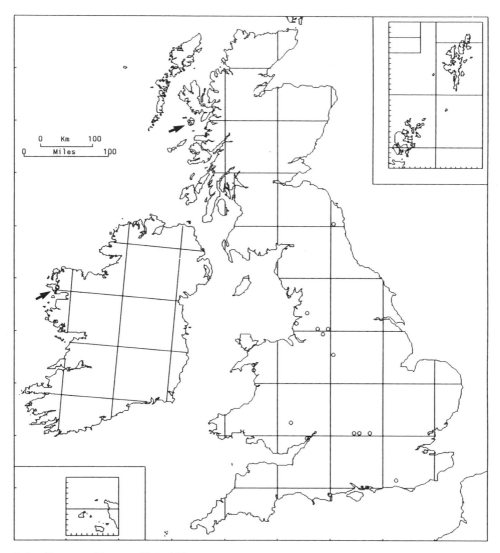

89/10. Bryum turbinatum (Hedw.) Turn.

As green to pinkish patches on damp basic soil and in dune-slacks. Lowland. GB 16*, IR 1*.

Dioecious; capsules ripe in spring and early summer. No gemmae known.

In Europe from Iceland and Scandinavia southward. Morocco, Algeria, Ethiopia, S. Africa; Asia south to Kashmir and Sikkim; doubtfully present in N. America, but known from the mountains of Peru and Ecuador and from Argentina and Chile.

Found only three times since 1930, most recently in 1947; now apparently extinct in Britain and Ireland.

A. C. CRUNDWELL

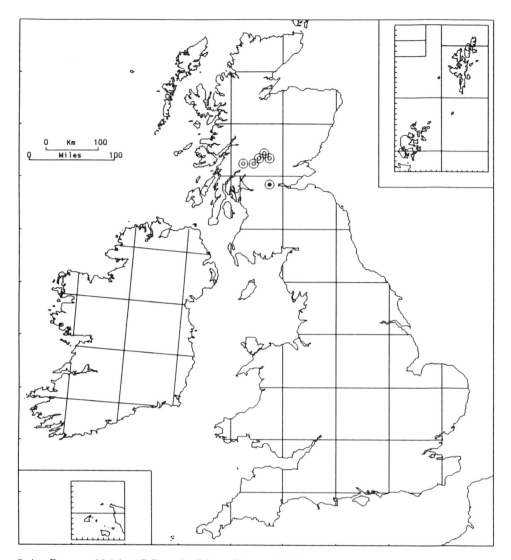

89/11. **Bryum schleicheri** DC. var. **latifolium** (Schwaegr.) Schimp.

A handsome plant forming yellow-green tufts, sometimes tinged with a copper colour, in upland flushes. Now known from only one locality (Touch Muir) at about 300 m. GB 1+5*.

Dioecious; capsules not known in Britain. No gemmae reported.

Frequent in flushes and by streamlets at about 1500–2000 m in C. Europe. Its distribution elsewhere is uncertain because of confusion with *B. schleicheri* var. *schleicheri* and *B. turbinatum*. The taxonomy of these plants needs further study.

A. C. Crundwell

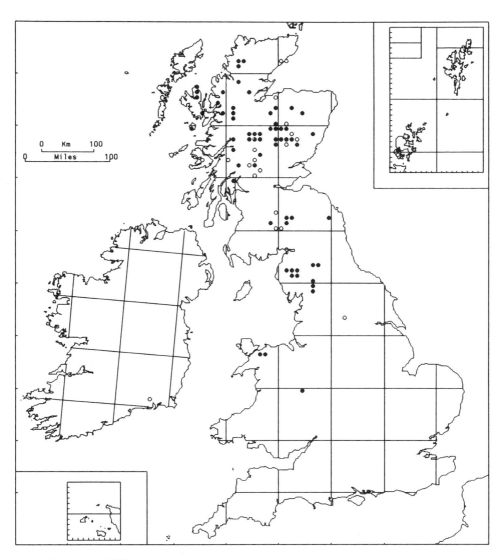

89/12. **Bryum weigelii** Spreng.

As pink or greenish-pink patches in mountain springs and flushes, by snow-patches and streams, and on the shores of Highland lochs. It avoids lime and its commonest associates are *Scapania undulata*, *Philonotis fontana* and *Pohlia wahlenbergii* var. *glacialis*. 230 m (Cairnacay) to 1070 m (Carn Eige). GB 62+21*, IR 1*.

Dioecious; capsules unknown in Britain. No gemmae.

Circumboreal with a Southern Hemisphere disjunction. In Europe from Iceland, Svalbard and N. Russia (Kola Peninsula) south to the Pyrenees, Alps and Carpathians. Kerguelen Island.

A. C. CRUNDWELL

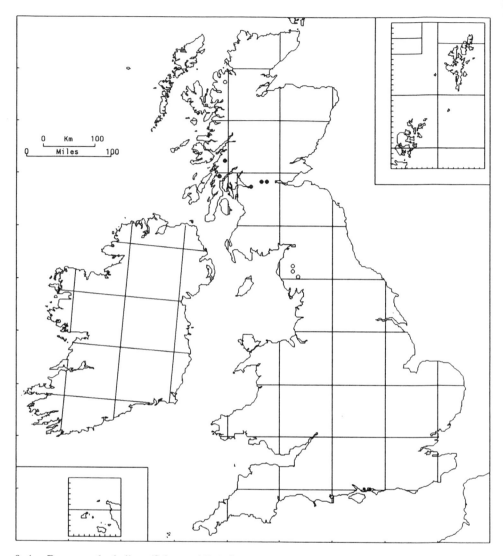

89/13. **Bryum cyclophyllum** (Schwaegr.) B. & S.

As loose, pale green patches on wet soil by streams and lakes. Lowland. GB 5+4*.

Dioecious; capsules not known in the British Isles. Filamentous gemmae sometimes abundant in the leaf axils. From Scandinavia south to C. Europe. Widespread in Asia east to Japan and Korea; N. America.

It is difficult to be quite sure of the extra-European range of this species because it has been confused in the literature and in herbaria with the arctic *B. cryophilum* Mårt. The two species are not like each other but the illegitimate name *B. obtusifolium* has been applied to both.

A. C. CRUNDWELL

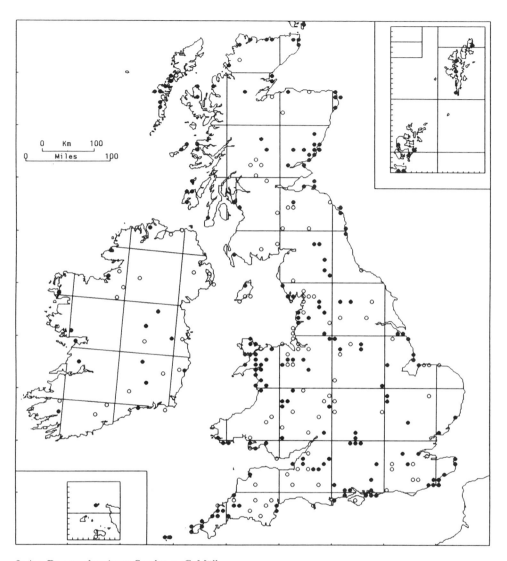

89/14. **Bryum algovicum** Sendtn. ex C. Müll.

As green patches or tufts on basic sand-dunes, less often on sandy soil by roadsides and on basic cliff-ledges; also on limestone dumps, in old quarries, on walls and old buildings. 0–460 m (Skye). GB 192+90*, IR 15+20*.

Usually synoecious; capsules common (the species is not identifiable without them), summer. No gemmae.

Circumboreal and occurring widely in other cool regions of the world. Europe from Svalbard south to Albania. Asia east to Japan; N. America and Greenland; mountains of Ethiopia, Tanzania and S. America; Australia, New Zealand, Falkland Islands, S. Georgia, Kerguelen Island.

Nearly all British plants belong to the synoecious var. *rutheanum* (Warnst.) Crundw. There are dioecious populations on the sand-dunes of northern Aberdeenshire. The autoecious var. *algovicum* has not been reported in Britain or Ireland.

A. C. CRUNDWELL

93

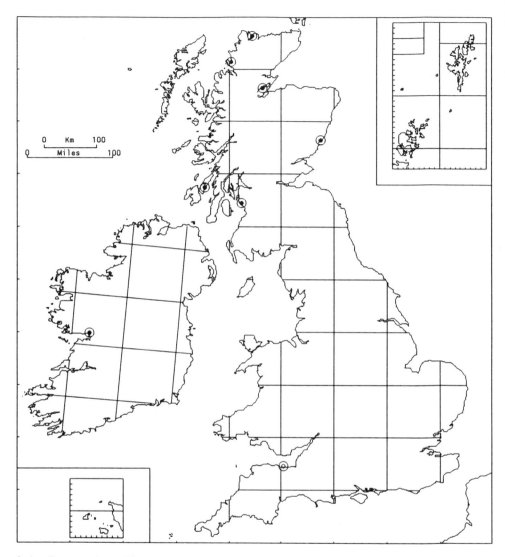

89/15. Bryum salinum Hagen ex Limpr.

Forms bright green patches in the upper parts of salt-marshes and on damp sandy ground where there is some saline influence. It is probably an obligate halophyte restricted to regions where the summer rainfall is sufficient to prevent excess concentration of salt from building up in the soil. Lowland. GB 6+1*, IR 1.

Synoecious; only identifiable with ripe capsules, maturing in spring and summer. Without gemmae.

Coasts of N. Europe from Denmark and Poland north to Iceland and Spitsbergen. Alaska, Canada, Greenland.

A species easily overlooked, first collected in England in 1925 but not distinguished in the British Isles until it was found in Ireland in 1957 (Nyholm & Crundwell, 1958). It is hardly to be distinguished in the field from *B. algovicum*, a much commoner maritime species. It is usually present only in small quantity and often on ground grazed by cattle, who do not improve the quality of the material.

A. C. CRUNDWELL

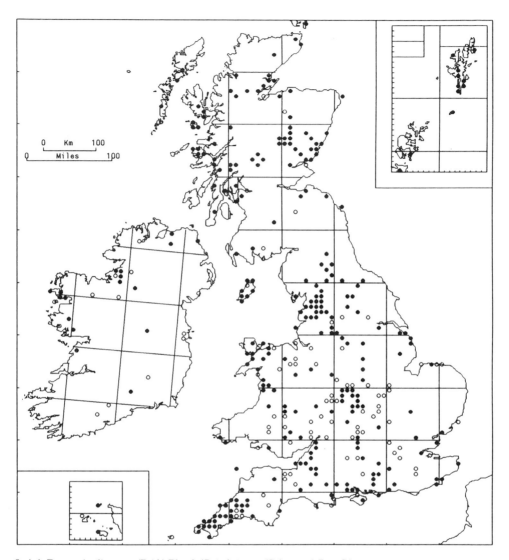

89/16. **Bryum inclinatum** (Brid.) Bland. (*B. imbricatum* (Schwaegr.) B. & S.)

As green patches or tufts on sandy banks and bare earth by roadsides and in dunes and waste places; also on basic rocks on mountains and cliffs, in quarries and occasionally on walls and old buildings. A mainly lowland species, ascending to 650 m (Snowdon). GB 260+84*, IR 17+12*.

Synoecious; capsules common, ripe in summer. Without gemmae.

Present through the frigid and temperate zones of both the Northern and Southern Hemispheres.

A not very interesting-looking species, probably under-recorded because the inner peristome must be examined to identify it.

A. C. Crundwell

95

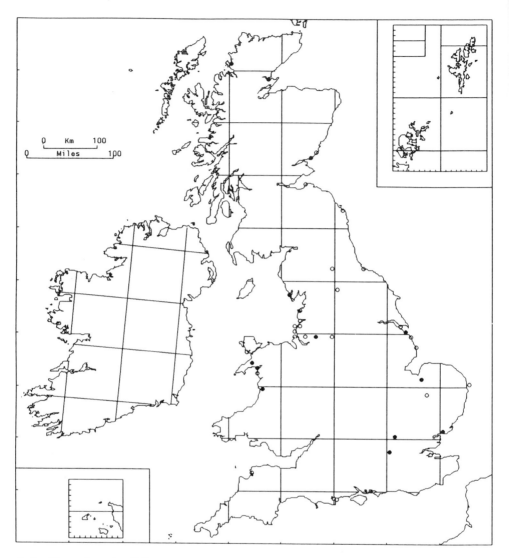

89/17. **Bryum knowltonii** Barnes

In green to reddish tufts in calcareous dune-slacks, less frequently in quarries and in moist sandy places inland. Lowland. GB 13+23*.

Synoecious; capsules common, ripe in late spring. Without gemmae.

Circumpolar. In Europe from Svalbard, Iceland, the Faeroes and Scandinavia south to N. France and Germany. Siberia, Himalaya, Alaska, Canada, Greenland.

A. C. CRUNDWELL

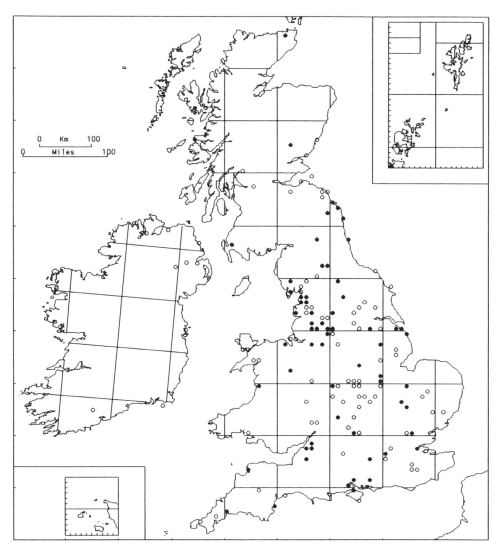

89/18. **Bryum intermedium** (Brid.) Bland.

A calcicolous plant forming green tufts on bare damp soil in sand-dunes and quarries, on roadsides and waste ground, sometimes also on rock-ledges and walls. Mainly lowland, to 310 m (Kielder Forest). GB 65+80*, IR 10*.

Synoecious; capsules maturing from February through to autumn, within each tuft successively, not simultaneously. Without gemmae.

In Europe from Scandinavia southwards. Siberia, Canada (Alberta), Greenland.

B. creberrimum and other species have sometimes been mistaken for *B. intermedium*, leading to erroneous records in the literature, which consequently gives no consistent picture of its world distribution. The map is likely to contain some errors, as is the published vice-county distribution in the most recent *Census Catalogue* (Corley & Hill, 1981).

A. C. CRUNDWELL

97

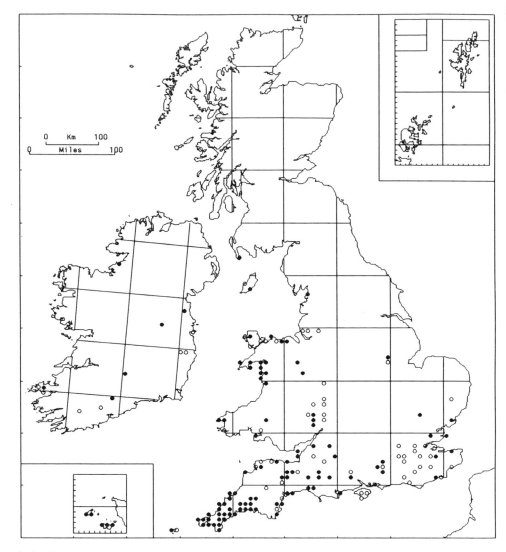

89/19. **Bryum donianum** Grev.

Occurs as dark green tufts on light slightly acid to slightly calcareous soil on banks by roadsides and in woodlands and quarries, occasionally also on stone bridges and derelict walls. It is especially frequent near the coast. Lowland. GB 95+52*, IR 6+7*.

Dioecious; capsules rare, ripe in spring and summer. No gemmae known.

France, Netherlands and S. Europe. Macaronesia, N. Africa, S.W. Asia. Also known from S. Africa (Cape Province).

A. C. CRUNDWELL

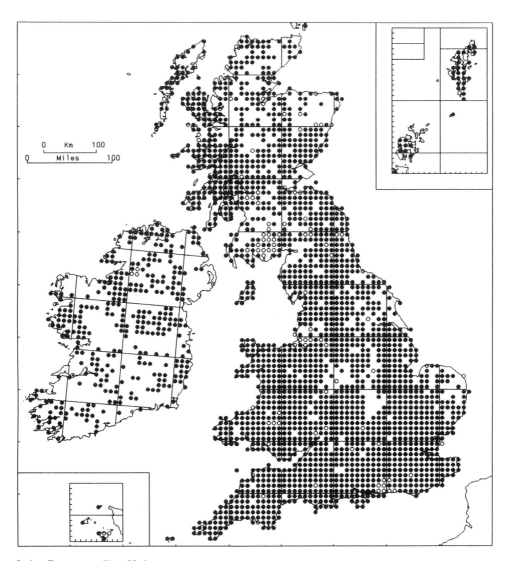

89/20. **Bryum capillare** Hedw.

In green tufts, sometimes tinged with red, on trees, fences, rotting wood, rocks, masonry and soil, especially where there is a little shelter. As an epiphyte it occurs on most species of tree. On the ground it grows on banks, tracksides, woodland rides and waste places, both on basic and slightly acid soils; also in turf. 0–610 m (Ben More). GB 2051+93*, IR 325+5*.

Dioecious; capsules frequent, ripening in spring and summer. Rhizoidal tubers frequently present. Gemmae develop on the protonema in culture (Duckett & Ligrone, 1992).

Found almost throughout the temperate and frigid zones of the Northern and Southern Hemispheres. There are also records from the tropics, but most of these are probably erroneous, being based on *B. torquescens*.

Var. *rufifolium* (Dix.) Podp. is, at its best, a very striking plant of dry limestone rocks, and some authors have been so impressed by the type specimen that they have wished to give it higher rank; but a continuous range of intermediates connects it to typical var. *capillare*.

A. C. CRUNDWELL

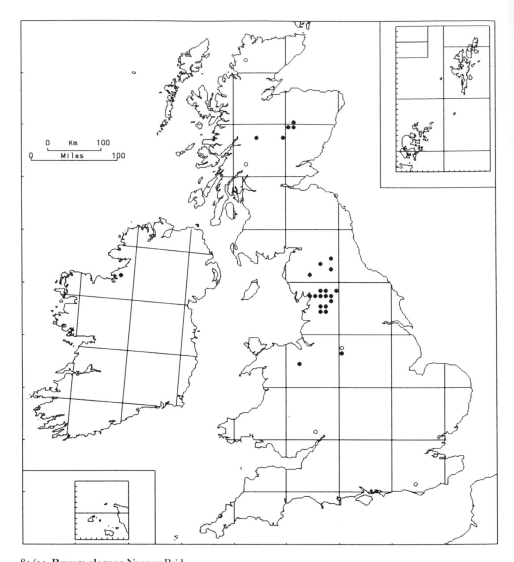

89/21. **Bryum elegans** Nees ex Brid.

In dense tufts, reddish or green tinged with red, on limestone rocks and in rock crevices, occasionally in thin limestone turf and on calcareous walls. It reaches 910 m on the Ben Alder group of mountains. GB 24+7*, IR 1.

Dioecious; capsules very rare. Brown tubers occur on the rhizoids but only rarely.

From Iceland and Arctic Scandinavia south to the mountains of C. Europe.

This species was frequently misunderstood by both British and continental authors before the paper of Syed (1973). Confusion with small forms of *B. capillare* has led to erroneous records and to treatment of *B. elegans* as a variety of *B. capillare* or even as a synonym of it. It may well have a wider distribution than is at present realized.

A. C. CRUNDWELL

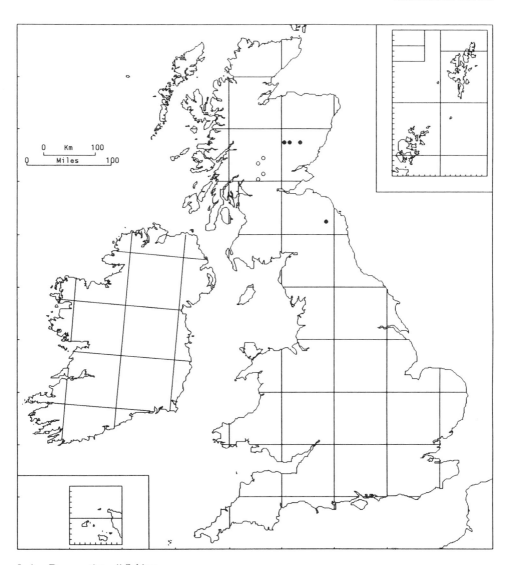

89/22. **Bryum stirtonii** Schimp.

As dense or loose tufts on soil, in springs and rock crevices, rarely on rocks. Certainly lime-tolerant, possibly a calcicole. A mountain plant reaching 1170 m on Ben Lawers and possibly not descending below about 500 m. GB 4+4*.

Dioecious; capsules uncommon; no gemmae reported.

Widely distributed in Scandinavia, also in Latvia and the mountains of Switzerland, Czechoslovakia and Bulgaria. Siberia, northern N. America (Alaska, Canada, Michigan).

This was not generally accepted as a good species before the publication of Syed's paper (1973). Several of the British records are old, but there is no particular reason to suppose that the species is becoming rarer.

A. C. CRUNDWELL

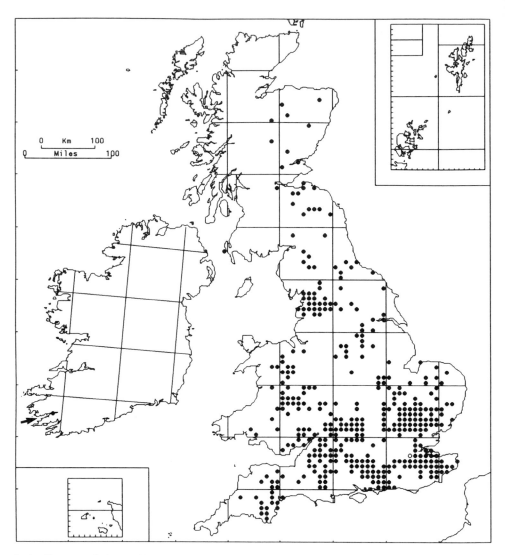

89/23. **Bryum subelegans** Kindb. (*B. flaccidum* auct. non Brid., *B. laevifilum* Syed)

Forms light green patches on trunks and branches of elder, ash, sycamore, maple and other trees. It occurs more rarely on stumps, rotten logs, rocks or soil. Lowland. GB 390+1*, IR 1.

Dioecious; fruit not known in Britain. Filamentous gemmae occur in the leaf axils, globular tubers on the rhizoids. Protonema-gemmae, resembling the axillary gemmae, are produced in culture on agar (Whitehouse, 1987).

From C. Scandinavia south to Spain and C. Europe. Turkey, N. America.

Not published as a British taxon until the revision of the *Bryum capillare* group by Syed (1973). Outside the British Isles, it is often not distinguished from *B. capillare*. The distribution indicated here is thus likely to be very incomplete. However, there are few records from well-recorded western areas such as Cornwall and N. Wales, suggesting that the species has a somewhat easterly distribution like that of *Aulacomnium androgynum*.

A. C. CRUNDWELL

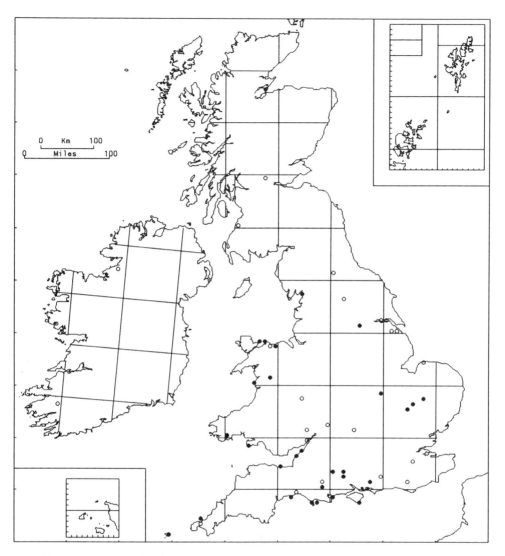

0 Km 100

0 Miles 100

89/24. Bryum torquescens Bruch ex De Not.

In green tufts sometimes tinged with red. It occurs on basic soil in grassland, on banks, at roadsides and on dunes, less commonly on rocks and walls, never on trees. Lowland. GB 29+22*, IR 2*.

Synoecious, rarely autoecious or dioecious; capsules frequent, ripening in spring. Large red tubers usually present on the rhizoids.

Common in S. Europe and rare in the north, though known from Gotland in Sweden. Known from every continent but tending to be in the warmer regions.

It is often confused with *B. capillare* and the details of its world distribution are thus imperfectly known. It was generally overlooked in Britain and Ireland until Syed (1973) published his revision; it must be under-recorded for the map.

A. C. CRUNDWELL

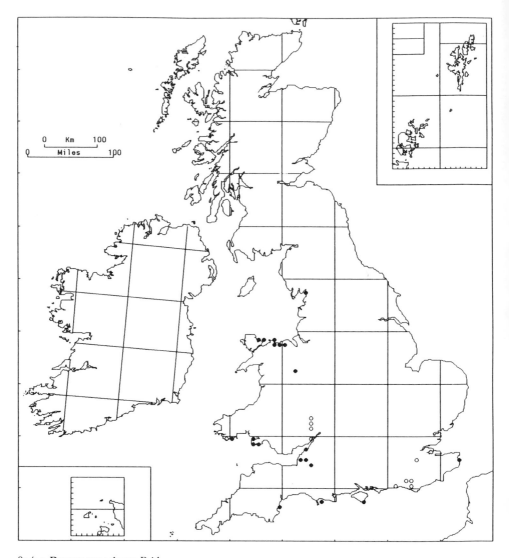

89/25. **Bryum canariense** Brid.

Dark green or reddish-green tufts on soil in crevices in dry limestone rocks, rarely on sand-dunes. Lowland. GB 20+10*.

Autoecious or occasionally synoecious; capsules occasional, ripe in spring. Red rhizoidal tubers usually present.

Common around the coasts of the Mediterranean and up the west side of Europe to Belgium. Macaronesia, W. and S. Africa, Pacific coasts of N. and S. America.

The world distribution given here is based on the generally accepted assumption that the autoecious and synoecious plant (*B. provinciale* Philib.) is synonymous with the dioecious non-British *B. canariense*. The latter is a smaller plant in which tubers are absent or rare and with some other reported vegetative differences (Pierrot, 1983). The status of *B. provinciale* must be re-examined.

A. C. CRUNDWELL

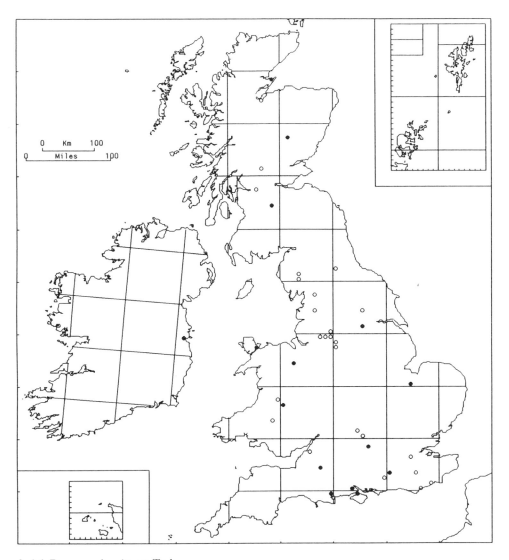

89/26. Bryum creberrimum Tayl.

In dense green tufts, reddish brown below, on soil on roadside banks and in waste places, occasionally on dunes and in crevices in rocks and walls. Lowland. GB 14+25*, IR 1.

Synoecious; capsules common, ripe in summer. Without gemmae.

Generally distributed in Europe from Svalbard southwards, and apparently a common weed in N. America. Asiatic and Australasian records are in need of confirmation.

Synoecious specimens of *B. pallescens* have sometimes been called *B. creberrimum* and there have consequently been some erroneous records. British and Irish herbarium specimens of these species were revised by Smith (1973, 1974a). The mapped distributions of both species are based on Smith's revision and on records made subsequent to his paper; they should not contain many errors.

A. C. CRUNDWELL

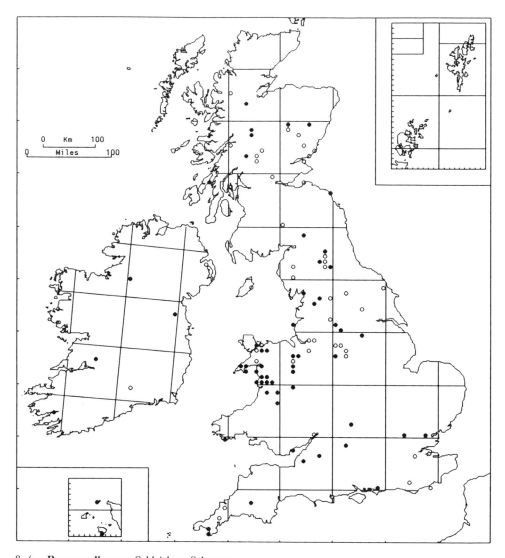

89/27. Bryum pallescens Schleich. ex Schwaegr.

In dense tufts, green above and reddish brown below, on soil in dunes and quarries, in crevices in rocks and walls, and on concrete and mine-waste. It is tolerant of heavy-metal contamination. 0–1205 m (Ben Lawers). GB 54+43*, IR 4+3*.

Synoecious or autoecious; capsules ripe in summer. Without gemmae.

Europe from Svalbard southward. W., C. and E. Asia, N. America, Greenland. C. and S. America, Falkland Islands, New Zealand.

A. C. CRUNDWELL

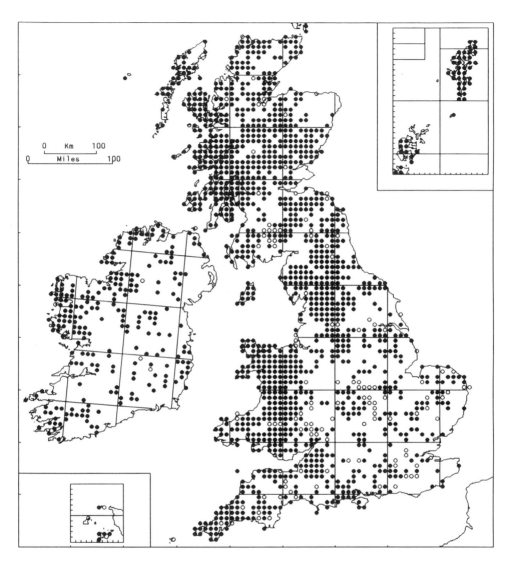

89/28. **Bryum pseudotriquetrum** (Hedw.) Schwaegr.

As green to reddish tufts and patches on wet soil in marshes, fens and flushes, in dune-slacks, by streams and pools, and on dripping rocks. It avoids extremely acid substrates, demands at least a little mineral content and tolerates lime. 0–1170 m (Ben Lawers). GB 1333+123*, IR 231+6*.

Dioecious or synoecious; fruit occasional, ripening in summer and autumn. Axillary filamentous gemmae are reported occasionally by European and American authors and have been found several times in Cambridgeshire (Whitehouse, 1983); similar gemmae are produced on the protonema in culture (Whitehouse, 1987).

Occurs almost throughout the temperate and frigid zones of both the Northern and Southern Hemispheres.

The species comprises var. *pseudotriquetrum*, which is dioecious, and var. *bimum* (Brid.) Lilj., which is synoecious and apparently has twice as many chromosomes. There seems to be little consistent difference in their frequency, in the frequency of sporophytes or in their ecology. Most plants are sterile and are thus not determinable to variety; in Skye (Birks & Birks, 1974) and North Wales (Hill, 1988), a large majority of fertile plants are dioecious.

A. C. CRUNDWELL

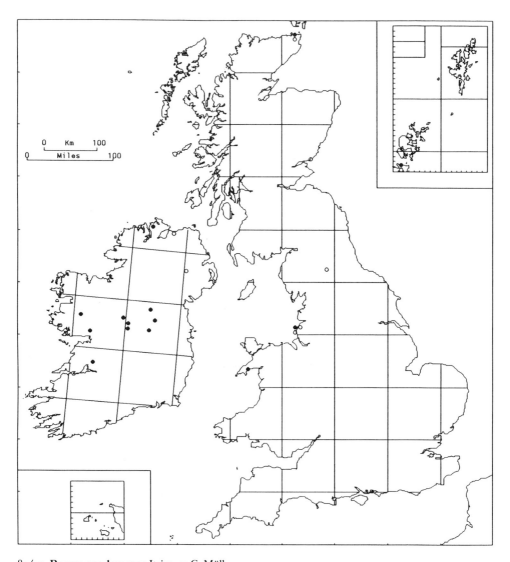

89/29. Bryum neodamense Itzigs. ex C. Müll.

As green or reddish-green, loose tufts or patches in calcareous fens, on wet limestone and in dune-slacks. Lowland. GB 3+5*, IR 10+2*.

Dioecious; fruit very rare, summer. Without gemmae.

From Iceland and N. Scandinavia south to the Pyrenees and mountains of C. Europe. Siberia, Altai, Alaska, Canada, Greenland.

A very distinctive species.

A. C. CRUNDWELL

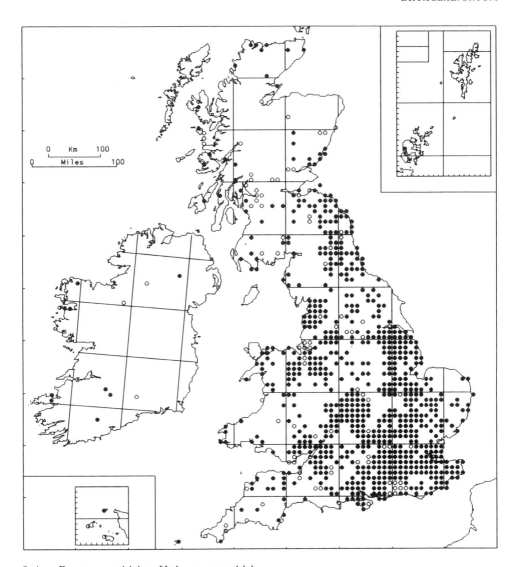

89/30a. Bryum caespiticium Hedw. var. **caespiticium**

A weedy species that grows in green or yellow-green tufts or patches on basic or neutral soil in dune-slacks, on waste ground, roadside banks and quarries, and on old walls and rocks. Lowland. GB 743+91*, IR 9+4*.

Dioecious; capsules common, ripe in summer. Mostly without gemmae but occasional gatherings have rather sparse inconspicuous chocolate-brown rhizoidal tubers.

An almost cosmopolitan species known from the frigid and temperate zones of both Hemispheres and from tropical mountain areas in Africa and America.

The above remarks apply to typical *B. caespiticium*, but there is another plant, probably specifically distinct, with small red tubers on the rhizoids. It remains distinct in culture. It was first found by Elsa Nyholm on waste ground in Stockholm and has since been found in chalk grassland in Suffolk, on a shady bank by a disused chalkpit in Cambridgeshire and in a stubblefield in Brittany. It is perhaps more strongly calcicolous than the typical plant.

A. C. CRUNDWELL

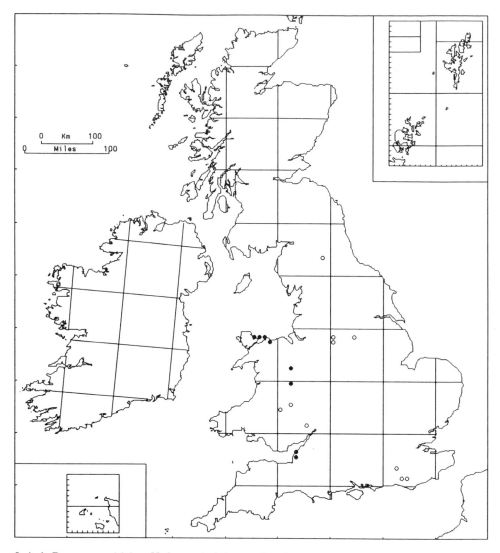

89/30b. **Bryum caespiticium** Hedw. var. **imbricatum** B. & S.

As pale green patches on calcareous rock-ledges and in thin calcareous turf in sunny places. Lowland. GB 8+10*.

Dioecious; fruit unknown in Britain. No gemmae reported.

Locally frequent on limestone in Provence (France) and probably widespread and frequent in S. Europe. Records from N. Africa and S.W. Asia need confirmation.

As pointed out by Hill (1988) this taxon is certainly worthy of specific rank, has little resemblance to var. *caespiticium* and is more likely to be confused, especially in the field, with *B. elegans*. Its name at the species level may be *B. kunzei* Hornsch., but this should not be used without study of the type, as many specimens so named are something else.

A. C. CRUNDWELL

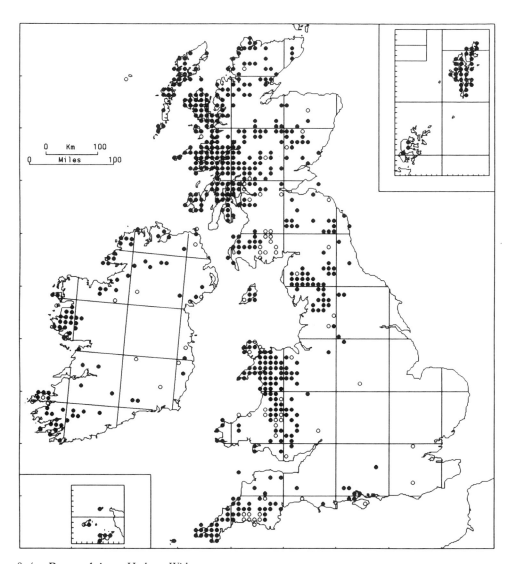

89/31. **Bryum alpinum** Huds. ex With.

In purple, red or, more rarely, green or variegated shining tufts on wet rocks and on bare acid or slightly basic soil by paths, on roadside banks, on lake shores and on peaty moorland. Found mainly at low elevations, ascending to 920 m (Coire an t-Sneachda). GB 533 + 63*, IR 82 + 14*.

Dioecious; fruit rare, summer. Red rhizoidal gemmae usually present in the tomentum; green protonemagemmae are produced in culture (Whitehouse, 1987).

Europe north to C. Norway and S. Sweden, with a few scattered records to 70°N on the Norwegian coast. Macaronesia, N. Africa, Asia Minor, C. Asia, Himalaya, N. America; also in Uganda, S. Africa, Madagascar, the mountains of Mexico and Peru, and in Argentina.

A badly-named plant, for though much commoner in the more mountainous parts of Ireland and Britain, it is most abundant at low altitudes. Green forms of *B. alpinum* have sometimes been given the name var. *viride* Husn. Most British specimens so called seem to be mere environmental modifications, mainly shade forms. In S. Europe one can find vigorous plants exposed to full sunlight and yet without red pigmentation. These may be worth recognition as a variety and this may be present in the British Isles.

A. C. Crundwell

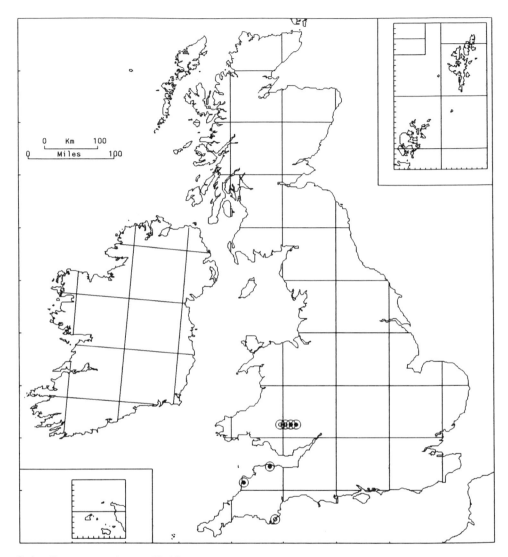

89/32. **Bryum gemmiparum** De Not.

In dense tufts, yellowish green above and reddish below, in rock crevices in the beds of streams and rivers. Lowland. GB 4+3*.

Dioecious; fruit unknown in Britain. Vegetative reproduction by bulbils, conspicuous in the axils of the upper leaves, and also by pinkish rhizoidal gemmae.

S. Europe. N. Africa, Cyprus, Turkey. Reported also from Belgium, Holland and S. Germany but these records need confirmation as do those from N. America.

Easily recognized when typical but there are problems. Plants without bulbils should not be given this name, nor plants with the bulbils atypical in form. Some specimens without bulbils have plentiful rhizoidal gemmae; these are not always present in bulbiliferous plants. The relationship of *B. gemmiparum* to the N. African *B. tophaceum* Dur. & Mont. needs elucidation.

A. C. CRUNDWELL

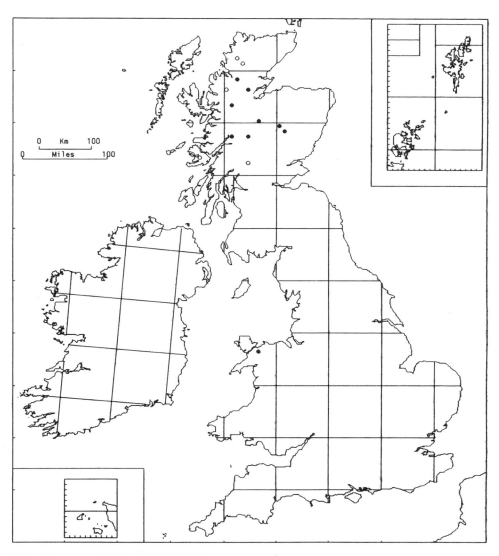

89/33. **Bryum muehlenbeckii** B., S. & G.

As dark greenish-red tufts on rocks in and by water in mountainous regions. Ascends to 800 m (Coire Cheap). GB 9+5*.

Dioecious; fruit unknown in Britain. Orange-red rhizoidal gemmae occur in the tomentum.

In Europe from N. Scandinavia south to C. Europe and the mountains of Spain and Corsica. Caucasus, N. America; also in S. Africa, Australia and Chile.

A. C. CRUNDWELL

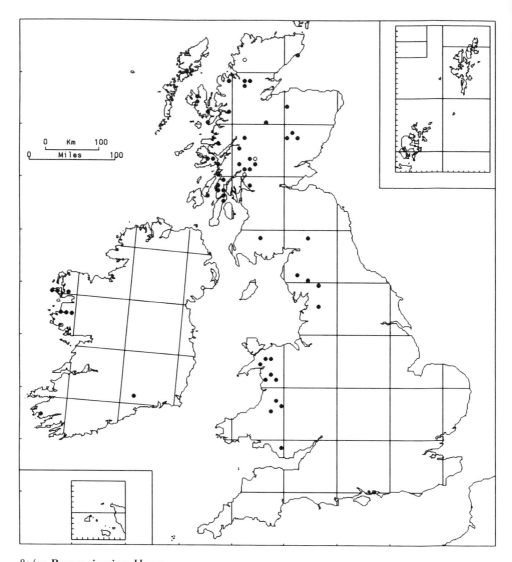

89/34. **Bryum riparium** Hagen

Occurs in a range of intermittently wet non-calcareous habitats, including crevices in rocks in stream-beds and beside streams where kept moist by seepage from sphagnum, on bare peat, on the sides of ditches, and in arable fields within the flood zone of streams. It appears to be a poor competitor with other plants, but takes advantage of any open acid habitat to which the rhizoidal tubers happen to be washed. From near sea-level (Lough Currane) to 760 m (Y Garn). GB 52+4*, IR 8+1*.

Only female plants known. Rhizoidal tubers abundant.

Known in Europe outside Britain and Ireland only from western Norway. There is a single unsexed record from eastern N. America (N. Carolina). Other species with flattened angular rhizoidal tubers occur in the Himalaya and Malaysia.

Although collected in Ireland in 1885, *B. riparium* was not generally recognized as distinct from *B. mildeanum* until revised by Whitehouse (1963). Like other tuberous moss species that were not distinguished till after the start of the Mapping Scheme, it must be somewhat under-recorded.

H. L. K. WHITEHOUSE

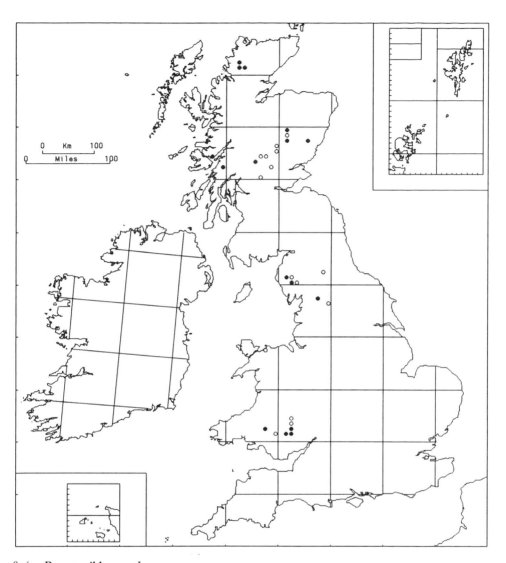

89/35. **Bryum mildeanum** Jur.

Found on rocks or sand by streams, and also away from rivers in limestone rock crevices, on earth-covered limestone, and on soil below basalt cliffs. Many of the habitats, perhaps all, appear to be on base-rich substrates. A montane species ranging from 70 m (bank of R. Wye, Whitney) to 850 m (Mickledore). GB 16+16*.

Dioecious; most British plants are female, males being known only from Ben Lawers and Crickhowell. Sporophytes are unknown in Britain and the plant has no obvious means of vegetative propagation.

Widespread in Europe in montane areas, with sporophytes occurring in C. Europe and Scandinavia. N. Africa, S.W. and S.E. Asia.

Many older records were erroneous. The map is based largely on specimens checked by Whitehouse (1963) after it was realized that *B. mildeanum* and *B. riparium* are distinct species.

H. L. K. WHITEHOUSE

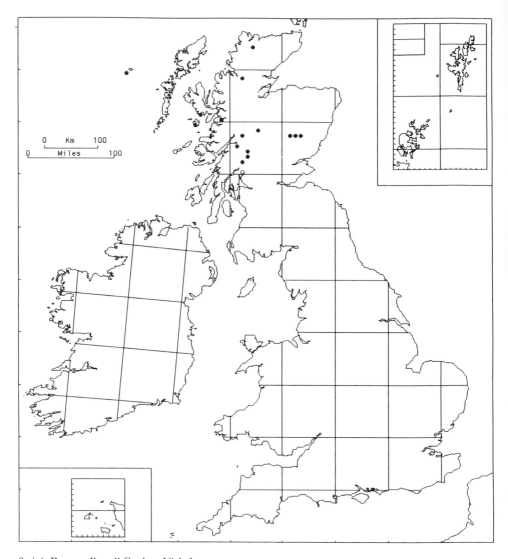

89/36. **Bryum dixonii** Card. ex Nicholson

It occurs in dense patches or small tufts among other bryophytes on damp basic montane rocks, usually schistose, and on basic schistose soil. It has been recorded on rocks in the bed of a stream, on rocks in flushed ground, on cliffs and rock scars, and in gullies. Recorded associates have included *Anomobryum filiforme, Bryum riparium, Dichodontium pellucidum, Diphyscium foliosum* and *Racomitrium ellipticum*. 30 m (St Kilda) to 850 m (Glen Coe). GB 15+1*.

Gametangia and sporophytes unknown. Axillary propaguliferous shoots sometimes present. In cultivation it produces tubers, but these have not been found in wild material (Whitehouse, 1992).

Possibly endemic to Scotland. According to Duell (1992) it has also been reported from Switzerland.

B. dixonii has been discovered in several new localities in recent years and it probably occurs in other sites in the Scottish mountains.

M. J. WIGGINTON

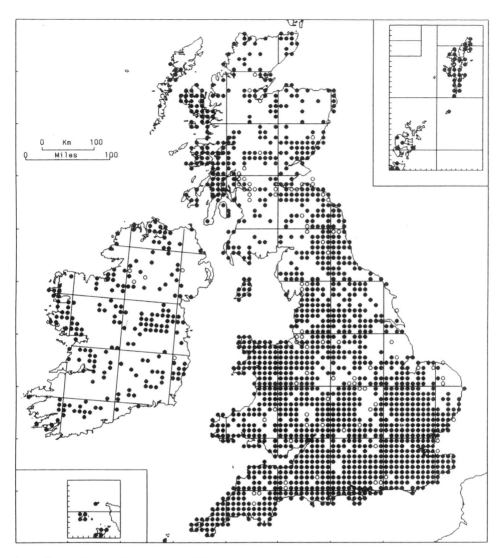

89/37. **Bryum bicolor** Dicks. (*B. barnesii* Wood)

A colonist in a wide variety of habitats. It occurs on compacted soil by paths and on roadsides, on tarmac, and on waste ground, often with *B. argenteum*. It grows in gardens, in fields, on walls, in quarries, on stonework, on cliffs, on periodically wet rocks by watercourses, on soil-banks by rivers and streams, on dunes, and on soil over limestone pavement. 0–490 m (Glen Coe). GB 1569+87*, IR 206+5*.

Dioecious; sporophytes occasional to frequent, maturing mainly in autumn and winter. Propagation is usually by caducous axillary bulbils, and, to a lesser extent, by tubers. Gemmae and lipid-packed bulbils develop on the protonema in culture (Duckett & Ligrone, 1992).

C. and N. Europe from Germany and Switzerland north to Norway and Iceland. Tunisia. Very widespread in other parts of the world but available data refer to the *Bryum bicolor* aggregate.

The map shows all records signified as *Bryum bicolor* on mapping cards but should contain very few errors, because other segregates recognized by Smith & Whitehouse (1978) are much less common. Several European authors (e.g. Wilczek & Demaret, 1976) treat *B. barnesii* as a distinct species.

M. J. WIGGINTON

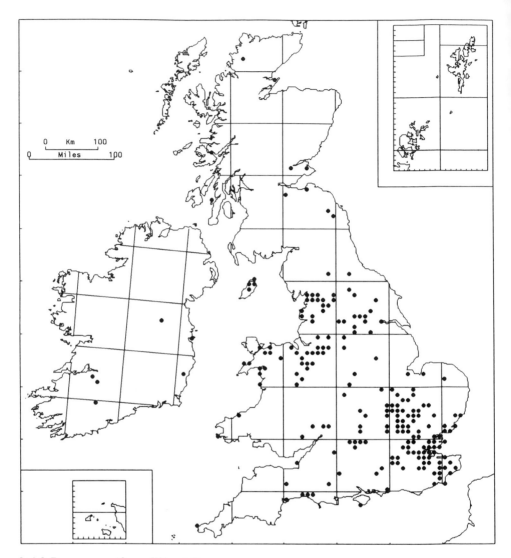

89/38. **Bryum gemmiferum** Wilcz. & Dem.

This species prefers a sandy, loamy, gravelly or rubbly substratum, but also grows on clayey or, rarely, chalky soil. It is found on banks, by ponds and ditches, in gravel workings and sand-pits, on waste ground, on cliffs by the sea, and on reservoir mud. Less frequent habitats include woodland rides, dune-slacks, chalkpits, railway ballast, cultivated ground and greenhouses. It is characteristic of eroding banks by rivers, where associates may include *Bryum bicolor*, *Ditrichum cylindricum*, *Pleuridium acuminatum*, *Pohlia annotina* and *Pseudephemerum nitidum*. In ruderal habitats recorded associates have included *Barbula convoluta*, *Bryum argenteum*, *Ceratodon purpureus* and *Funaria hygrometrica*. Mainly lowland, to 350 m (Llyn Brenig). GB 202, IR 6.

Dioecious; sporophytes occasional, ripening in spring or summer. Propagation mostly by axillary bulbils; tubers are sometimes produced in culture but are rarely if ever found in the wild (Smith & Whitehouse, 1978).

Western Europe east to Germany and the Netherlands. Canary Islands.

It was collected at Cassington near Oxford in 1948 but was not generally recognized as distinct until about 1965. Smith & Whitehouse (1978) published the notice of its occurrence in Britain and Ireland. It is still under-recorded.

M. J. WIGGINTON

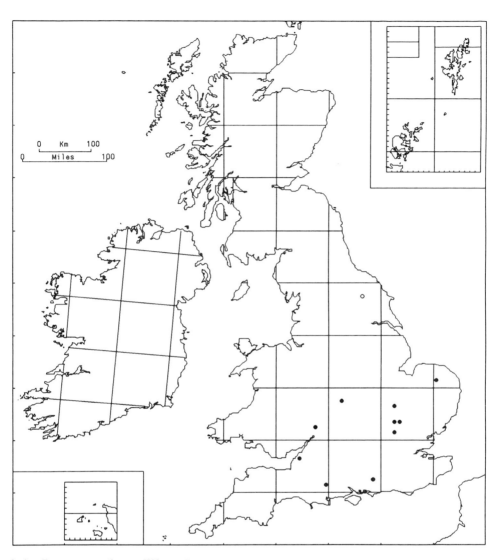

89/39. **Bryum gemmilucens** Wilcz. & Dem.

It has been found in arable and stubble fields, and in a non-calcareous loamy woodland ride. In an arable field on Oxford Clay in Cambridgeshire, it had a rich list of associates, including *Riccia* spp., *Acaulon muticum*, *Barbula unguiculata*, *Dicranella staphylina*, *D. varia*, *Ephemerum serratum* var. *minutissimum*, *Phascum cuspidatum* and *Pottia truncata*. These indicate a slightly acid soil. At another site, on a woodland ride, *Phascum cuspidatum* was also noted as an associate. Lowland. GB 10+1*.

Dioecious; gametangia not known in Britain, sporophytes unknown. Propagation by axillary bulbils.

Spain, France, Belgium, Hungary. Turkey, western N. America (California).

Although clearly rare in Britain it was not generally recognized as a distinct species until the notice of its occurrence by Smith & Whitehouse (1978). It is doubtless under-recorded both in Britain and in the rest of the world.

M. J. WIGGINTON

119

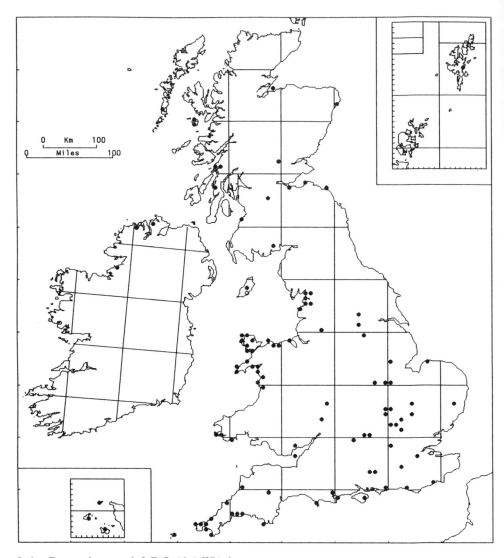

89/40. Bryum dunense A. J. E. Smith & Whitehouse

A species mainly of open, sandy or gravelly ground in coastal regions, sometimes where distinctly brackish, but also occurring on a wider range of soils including those that are strongly calcareous, and, more rarely, inland. It occurs on foreshores and dunes, on sandy cliffs and on waste ground, with associates such as *Barbula unguiculata*, *Bryum algovicum*, *B. bicolor*, *Ceratodon purpureus* and *Pottia heimii*; and on damp sandy clay with *Barbula tophacea*, *Pohlia annotina* and *P. carnea*. Other habitats include moorland near the coast, wet ground by an old railway, walls, a calcareous quarry floor, and a spoil-heap in a brickworks. Lowland. GB 98+1*, IR 3.

Dioecious; sporophytes occasional. Propagation commonly by deciduous axillary bulbils; tubers are produced in culture but are rarely if ever seen in wild plants (Smith & Whitehouse, 1978).

Spain, Italy and Greece north to Ireland and Sweden. Turkey.

First described by Smith & Whitehouse (1978), it is still somewhat under-recorded. Its European distribution is poorly known.

M. J. WIGGINTON

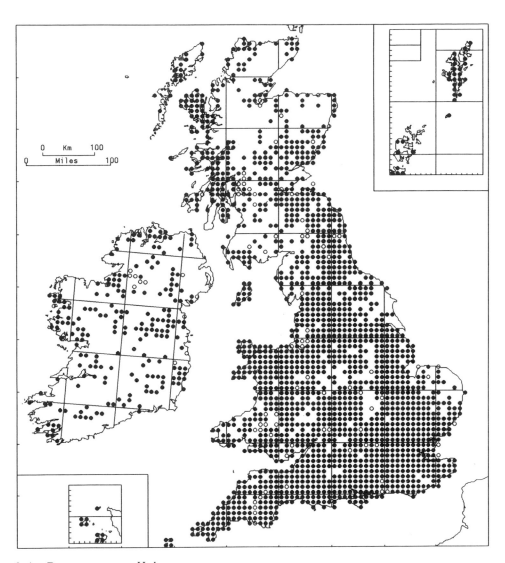

89/41. Bryum argenteum Hedw.

A pollution-tolerant ruderal, found in a wide range of habitats, particularly those that are rich in nitrogen. It grows on dry or moist sandy, silty or, less commonly, clay soils, often where trampled, by paths, on roadsides, between paving-stones, on waste ground and on railway ballast, where associated species may include *Barbula* spp., *Bryum bicolor*, *B. caespiticium*, *B. rubens* and *Ceratodon purpureus*. It occurs on stone walls, mortar of buildings, roofs and concrete slabs with such species as *Grimmia pulvinata* and *Tortula muralis*, and on asphalt roads and cultivated ground. It is not uncommon in natural habitats, including steep eroding banks by rivers and streams, shaly rock-scars, sand-dunes and cliff-slopes. It is drought-tolerant but grows luxuriantly on periodically wet rocks by rivers. 0–490 m (Ben Lawers). GB 1761+80*, IR 207+12*.

Dioecious; sporophytes occasional, mainly autumn to spring. It sometimes produces axillary bulbils.

Cosmopolitan; one of the most widely distributed species, occurring in all continents, including Antarctica, and extending even to remote oceanic islands.

Var. *lanatum* (P. Beauv.) Hampe is of doubtful taxonomic value (Longton, 1981) and is not mapped separately.

M. J. WIGGINTON

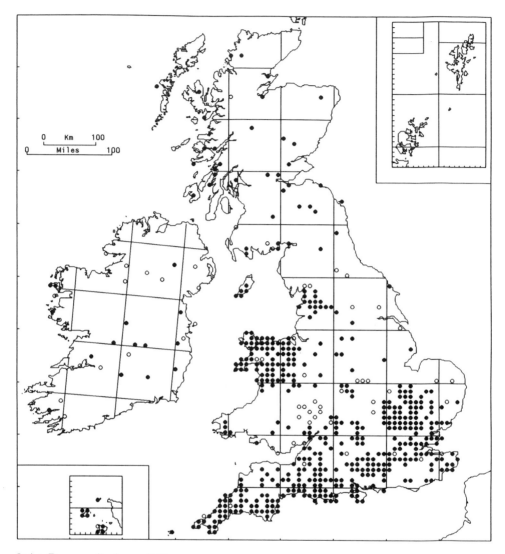

89/42. **Bryum radiculosum** Brid.

Occurs on limestone rocks, hard calcareous earth and the mortar of old walls. Lowland. GB 457+47*, IR 15+15*.

Dioecious; both sexes are widespread and sporophytes are frequent. Rhizoidal tubers are abundant and evidently provide a means of survival and vegetative propagation. Protonema-gemmae are produced in culture (Whitehouse, 1987).

S. Europe north to Scotland, Denmark and Czechoslovakia. Outside Europe it is recorded from the Canary Islands, Madeira, N. Africa, Palestine, Turkey, the Caucasus, S.E. Asia and Japan, and it is reported from Bermuda.

Although known in Britain since the nineteenth century under the name *Bryum murale* Wils., it was not recognized as a common plant until Whitehouse discovered that it produces tubers (Crundwell & Nyholm, 1964a). It is still under-recorded in many areas.

H. L. K. WHITEHOUSE

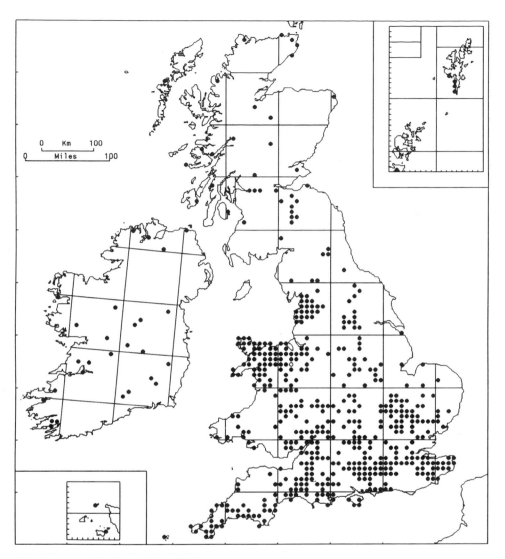

89/43. **Bryum ruderale** Crundw. & Nyh.

Occurs on trampled ground such as field entrances, roadsides and paths. It is not usually found in regularly disturbed ground such as arable fields. 0–310 m (near Colwyn Bay). GB 498+2*, IR 27.

Dioecious; male plants are widespread, females apparently less common. Sporophytes are very rare and have not been found in Britain or Ireland. Rhizoidal tubers frequent. Protonema-gemmae are produced in culture (Whitehouse, 1987).

Widespread in Europe but extending less far north and more frequent in the south than related species except *B. radiculosum*. Canaries, Egypt, Azerbaijan, N. America (Missouri, Texas, Ontario).

B. ruderale was rather seldom collected in Britain and Ireland before the revision of the *Bryum erythrocarpum* complex by Crundwell & Nyholm (1964a).

H. L. K. WHITEHOUSE

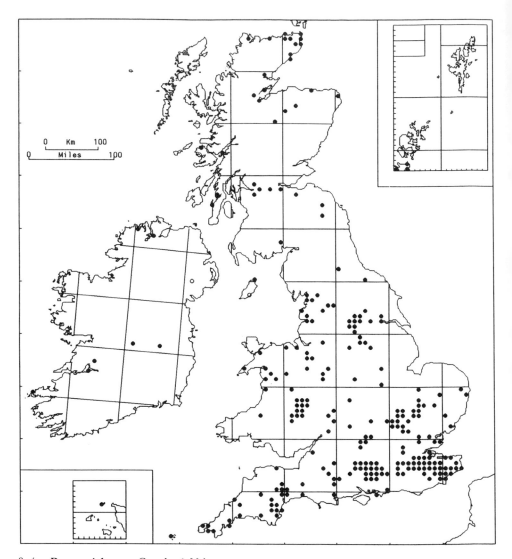

89/44. **Bryum violaceum** Crundw. & Nyh.

Occurs on calcareous to slightly acid, disturbed soil in temporarily open habitats such as arable fields, earth-banks, roadsides and waste ground. It is evidently an early colonist after disturbance. Lowland. GB 242+1*, IR 7+1*.

Dioecious; male plants are rare and sporophytes not known in Britain or Ireland. Rhizoidal tubers abundant.
Widely distributed in Europe. Canary Islands (Tenerife), Kashmir, Canada, U.S.A., Argentina (Patagonia).
An inconspicuous plant, hardly ever collected in Britain and Ireland before 1960, when interest in the *Bryum erythrocarpum* complex began to increase (Crundwell & Nyholm, 1964a). It is doubtless still often overlooked.

H. L. K. Whitehouse

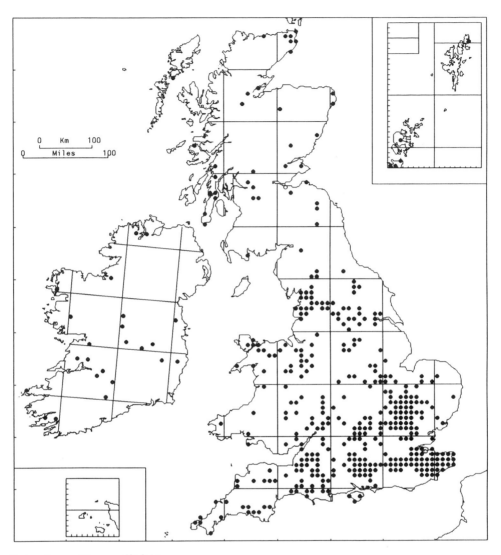

89/45. **Bryum klinggraeffii** Schimp.

Occurs on slightly acid to highly basic disturbed soil such as in arable fields. It is particularly frequent on bare ground beside pools and has often been detected in specimens of *Physcomitrella patens*. 0–350 m (beside Llyn Brenig). GB 395+1*, IR 23.

Dioecious; both sexes are widespread in Britain, but sporophytes have not been found in Britain or Ireland. They were produced, however, in a mixed culture on sterilized soil of plants grown from single tubers from material collected at Comberton, Cambridgeshire (female) and Crewe, Cheshire (male). Rhizoidal tubers abundant. Protonema-gemmae are produced in culture (Whitehouse, 1987).

Widespread in Europe. China, Japan, Canada, U.S.A., Argentina (Patagonia).

First reported from Britain by Crundwell (1962), but English specimens dating back to the nineteenth century have been detected in herbaria.

H. L. K. WHITEHOUSE

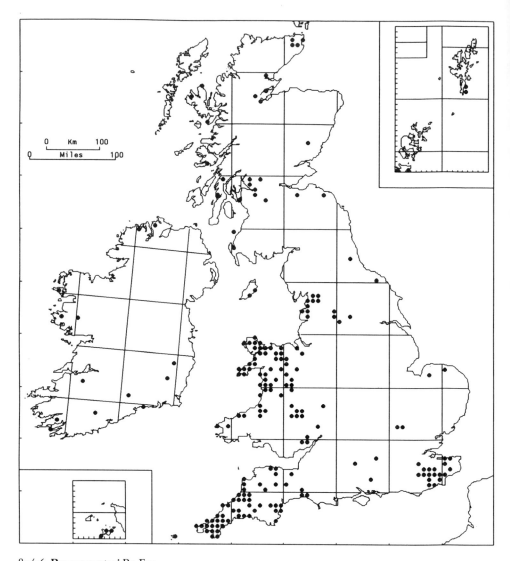

89/46. **Bryum sauteri** Br. Eur.

A plant of disturbed non-calcareous ground such as roadsides, banks, molehills and arable fields. Lowland. GB 192, IR 13.

Apparently dioecious in Britain and Ireland, with male plants and sporophytes not recorded. Synoecious or autoecious in C. Europe, where it is a montane plant. Rhizoidal tubers abundant. Protonema-gemmae are produced in culture (Whitehouse, 1987).

Outside Britain and Ireland the non-fruiting plant is recorded from western France, Canary Islands (Tenerife) and New Zealand. The fruiting plant is known from C. Europe, Norway and the Caucasus.

Reported as new for Britain by Crundwell (1962). An inconspicuous plant, it was very rarely collected in Britain before 1960 and is still under-recorded.

H. L. K. Whitehouse

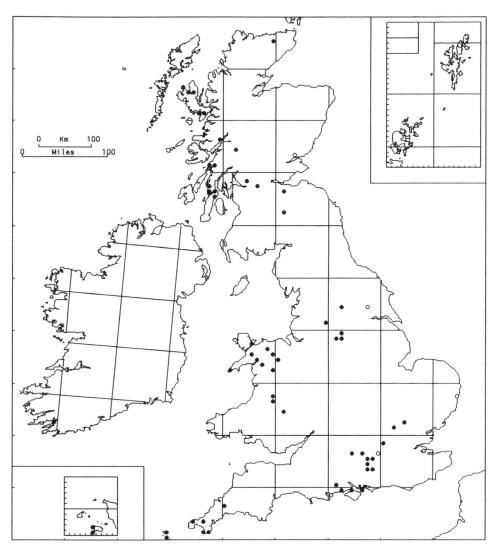

89/47. **Bryum tenuisetum** Limpr.

Occurs on acid sandy or peaty soils, often on heathland or at the margin of upland reservoirs, very rarely in acid arable fields. 0–350 m (beside Llyn Brenig). GB 56+5*, IR 1.

Dioecious or occasionally synoecious; both sexes and sporophytes occur in Britain. Rhizoidal tubers abundant. Gemmae develop on the protonema in culture (Duckett & Ligrone, 1992).

Widespread in C. and N. Europe. N. America.

It was rarely collected before being distinguished, new for Britain, in the revision of the *Bryum erythrocarpum* complex by Crundwell & Nyholm (1964a).

H. L. K. Whitehouse

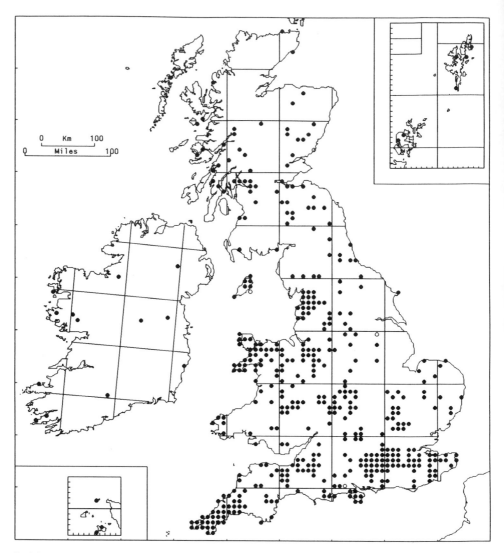

89/48. Bryum microerythrocarpum C. Müll. & Kindb. (*B. subapiculatum* Hampe)

Occurs on heathland and other sandy or peaty non-calcareous bare ground such as on molehills, the sides of paths and in arable fields. It is absent from large areas of moorland and generally avoids peaty soils unless they are subject to major disturbance such as forestry operations, ploughing, fire or turf-stripping. 0–310 m (Foel Gasyth). GB 438+4*, IR 13.

Dioecious; sporophytes frequent. Rhizoidal tubers abundant.

Widespread in C. and N. Europe. N. America, New Zealand.

A member of the *Bryum erythrocarpum* group, not clearly distinguished from *B. rubens* in Britain and Ireland until the revision by Crundwell & Nyholm (1964a).

H. L. K. Whitehouse

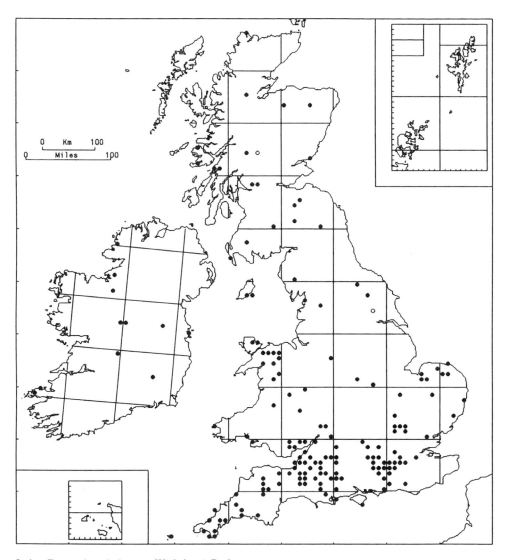

89/49. **Bryum bornholmense** Winkelm. & Ruthe

Occurs on sandy or peaty soil where relatively undisturbed. It tolerates shade, so is sometimes found under trees. There are no records from arable fields. 0–400 m (near source of R. Teme). GB 158+3*, IR 11.

Dioecious; sporophytes uncommon. Rhizoidal tubers abundant.

Widely distributed in Europe. One record from N. America (Texas).

This species was distinguished at a relatively late stage during studies of the *Bryum erythrocarpum* group by Crundwell & Nyholm (1964a). It lacks the experimental backing that other members of the complex received through being grown in pure cultures on agar. Records mapped here are probably based on more than one taxon.

H. L. K. Whitehouse

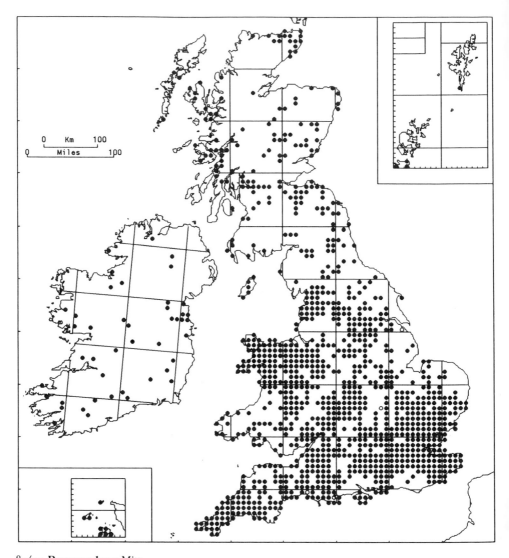

89/50. **Bryum rubens** Mitt.

Occurs on slightly acid to highly basic disturbed ground such as arable fields, molehills, bare patches at roadsides, in woodland rides, and in turf. It is usually found in eutrophic sites and can be very abundant in stubblefields where nutrient levels are high. 0–450 m (Snowdon). GB 1071+4*, IR 44.

Dioecious; both sexes are widespread in Britain and Ireland and sporophytes are occasional. Rhizoidal tubers abundant.

Widespread in Europe. Azerbaijan, India, Japan, N. America (rare, presumably introduced), New Zealand (presumably introduced).

Although described on the basis of English material by Mitten in 1856, it was not distinguished clearly from *B. microerythrocarpum* until the revision of the *Bryum erythrocarpum* complex by Crundwell & Nyholm (1964a).

H. L. K. WHITEHOUSE

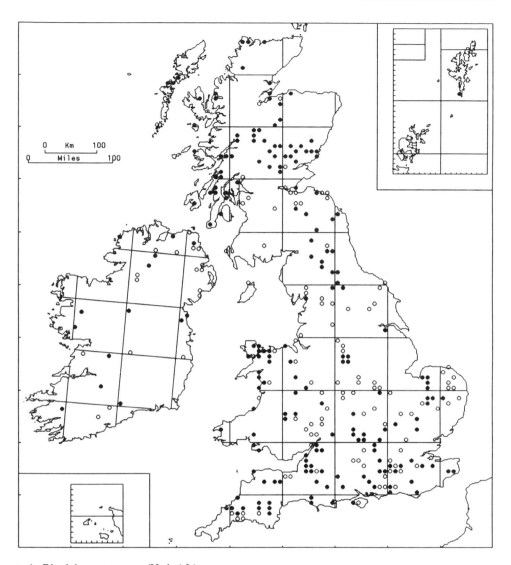

90/1. Rhodobryum roseum (Hedw.) Limpr.

This large, handsome moss is a calcicole of well-drained, sheltered ground in light shade. It is found most frequently in the shelter of higher plants on ant-hills in chalk and limestone grassland, especially where there is a thin layer of drift, and on calcareous dunes; less often it occurs in open woodland on light soils. In eastern England it is noted mainly from sandy ground, and can be locally plentiful in calcareous turf in Breckland. In the north and west it occurs also on lightly wooded banks, in gullies, on rock-ledges, and among stones on grassy hillsides. 0–550 m (The Storr). GB 184+102*, IR 14+17*.

Dioecious; capsules rare, ripe in winter. Colonies spread clonally by underground stolons.

Widespread in boreal, cool-temperate and montane parts of the Northern Hemisphere but apparently absent from eastern N. America. Most of Europe north to Iceland and N. Scandinavia.

The closely related *R. ontariense* (Kindb.) Kindb., which might possibly occur in Britain, is more strongly calcicolous and has a more southern distribution in Europe (Orbán & Pócs, 1976).

M. O. Hill

131

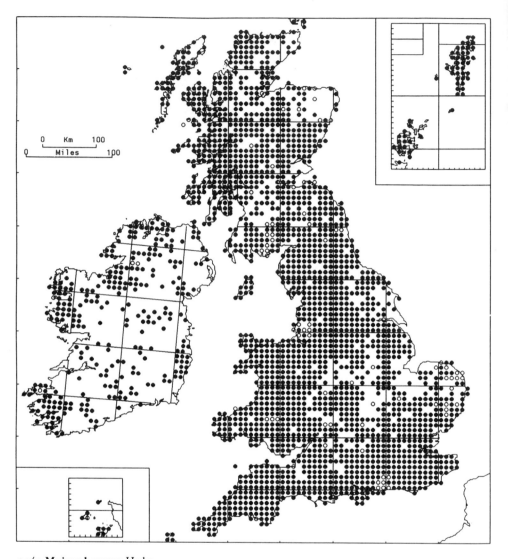

91/1. Mnium hornum Hedw.

As dark green or yellow-green tufts or patches, sometimes tinged with red in exposed situations, on soil in woods, on shady banks, on boles and roots of trees, on stumps and on rotting logs; also on rocks and in rock crevices. Calcifuge. It is mainly a lowland plant but is also quite frequent on mountains, ascending to 970 m on Ben Lawers. GB 2112+77*, IR 283+5*.

Dioecious; capsules common, ripe in early summer.

In Europe from Iceland and Scandinavia, where it is uncommon in the north, southwards. Algeria, Turkey, the Far East (Sakhalin, Japan), eastern N. America.

A. C. CRUNDWELL

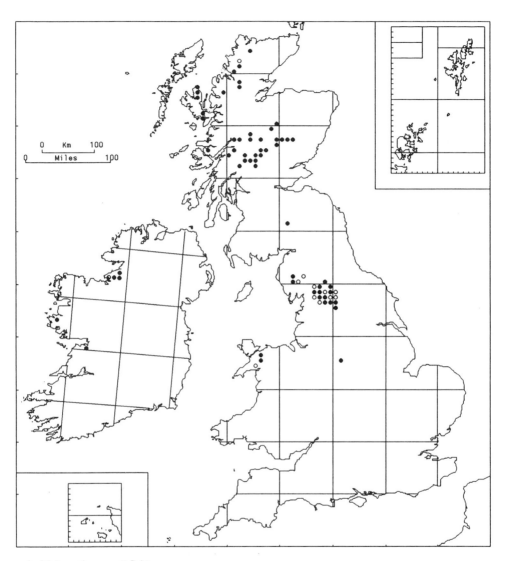

91/2. Mnium thomsonii Schimp.

In dense, pale green tufts or patches, reddish below, on basic rock-ledges and in crevices in basic rocks. A mountain plant reaching 900 m (Bidean nam Bian), but descending to 60 m on Skye. GB 54+11*, IR 5+1*.

Dioecious; capsules very rare, ripe in summer.

From Svalbard south to the mountains of C. Europe and Morocco. Iran, Caucasus, Urals, Arctic Siberia and C. Asia to Japan; N. America, Greenland.

M. thomsonii, M. ambiguum and *M. marginatum* are closely related species with similar ecology but different, though overlapping, altitudinal ranges. *M. marginatum* is found mainly at the lower levels, *M. ambiguum* at the highest, *M. thomsonii* at intermediate ones.

A. C. CRUNDWELL

133

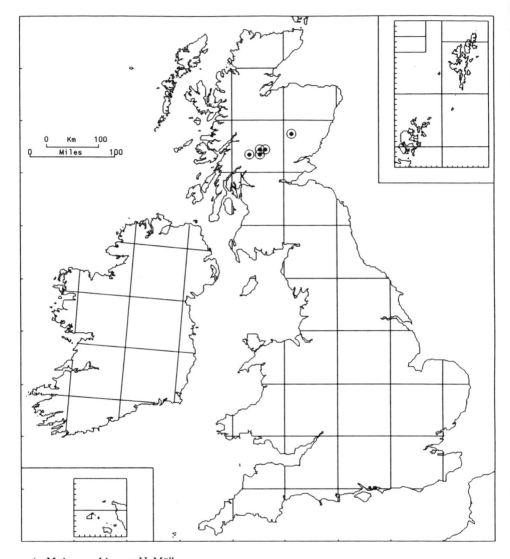

91/3. Mnium ambiguum H. Müll.

As green tufts, often tinged with red, in rock crevices and on ledges of damp mountain rocks. Calcicolous. Known only from a few localities at about 900 m; on Ben Lawers it occurs in the altitudinal range 800–1205 m. GB 5.

Dioecious; capsules very rare, summer.

From Iceland and northern Scandinavia south to the Pyrenees and the mountains of C. Europe. Himalaya east and north to China and Japan, N. America.

In N. America it is widespread, extending south to warmer localities than in the rest of its range and occurring sometimes on rotten wood and humus-rich woodland soils.

A. C. CRUNDWELL

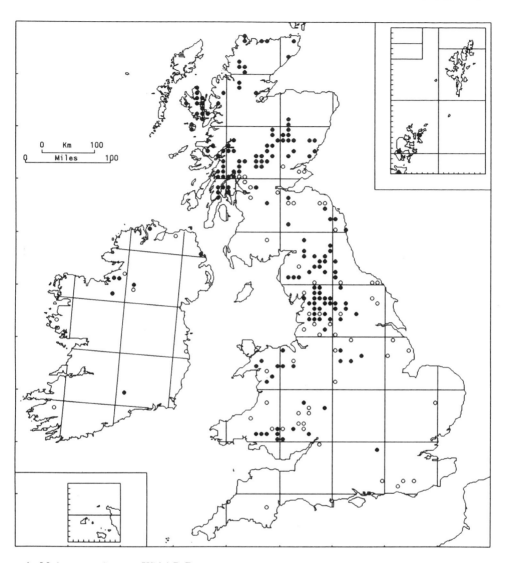

91/4. Mnium marginatum (With.) P. Beauv.

In green tufts, usually tinged with red, on sheltered damp (but not wet) soil and on rocks and in rock crevices in lightly shaded, sheltered situations. Calcicolous. A species of mainly lowland habitats, but recorded at 1170 m on Ben Lawers. GB 166+60*, IR 7+6*.

Usually synoecious, rarely dioecious; capsules frequent, ripe in early summer.

In Europe from Iceland and northern Scandinavia southwards. Siberia, Himalaya, China, Hawaii, N. America, Mexico and Guatemala.

Dioecious plants (var. *dioicum* (H. Müll.) Crundw.) are very rare and do not appear to differ from the synoecious var. *marginatum* in any morphological character except the inflorescence, nor in ecology or geographical distribution. A chromosome count would be interesting.

A. C. CRUNDWELL

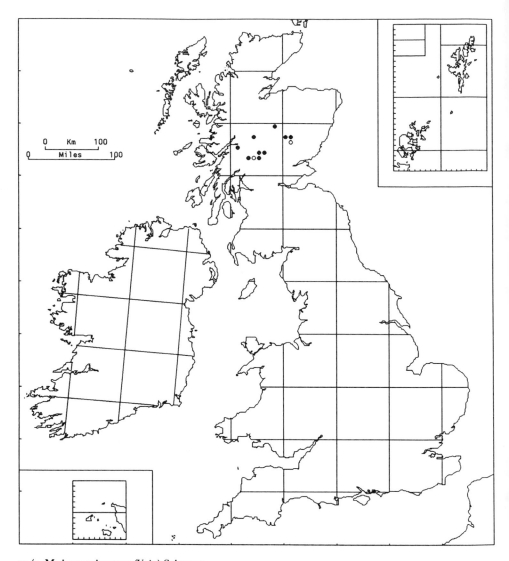

91/5. Mnium spinosum (Voit.) Schwaegr.

As dull green tufts or patches in turf among rocks, in rock crevices and on rock-ledges. Calcicolous. Mainly at higher altitudes, ascending to 1100 m (Ben Lawers). GB 9+2*.

Dioecious; fruit very rare, late summer.

In Europe from Svalbard and northern Scandinavia south to the Pyrenees and the mountains of C. Europe. Turkey, Caucasus, Iran, Himalaya, Siberia, E. Asia, western N. America.

A. C. CRUNDWELL

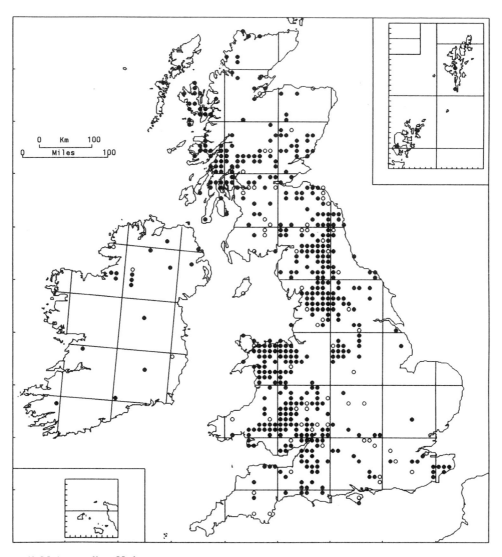

91/6. Mnium stellare Hedw.

A calcicole of well-drained substrata in shady situations, growing in dense bright or bluish-green tufts on sheltered woodland banks, in turf, on old walls and rocks, in rock crevices and on rock-ledges. Most frequent at low altitudes but reaching 490 m in Skye. GB 462+66*, IR 16+3*.

Dioecious; capsules very rare, ripe in spring.

From Iceland and northern Scandinavia southward. Morocco, Algeria, Turkey, Caucasus, Iran, Himalaya, E. Asia, eastern N. America.

A. C. CRUNDWELL

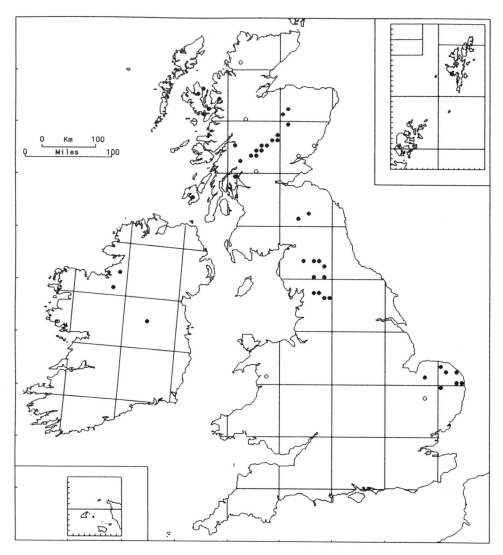

92/1. Cinclidium stygium Sw.

In densely tomentose, green or reddish tufts or patches in calcareous fens and marshes and in basic springs and flushes. It ascends to 970 m on Ben Lawers. GB 41+7*, IR 3.

Synoecious; capsules uncommon, ripe in summer.

From northern Scandinavia south to the Alps. Siberia, N. America south to Michigan, Greenland, southernmost S. America.

A boreal and Low-Arctic species that was more widespread in Late Glacial times than now. The East Anglian records may be regarded as relict.

A. C. CRUNDWELL

138

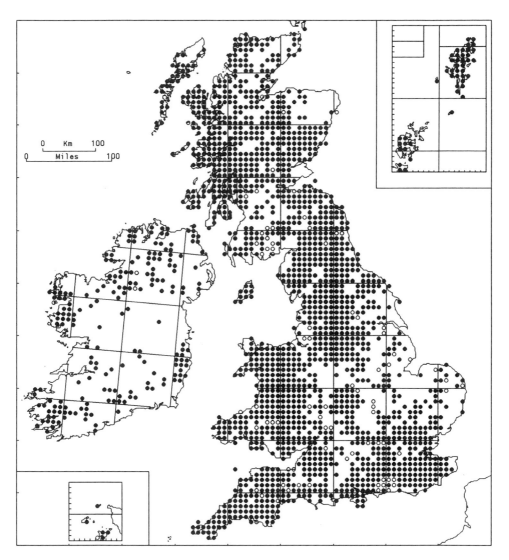

93/1. Rhizomnium punctatum (Hedw.) Kop.

As loose tufts, dark green above, deep brown below, on soil in fens and marshes and damp woodland, and on rotting logs; also in springs and mountain flushes, beneath dripping rocks and on wet rock-faces. Frequent also as a persistent chocolate-brown protonema, sometimes with a very few depauperate leafy shoots, on stones, damp rocks and walls in deep shade. Found in slightly acidic to strongly basic habitats. Equally frequent in lowland habitats as on the mountains, reaching 1165 m on Ben Lawers. GB 1713+83*, IR 193+4*.

Dioecious; capsules frequent, ripening in spring.

Almost throughout Europe, but rare in the extreme north and the extreme south. Macaronesia, N. Africa, Turkey, Iran, Siberia (Baikal). In N. America represented by ssp. *chlorophyllosum* (Kindb.) Kop.

A. C. CRUNDWELL

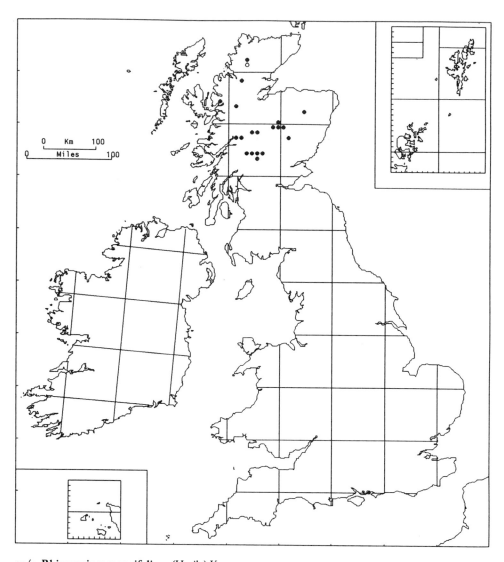

93/2. **Rhizomnium magnifolium** (Horik.) Kop.

As dull green patches on wet ground. In high mountain habitats a calcifuge of areas of late snow-lie, of sheltered gullies and of stony ground below dripping rocks. In flushes, sometimes at lower altitudes, tolerant of mildly basic conditions. 420 m (Allt Mhainisteir) to 1020 m (Feith Buidhe). GB 19+1*.

Dioecious; capsules known in Britain.

A boreal species known from Fennoscandia, the adjacent regions of Russia, the Pyrenees and the mountains of C. Europe. Siberia, Himalaya, E. Asia, N. America from Alaska to New Mexico in the west and from Labrador to Michigan in the east, Greenland.

Not reported as a British species until after the beginning of the BBS Mapping Scheme (Crundwell, 1978) and probably under-recorded owing to past confusion with *R. punctatum*. A fine account of its ecology and distribution in Britain was published by Long (1982a). Its foreign distribution suggests that it may be found in some more southern and less montane habitats than those from which it is at present known.

A. C. CRUNDWELL

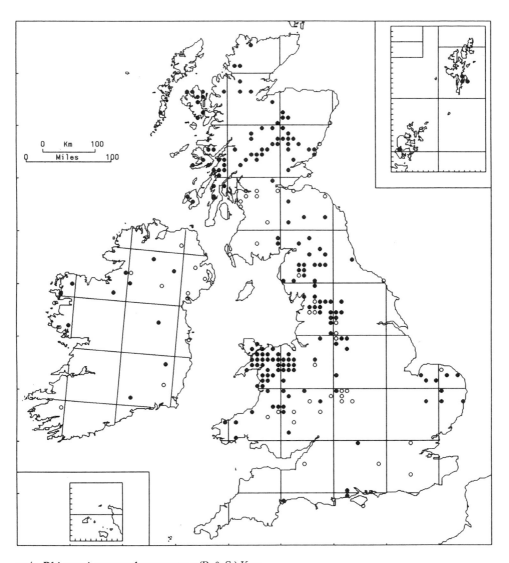

93/3. **Rhizomnium pseudopunctatum** (B. & S.) Kop.

In dark green patches in fens, marshes and basic flushes, often associated with *Drepanocladus revolvens* and *Scorpidium scorpioides*. It reaches 970 m on Ben Lawers. GB 187+43*, IR 11+11*.

Synoecious; capsules frequent, ripening in winter.

From northern Scandinavia and neighbouring regions of Russia to the lowlands of C. Europe and the S. European mountains. Siberia, N. America from Alaska and Labrador south to Colorado and Michigan, Greenland.

Easily confused with other species of *Rhizomnium*, but the map probably contains very few errors.

A. C. CRUNDWELL

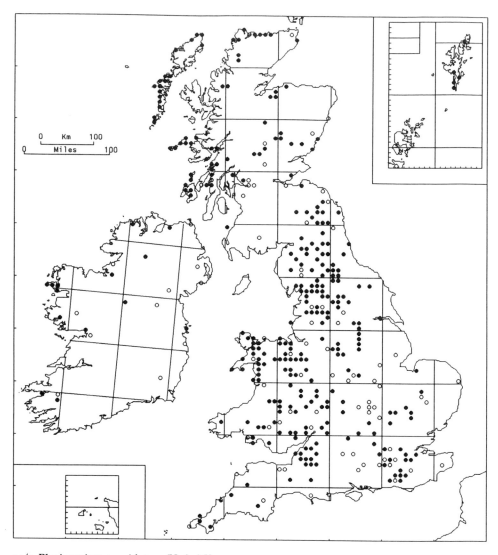

94/1. Plagiomnium cuspidatum (Hedw.) Kop.

As dark green tufts or patches on stumps and boles of trees, on shaded rocks and walls, and in damp basic turf. A mainly lowland species, reaching 650 m near Blair Atholl (Meall Breac). GB 289+73*, IR 14+11*.

Synoecious; capsules frequent, ripe in spring.

From the Pyrenees to northern Scandinavia. N. Africa, Uganda, Himalaya, Siberia, E. Asia, N. America, Mexico, Cuba, Greenland.

A declining species, now uncommon in most of its British range though remaining frequent on the Magnesian limestone of N. England. Destruction of habitats can only partially explain the decline. It was probably never as common in the British Isles as in continental Europe or in N. America.

A. C. CRUNDWELL

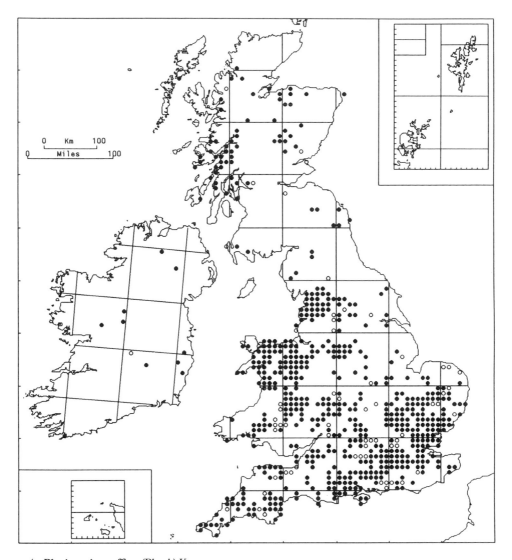

94/2. Plagiomnium affine (Bland.) Kop.

In loose green patches on damp basic or slightly acid ground in woods, on shady tracksides, on logs, in turf and sometimes on earth-covered rocks. Lowland. GB 572+63*, IR 10+1*.

Dioecious; fruit very rare, ripe in spring.

From Madeira and the mountains of southern Europe north to southern Scandinavia. Turkey, Iran.

The name *Mnium affine* was formerly used in an aggregate sense, including also *Plagiomnium elatum*, *P. ellipticum* and *P. medium*. In the early stages of the BBS Mapping Scheme, this confusion had not completely been eliminated, and some mapped records of *P. affine*, especially those from marshes, must be erroneous. Nowadays *P. affine* is not very often confused with these species, but in the field it is very easily overlooked as stunted *P. undulatum*.

A. C. CRUNDWELL

143

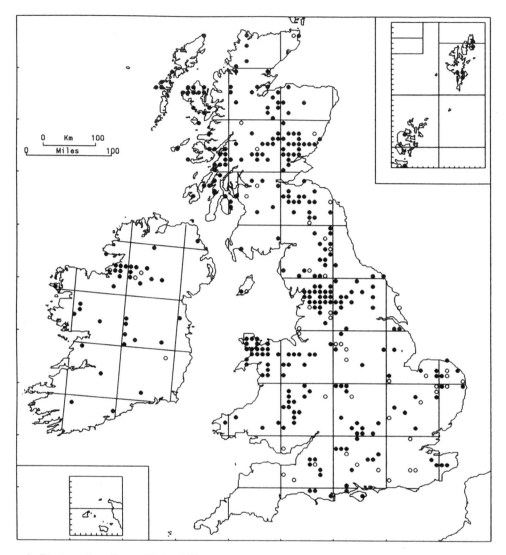

94/3. Plagiomnium elatum (Br. Eur.) Kop.

Forming patches, sometimes extensive, in very wet calcareous fens and flushes, beside ditches and streams, and in reed-swamps, especially with *Phragmites* or *Juncus subnodulosus*. A mainly lowland species, but recorded from 850 m in the eastern Highlands of Scotland (Caenlochan Glen). GB 294 + 48*, IR 35 + 4*.

Dioecious; fruit not known with certainty in Britain or Ireland.

From the Alps to Iceland, northern Scandinavia and adjacent parts of Russia.

Formerly confused with *P. affine* and so probably under-recorded.

A. C. CRUNDWELL

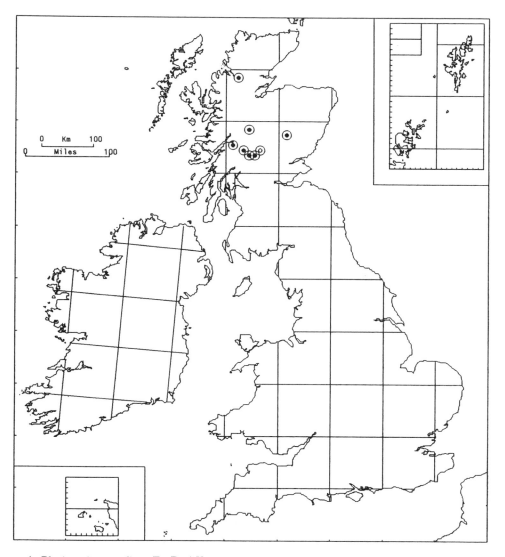

94/4. **Plagiomnium medium** (Br. Eur.) Kop.

In lax, green tufts in basic flushes and on damp calcareous rock-ledges at high altitudes. Associated species include *Saxifraga hypnoides* and *Hylocomium splendens*. It is at 1070 m on Bidean nam Bian, and is not known below 800 m. GB 7+1*.

Synoecious; capsules unknown in Britain.

Generally distributed in Europe from the Pyrenees northwards. Morocco, Caucasus, Siberia, E. Asia, N. America, Greenland.

Though discovered in Scotland in 1899 this species was omitted from British lists for many years and was consequently passed by as *P. affine* (Crundwell, 1957). Outside Britain *P. medium* is mainly a plant of boreal woodland, where it grows around springs and on stream-banks.

A. C. CRUNDWELL

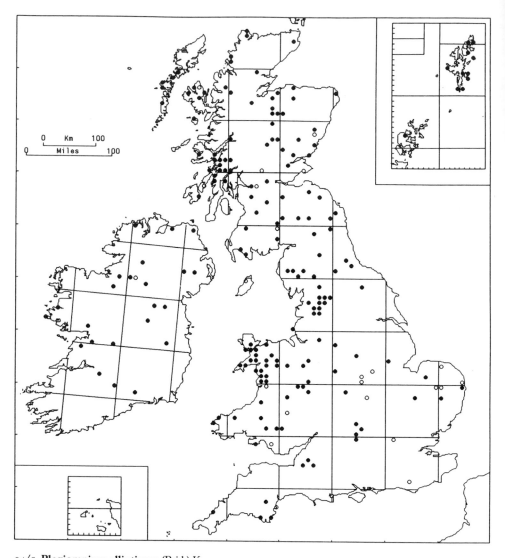

94/5. **Plagiomnium ellipticum** (Brid.) Kop.

In green patches or tufts in flushes and by springs, on stream-banks, in damp grassland and in wet woodlands. It avoids very acid habitats and does not demand such a high base-status nor quite so much moisture as *P. elatum*, though their habitats overlap. Mainly at low altitudes, reaching 550 m in Skye. GB 167+22*, IR 23+1*.

    Dioecious; capsules very rare, ripe in summer.

    Circumboreal; ranging in Europe from Svalbard and Iceland south to the Alps, and common in N. America. Known also from Argentina and Chile.

    Probably under-recorded in the British Isles because of past confusion with *P. affine*.

A. C. CRUNDWELL

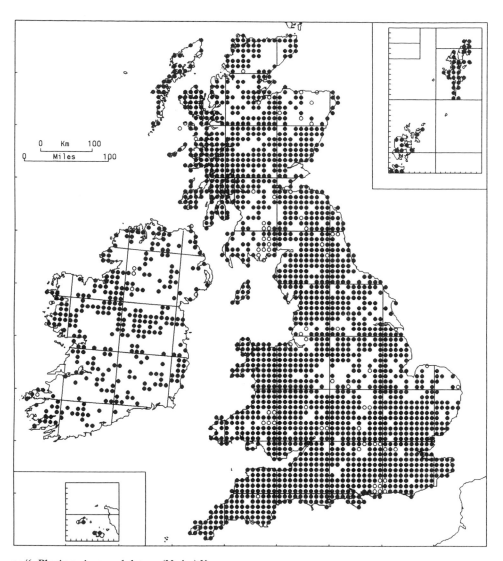

94/6. **Plagiomnium undulatum** (Hedw.) Kop.

Abundant in woodland on damp basic or neutral soils, on shady lane-banks and on dripping rocks in sheltered glens. Frequent also in damp grassland, especially in churchyards and damp lawns, where it is often very stunted. A mainly lowland plant, but it reaches 850 m on Creag Meagaidh. GB 1971 + 72*, IR 308 + 7*.

Dioecious; capsules rare but often abundant when present, ripe in spring. Male plants are less frequent than female and produce gametangia less frequently (Newton, 1971).

In all Europe except Arctic Scandinavia. Morocco, Algeria, Tunisia, S.W. Asia.

A. C. CRUNDWELL

147

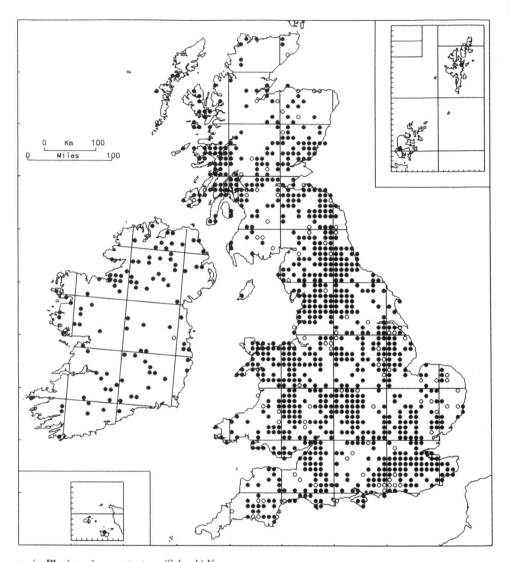

94/7. **Plagiomnium rostratum** (Schrad.) Kop.

As loose, green patches on shaded basic rocks and base-rich soil, and on damp walls. In the north of England it is especially characteristic of river-banks, occurring on gritty ground about tree-roots in places that are infrequently flooded. In the main a plant of low altitude, but ascending to 460 m on Snowdon. GB 934 + 101*, IR 100 + 2*.

Synoecious; fruit common, spring.

Europe south of the Arctic Circle except Iceland and the Faeroes. From Turkey, the Caucasus and Iran through to the Himalaya and N.W. China; temperate N. America, Mexico.

Sterile plants can be difficult to distinguish in the field from *P. affine*, and this may have led to some errors. Records from marshes are particularly suspect.

A. C. Crundwell

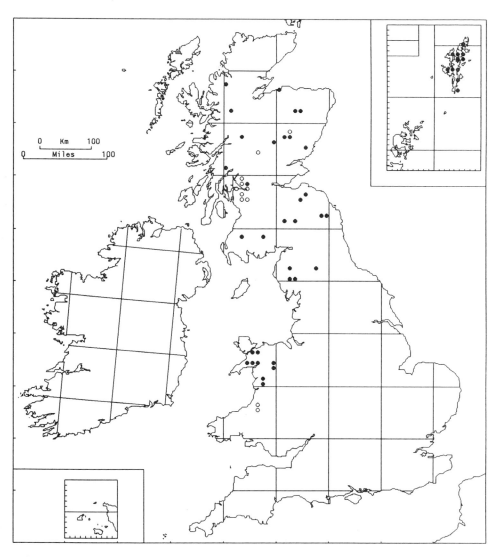

95/1. Pseudobryum cinclidioides (Hüb.) Kop.

As bright green patches or swards, deep brown below, in marshes, carr and wet woodland by the shores of lakes, and in flushes and springs and by peaty pools upon the hills. It ascends to 800 m (Glen Coe). GB 44 + 11*.

Dioecious; capsules very rare, ripe in early summer.

From the mountains of C. Europe to Iceland and the north of Scandinavia. Siberia, Himalaya, Sakhalin, China, Japan, N. America, Greenland.

A. C. CRUNDWELL

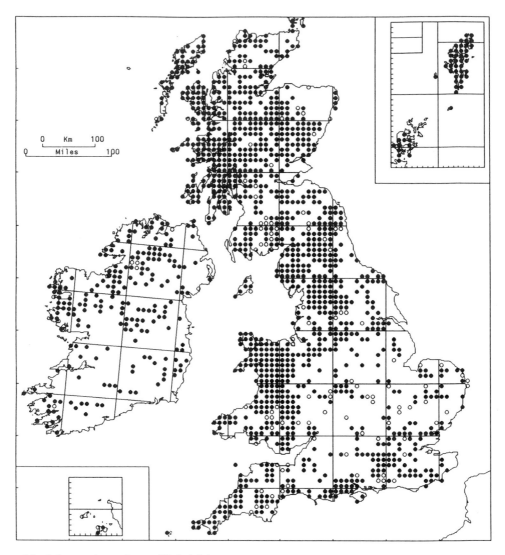

96/1. **Aulacomnium palustre** (Hedw.) Schwaegr.

A. palustre is found on damp, more-or-less acid, usually peaty ground in a wide range of habitats. These include moist grassland, heathland, moorland and meadows, communities dominated by *Carex rostrata*, valley-bogs, blanket-bogs, fen carr and open woodland. Although present in many base-poor habitats, it is most frequent in sites which receive a supply of base-enriched water. In lowland England *A. palustre* has been found in damp sand- and gravel-pits and disused railway cuttings. 0–975m (Lochnagar). GB 1202+101*, IR 207+7*.

Dioecious; sporophytes occasional, maturing in summer. Leaf-like gemmae are sometimes produced on elongated stem-apices, perhaps when conditions are unfavourable.

Widespread in Europe, extending north to Svalbard, but absent from the drier south. Found throughout the Northern Hemisphere and in S. America, Australia and New Zealand.

Habitat destruction has led to the loss of many sites for this species in S. England. It has, however, invaded some newly-available habitats, including brick-pits at Chawley (Jones, 1986) and carr at Wicken Fen (Lock, 1990).

C. D. Preston

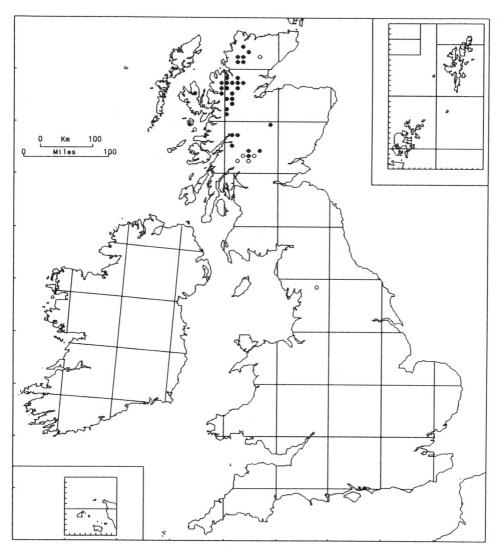

96/2. Aulacomnium turgidum (Wahlenb.) Schwaegr.

A montane species, found in short vegetation over base-rich rocks in sites where the development of a taller sward is prevented by exposure or grazing. It grows in *Racomitrium lanuginosum* heath, amongst dwarf herbs such as *Silene acaulis* and in *Agrostis capillaris-Festuca ovina* turf. These communities are usually species-rich. Some sites are subject to cryoturbation; *A. turgidum* is often found on the well-drained edges of soil-creep terraces. Usually between 700m and 1000m (Beinn Fhada), but descends to 430m (The Storr). GB 32+7*.

Dioecious; sporophytes have not been found in Britain and the species is not known to produce gemmae.

Circumpolar arctic-alpine; common in the Arctic and extending south on mountains to E. Africa, the Himalaya, Japan and Mexico.

Subfossil *A. turgidum* has been found in Britain at sites south of its current range (Dickson, 1973). In Europe it was widespread during the last glacial, and it has been detected in the stomach of a frozen woolly mammoth (*Mammuthus primigenius*) which roamed the Siberian tundra-steppe some 40,000 years ago (Farrand, 1961; Ukraintseva, 1986).

C. D. PRESTON

151

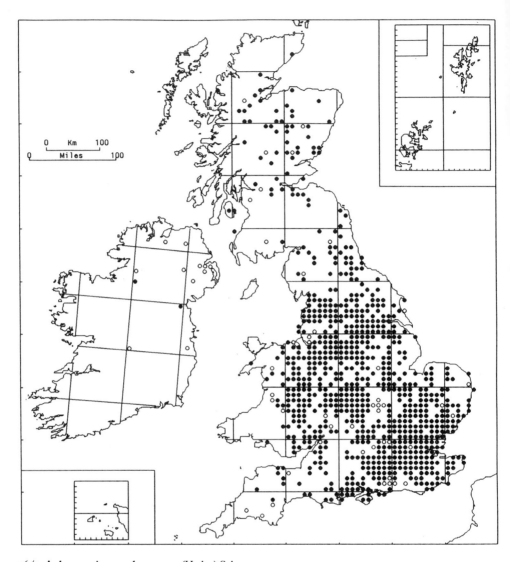

96/3. **Aulacomnium androgynum** (Hedw.) Schwaegr.

This species is usually found on decaying organic matter, being most frequent on rotting tree-stumps but also colonizing logs, twigs, old grass tussocks, fence-posts, pollard willows, peat-diggings, and banks in hedgerows and woods where the soil has a high humus content. Other habitats include the bases of acid-barked trees and the inclined trunks and branches of elders and willows. It is locally frequent on sandy banks and sandstone. Lowland. GB 775+45*, IR 2+11*.

Dioecious; capsules very rare, summer. Gemmae, in spherical clusters at the stem apices, abundant; morphologically different gemmae are produced on the protonema in culture (Duckett & Ligrone, 1992).

Widespread in W. and C. Europe, north to 63°N in Norway; rare in S. and E. Europe. Canaries, N. Africa, W. Asia, Japan, Korea, N. America, Argentina (Patagonia).

A. androgynum is known to have increased in frequency since the beginning of the century in some well-recorded counties. It has also spread to new habitats. In Oxfordshire it had only once been recorded on elders and had never been seen on willow before 1952; it is now frequent on both species (Jones, 1991).

C. D. Preston

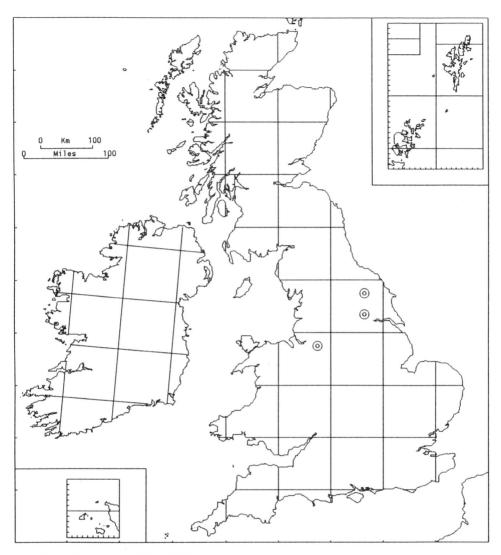

97/1. **Paludella squarrosa** (Hedw.) Brid.

In rich to intermediate fens, usually with cool, mineral-rich spring or seepage water percolating up but not inundating the bryophyte layer. Specimens from Terrington North Carr, a site described as a small swampy hayfield, include abundant *Homalothecium nitens*, with *Calliergon cuspidatum* and *Plagiomnium elatum*. Associates at the other British localities are unknown. Two of its three localities were shared with *Helodium blandowii* but the two apparently did not occur intermixed. Lowland. GB 3*.

Dioecious; sporophytes found only once, at Terrington in 1854 (specimen in BM).

Circumpolar, mainly in the Arctic and northern boreal zones, scattered south to the Alps, Carpathians, Kazakhstan, Mongolia, Japan and northern U.S.A. Considered a glacial relic in C. Europe, extinct in the Netherlands and rare and scattered in Germany.

Paludella was a common mire moss in northern Britain after the last Ice Age, forming extensive peat in some sites (Dickson, 1973). During the last 5000 years its habitats disappeared and it was latterly restricted to three sites: Knutsford Moor (found only in 1832), Terrington North Carr (drained 1861, a few plants remaining till 1868), and Skipwith Common (found only in 1916).

F. J. RUMSEY

153

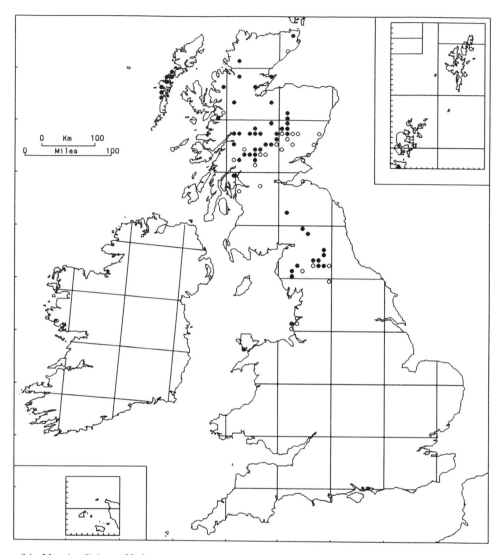

98/1. **Meesia uliginosa** Hedw.

This calcicole of wet, open ground occurs both in dune-slacks by the sea and on seeping rocks, gravel and tufa in the mountains. Characteristic associates in the mountains are *Carex demissa, Juncus triglumis, Saxifraga aizoides, S. oppositifolia, Blindia acuta, Ctenidium molluscum, Drepanocladus revolvens* sensu lato and *Scorpidium scorpioides*. 0–990 m (Ben Lawers). GB 56+29*.

Monoecious; capsules common, early summer.

Circumpolar; widespread in the boreal zone and common in the Arctic, reaching the High Arctic, extending south in Europe to the Pyrenees, Alps and Carpathians, in Asia to the Caucasus, Himalaya and Mongolia, and in N. America to California and New York State.

It is very similar in its requirements to *Amblyodon dealbatus* and *Catoscopium nigritum*, with which it sometimes grows. All three species have been found in the limestone flushes of Upper Teesdale.

M. O. HILL

154

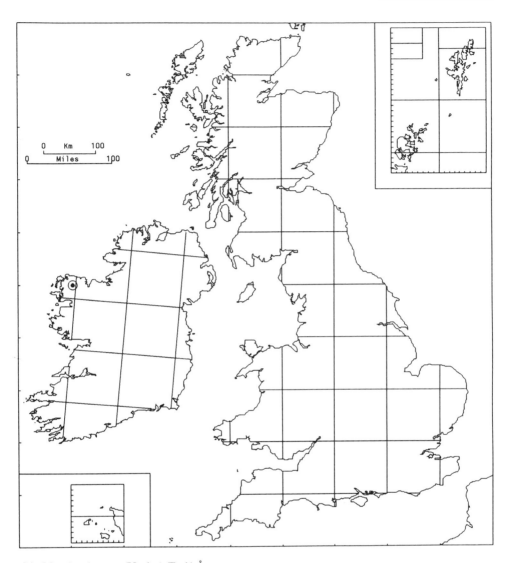

98/2. Meesia triquetra (Hook. & Tayl.) Ångstr.

It formerly grew in a gently sloping, iron-rich, calcareous flush within a vast tract of blanket-bog, where a stream called Sruffaunnamuingabatia is deflected by a low gravel ridge. *M. triquetra* was found near open water below the gravel ridge, in almost pure patches, with some *Sphagnum auriculatum* growing through them. This was the richest part of the flush, with luxuriant *Rhizomnium pseudopunctatum*. Lowland. IR 1.

Dioecious; sterile in Ireland.

Circumpolar, widespread in the boreal zone and Arctic, reaching the High Arctic; scattered in mountains further south. It also occurs in S.E. Australia.

This distinctive 'rich-fen' species has declined in the lowlands of N. Europe, along with several other species occurring in the same habitat. First found in Ireland in 1957 (Warburg, 1958), it was refound in 1958 (King & Scannell, 1960). The blanket-bog has been exploited for peat extraction and *M. triquetra* has not been seen in Ireland since 1958 in spite of careful searching. There are subfossil records from Britain (Dickson, 1973), from which it disappeared before bryophyte recording began.

M. O. Hill

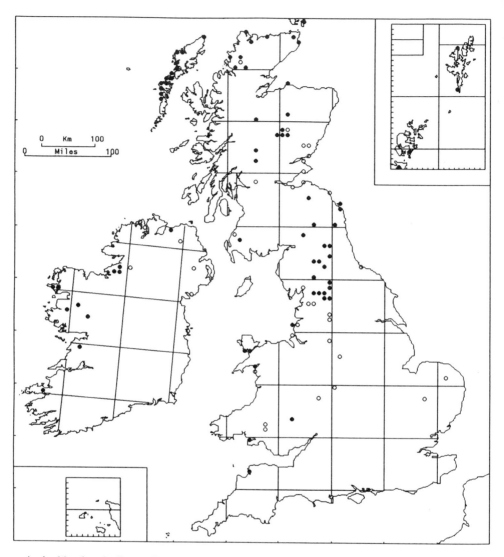

99/1. Amblyodon dealbatus (Hedw.) Br. Eur.

A calcicole of wet, open ground. In duneland it is found in slacks and on moist ditch-sides. On hills and mountains it occurs on tussocks in tufa-springs and on stream-banks, cliffs and gravel where flushed by calcareous water. 0–760 m (Creag an Lochain). GB 62+29*, IR 11+7*.

Autoecious; capsules common, summer.

Circumboreal, scattered widely in the boreal zone and on mountains, rare in the Arctic and above the tree-limit, extending south in Europe to Sierra Nevada, Pyrenees and Alps; in Asia to Caucasus, Iran and Altai Mts; and in N. America to Colorado and S.E. Canada.

Amblyodon is a member of a small but notable group of species that occur both at sea-level on calcareous sand and at higher elevations in calcareous springs. Other such species are *Equisetum variegatum*, *Catoscopium nigritum* and *Meesia uliginosa*, but, at the world scale, these are both commoner and more northerly in distribution.

M. O. HILL

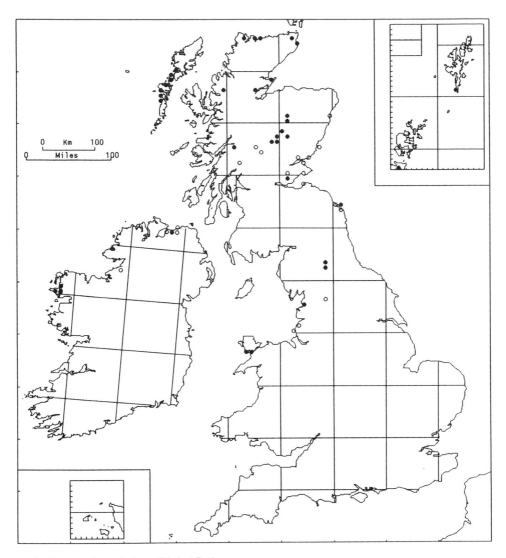

100/1. **Catoscopium nigritum** (Hedw.) Brid.

This species, like a number of montane calcicoles, occurs both on mountains and on sandy ground by the sea. In both types of habitat it forms low hummocks in open, very short, intensely nutrient-poor vegetation, where it is flushed or intermittently flooded by water that deposits calcium carbonate. Although generally rare, very locally in dune-slacks it can form quite extensive carpets. In montane flushes, it is characteristic of semi-open rocky ground with associates such as dwarf sedges, *Saxifraga aizoides*, *Aneura pinguis*, *Drepanocladus revolvens* sensu lato and *Scorpidium scorpioides*. In Upper Teesdale it has been noted in calcareous springs and on the sides of tussocks in calcareous flushes, forming small hummocks similar to those of *Gymnostomum recurvirostrum* and mixed with a range of northern calcicoles including *Primula farinosa* (Pigott, 1956). 0–550 m (Meall Breac). GB 31+17*, IR 5+4*.

Dioecious; capsules frequent, autumn.

Circumpolar arctic-alpine, widespread in the Arctic and occurring north to the High Arctic, extending south to the main European mountain ranges; to the Caucasus and Sayan Mts in Asia; and to British Columbia and the Great Lakes in N. America.

M. O. HILL

157

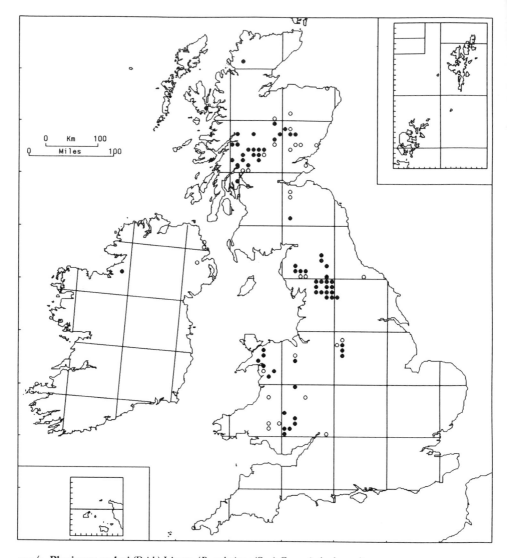

101/1. **Plagiopus oederi** (Brid.) Limpr. (*P. oederiana* (Sw.) Crum & Anderson)

Occurs on basic rock-ledges and in crevices, mostly on hard calcareous rocks. Habitats include gills, ravines and valley woods, wooded limestone scars and N.-facing crags. It extends to a few low-lying localities in rocky woods, but is mostly confined to upland regions and is rather common at moderate altitudes on the Carboniferous Limestone of northern England. On sufficiently basic rocks, such as calcareous schist, it ascends to high mountain-cliffs and gullies. *Ctenidium molluscum*, *Ditrichum flexicaule* and *Neckera crispa* are among its many associates. 0–730 m (Creag na Caillich). GB 66+30*, IR 1+4*.

Synoecious; capsules common, maturing spring and summer.

Circumboreal, north to the High Arctic; recently found in southern Africa. Widespread in mountains throughout Europe, including the Mediterranean mountains.

T. L. BLOCKEEL

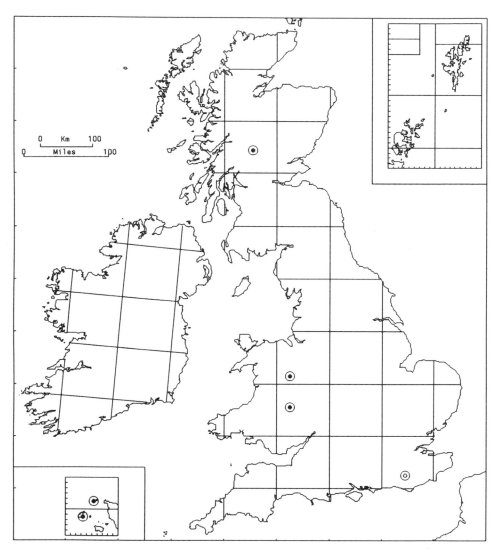

102/1. **Bartramia stricta** Hedw.

A xerophytic moss occurring on soil and in rock crevices on S.-facing slopes. The recorded sites in Wales and Scotland are all on basic igneous rock outcrops, the former site in Sussex was on calcareous sandstone, and the Channel Island localities receive basic mineral inputs from salt spray and blown calcareous dune sand. On Alderney and Guernsey it occurs in communities kept open by summer drought and erosion, and is associated with species such as *Anagallis arvensis, Plantago coronopus, Romulea columnae, Sedum anglicum, Campylopus polytrichoides* and *Scleropodium tourettii*. Lowland. GB 3+1*.

Synoecious; sporophytes are produced abundantly in most years.

Common in the Mediterranean region, extending north along the Atlantic seaboard to Britain. Macaronesia, N. and tropical Africa, S.W. Asia, western N. America (British Columbia to California), Australia.

Yarranton (1962) gives a description of communities containing the species on Breidden Hill, where it is believed to have been destroyed by recent quarrying. Its British populations are all small, and it is also rare in N.W. France, perhaps because it cannot tolerate high humidity.

J. W. BATES

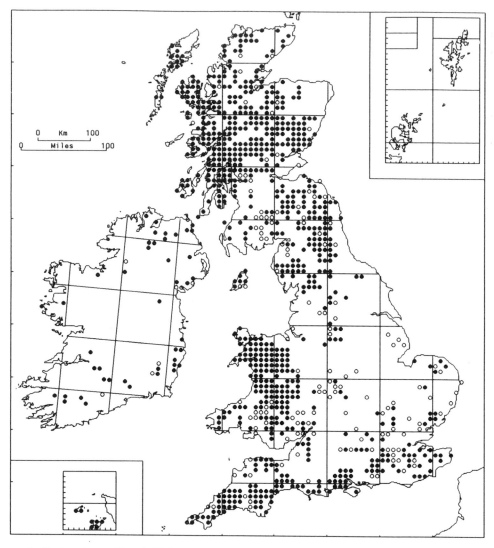

102/2. **Bartramia pomiformis** Hedw.

In C. and E. England this moss is found on shaded earth-banks, often amongst tree-roots or stones, and principally on strongly acidic gravels, sands and sandrocks. Elsewhere it is also common in acidic or mildly basic rock crevices, particularly in ravines and on cliffs, and on rocky banks in woodland. It occupies both dry and irrigated crevices, the production of epicuticular wax probably facilitating gas exchange in the latter situation by preventing the formation of a water film on the leaves. 0–900 m (Ladhar Bheinn). GB 766+141*, IR 43+9*.

Monoecious; sporophytes produced abundantly, ripe summer.

Circumboreal, from the Arctic south to Macaronesia, N. Africa, Himalaya and southern U.S.A. (Oregon and Arkansas). Reported from southern S. America and New Zealand.

The disappearance of *B. pomiformis* from many lowland localities is probably attributable to the effects of air pollutants and the cessation of traditional methods of hedgebank maintenance.

J. W. Bates

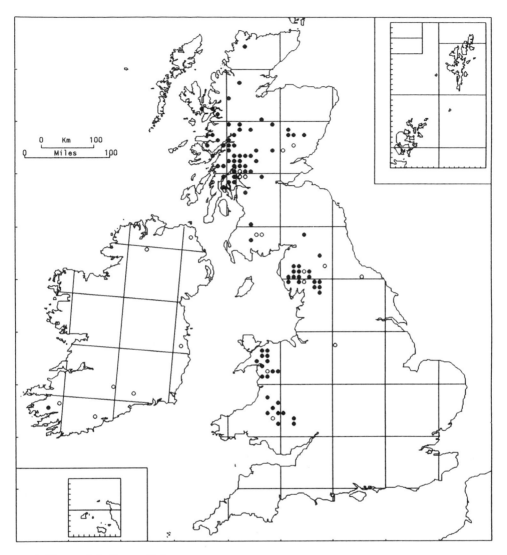

102/3. **Bartramia hallerana** Hedw.

On usually dry, moderately calcareous, rock crevices and overhangs in wooded ravines and on mountain-cliffs. Although restricted to the major mountain massifs it frequently favours low-lying situations within these. 60 m (Skye) to 730 m (Glen Coe). GB 91+15*, IR 1+7*.

Monoecious; sporophytes common, ripe summer.

Boreal and montane, from N. Russia (Kola Pensinsula) south to southern Europe. In Asia it extends south to the Caucasus, Himalaya, S. China (Yunnan, Jiangxi) and Japan; very rare in N. America (Canadian Rockies). Widespread on mountains in the tropics and Southern Hemisphere, including tropical Africa, New Guinea, Australasia, S. America and Hawaii.

J. W. Bates

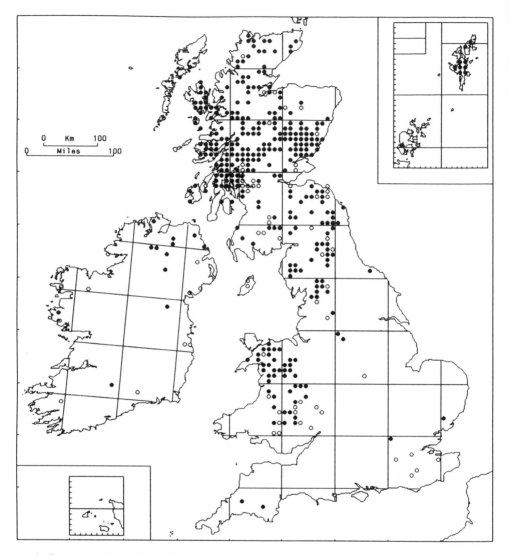

102/4. **Bartramia ithyphylla** Brid.

It is most frequently found in shaded crevices of mildly basic mountain-cliffs. Although generally more montane than *B. pomiformis*, it formerly occurred on sandy hedgebanks in Surrey and Sussex, and still persists on banks and walls in a few other low-lying areas. 0–1170 m (Ben Lawers). GB 323+59*, IR 12+9*.

Synoecious; sporophytes frequent or abundant, ripe summer.

Circumboreal and montane, from the Arctic (widespread) south to the mountains of N. and tropical Africa, Himalaya, S. China (Yunnan, Taiwan), C. and S. America.

It occurs on organic soils in some types of tundra.

J. W. BATES

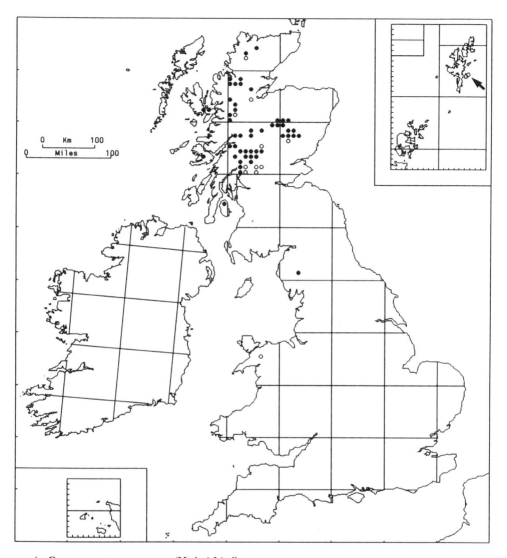

103/1. **Conostomum tetragonum** (Hedw.) Lindb.

It grows mainly on humus or undifferentiated acid soil subject to frost-heave in late-snow beds, with associates such as *Carex bigelowii*, *Salix herbacea*, *Anthelia juratzkana*, *Gymnomitrion concinnatum*, *Marsupella brevissima*, *Kiaeria falcata*, *K. starkei*, *Polytrichum alpinum* and *P. sexangulare*. It also occurs on wind-swept summits with *Juncus trifidus* and on humus over rock in N.E.-facing corries. Mainly above 800 m, to 1335 m on the summit of Ben Nevis, but found last century on Bressay, where it cannot have been above 230 m. GB 50+12*.

Dioecious; capsules occasional, summer.

A circumpolar arctic-alpine, widespread in the Arctic and reaching south to the Alps, C. Asia (Sayan Mts), Japan, and northern U.S.A. (Montana and New York State).

M. O. HILL

163

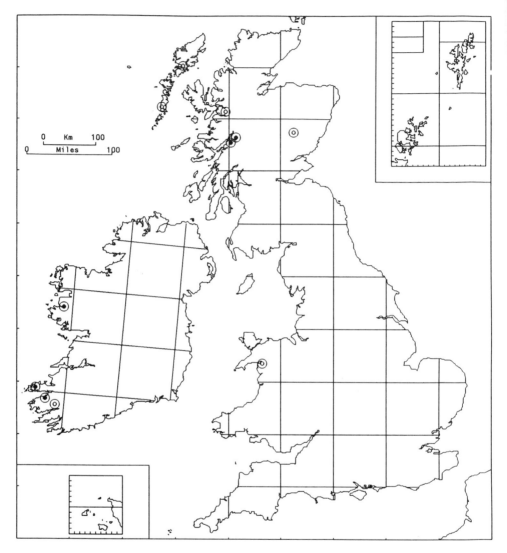

104/1. **Bartramidula wilsonii** Br. Eur. (*Philonotis cernua* (Wils.) Griffin & Buck)

Usually on damp, bare, previously burnt, acidic peat on heathy montane slopes, occasionally on unstable mineral soil amongst loose scree. Although rare and sporadic, it may occur in large patches. From near sea-level to 500 m (Glen Doll). GB 2+5*, IR 3+1*.

Synoecious; capsules normally abundant, late autumn. No specialized method of vegetative dispersal is known.

In Europe it is known with certainty only in Britain and Ireland. W. Africa (Bioko, formerly known as Fernando Po), China (Yunnan), eastern N. America (mountains of N. Carolina and Tennessee), Mexico, Costa Rica, Brazil. Probably more widespread in S. America but obscured in synonymy (Crum & Anderson, 1981).

It has not been seen for very many years at two of its best-known and first-discovered stations, in Wales and eastern Scotland. It was perhaps over-collected at both, but is a surprisingly persistent if intermittent species, and could conceivably recur following burning. It may have been overlooked in some localities because of its small size.

F. J. RUMSEY

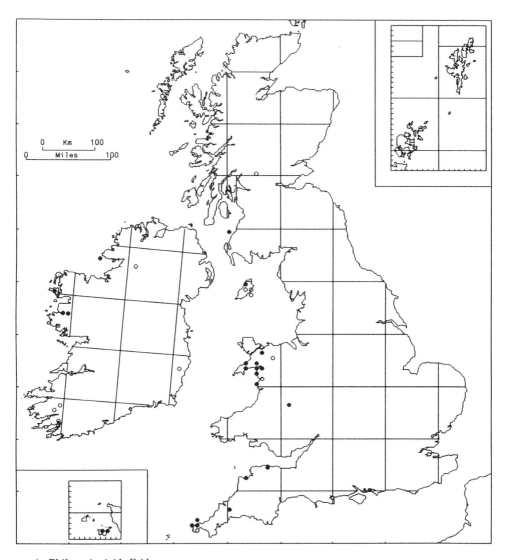

105/1. **Philonotis rigida** Brid.

On seeping cliffs, undercliffs, coastal slumps and river-banks, occurring both on soil and in rock crevices. Most occurrences are in natural habitats on the coast; it has also been found in quarries. The substratum is typically non-calcareous but with flushing by moderately basic water. Lowland. GB 17+8*, IR 4+6*.

Autoecious; capsules common, ripe in spring. Deciduous branchlets are frequently produced in leaf axils and presumably function as propagules.

W. and S. Europe north to Scotland, Normandy and the Alps. Macaronesia, N. Africa, S.W. Asia.

Raeymaekers (1983) has mapped its European distribution.

M. O. HILL

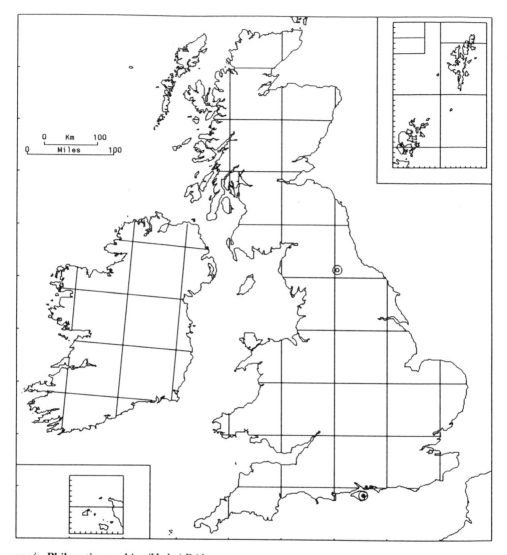

105/2. **Philonotis marchica** (Hedw.) Brid.

At its northern site it was found in crevices of wet shale rocks by a river. On the Isle of Wight it occurs in a coastal locality on moist soil slumps and on the surface of wet friable sandstone rocks, sometimes in considerable shade. Lowland. GB 1+1*.

Dioecious; capsules not found in Britain. In other parts of Europe, bulbils are commonly produced in the leaf axils and these are likely to occur in Britain.

S. Europe north to Britain, Denmark and Ukraine. Macaronesia, N. Africa, S.W. Asia, Caucasus, E. Asia, N. America, Colombia.

It was discovered on the Isle of Wight in 1869 but there was uncertainty about the identity of British plants until the matter was resolved by Smith (1974b).

M. O. HILL

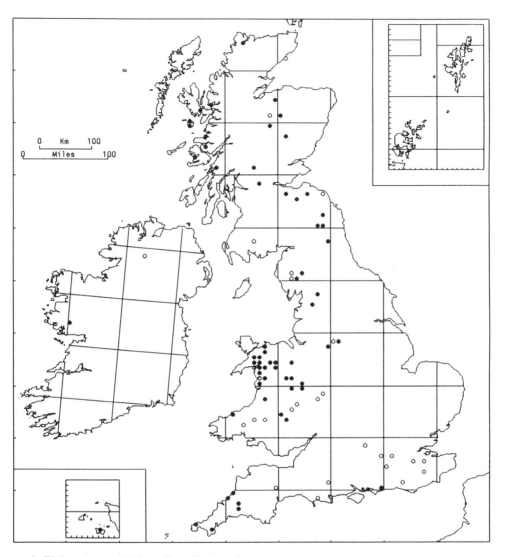

105/3. **Philonotis arnellii** Husn. (*P. capillaris* auct.)

On slightly acid to slightly basic non-calcareous soil and rock crevices in situations that are not permanently wet but where it would be intermittently flushed or flooded or where the ground is kept open by slippage. It occurs in a wide variety of habitats, including woodland rides, heathy tracks, earthy banks, streamsides, mud beside reservoirs, floors of quarries, and basic mountain-cliffs. It is especially characteristic of crumbling earth on rocky banks in light shade, where it typically occurs in rather small quantity mixed with common mosses such as *Bryum capillare*, *Eurhynchium praelongum* and *Rhytidiadelphus squarrosus*. From the lowlands to at least 640 m (Ingleborough) and probably higher on Snowdon and Scafell Pike. GB 59+27*, IR 1+1*.

Dioecious; capsules very rare. Axillary flagelliform shoots are produced in the leaf axils.

Widespread in Europe north to Iceland and S. Scandinavia, montane in the south. Turkey, western and eastern N. America, Greenland.

Most British and Irish plants formerly referred to *Philonotis capillaris* are *P. arnellii*. Older records mapped here were checked for a vice-county revision (Corley & Hill, 1981).

M. O. HILL

167

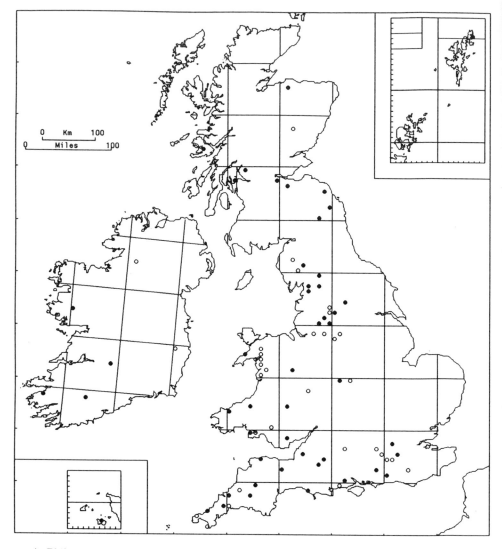

105/4. **Philonotis caespitosa** Jur. (*P. fontana* (Hedw.) Brid. var. *caespitosa* (Jur.) Schimp.)

On moist or wet non-calcareous soil and rocks, where flushed or flooded by slightly basic mineral water. Recorded habitats include springs, streamsides, ditches, marshes, boggy fields, margins of lakes and reservoirs, tracksides, quarries and dripping rocks. 0–300 m (Edale). GB 44+31*, IR 4+2*.

Dioecious; capsules found once, in Ashdown Forest in 1901. Vegetative propagules have not been recorded in Ireland or Britain.

Widespread in the Northern Hemisphere, especially in the boreal zone and mountains. European distribution from the Arctic south to the mountains of W., C. and S. Europe.

The map is based on a vice-county revision of British material (Corley & Hill, 1981) and is unsatisfactory because it includes relatively large plants with lax areolation and falcate leaves. Wilson's type material, from Cheshire, is small and male. Specimens with the characteristic small size and lax areolation have been seen from North Wales, the Lake District, Shetland and W. Ireland.

M. O. HILL

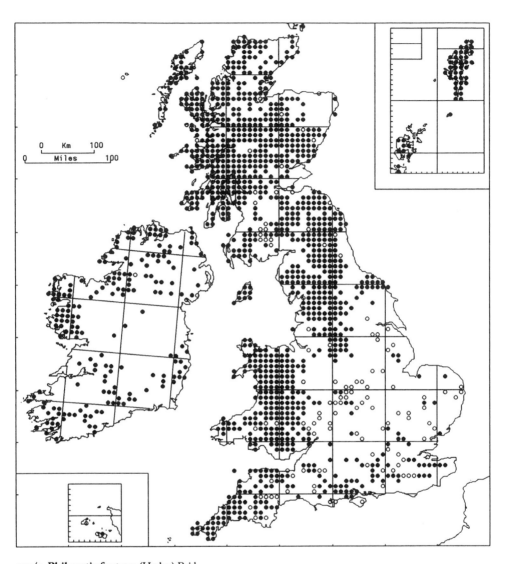

105/5. **Philonotis fontana** (Hedw.) Brid.

On moist or wet, open or slightly shaded soil and rocks, where flushed or intermittently flooded by non-calcareous but usually slightly basic water. It is especially prominent in springs, by moorland streams, in ditches, on dripping cliffs and in ruts on wet tracks. Other habitats include marshes, boggy fields, lakesides, reservoir margins, wet rocks and quarries. 0–1335 m (Ben Nevis). GB 1140+119*, IR 184+4*.

Dioecious; capsules frequent. Deciduous branchlets have been produced in culture (Field, 1986) but they are rarely found in wild material (Petit, 1976).

Widespread in sub-Arctic, boreal, cool-temperate and mountainous regions of the Northern Hemisphere. Distributed through most of Europe from N. Scandinavia southwards, becoming montane in the south.

Although eliminated by drainage from some former sites in S.E. and C. England, it persists widely in woodland rides and is unlikely to decrease further.

M. O. HILL

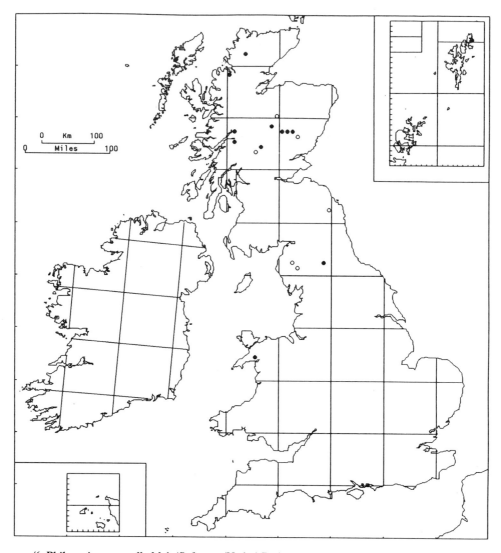

105/6. **Philonotis tomentella** Mol. (*P. fontana* (Hedw.) Brid. var. *pumila* (Turn.) Brid.)

On wet, basic cliffs and rocky slopes on mountains, with associates such as *Bryum pseudotriquetrum, Ctenidium molluscum, Ditrichum flexicaule* and *Plagiobryum zieri*. It has also been found on sandy ground beside a river and on flushed, open basic ground near a lake. A mountain plant, very rare below 400 m, ranging from 80 m (Shetland) to 1100 m (Ben Lawers). GB 13+6*.

Dioecious; capsules not found in Britain.

Circumpolar; widespread in the Arctic and in the mountains of Europe, Asia and America, extending south to the Pyrenees, Alps, Carpathians, Turkey, Iraq, Mongolia and southern U.S.A. (California and Tennessee). According to Steere (1978) it is much commoner than *P. fontana* in Arctic Alaska.

The map is based on a vice-county revision for the *Census Catalogue* (Corley & Hill, 1981). Up to that time *P. tomentella* had been very poorly understood by British bryologists, with only about one-third of specimens correctly named. There is a marked concentration of records from the Breadalbane Mountains; this does not show clearly on the map because five separate records fall in a single 10-km square.

M. O. Hill

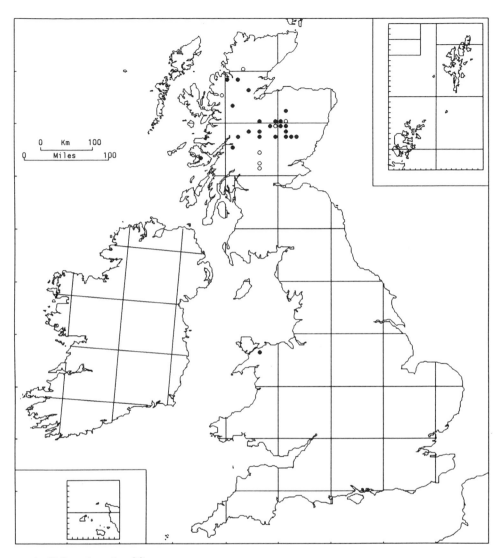

105/7. **Philonotis seriata** Mitt.

Most occurrences are in non-calcareous springs and streams on mountains, typically in sites with prolonged snow-lie, with associates such as *Saxifraga stellaris*, *Scapania undulata*, *Calliergon sarmentosum*, *C. stramineum*, *Drepanocladus exannulatus*, *Pohlia ludwigii* and *Sphagnum auriculatum*. A few records are from more calcareous waters, with *Cratoneuron* spp. and *Drepanocladus revolvens* sensu lato. 450 m (N. Wales) to 1000 m (Clova); according to Duncan (1966) it is best developed in the altitudinal range 850–1000 m. GB 23+7*.

Dioecious; capsules very rare.

Arctic-alpine, widespread in Europe from Iceland, Svalbard and Arctic Scandinavia south to the mountains of W., C. and E. Europe. Urals, Caucasus, S.W. and C. Asia, Greenland.

M. O. HILL

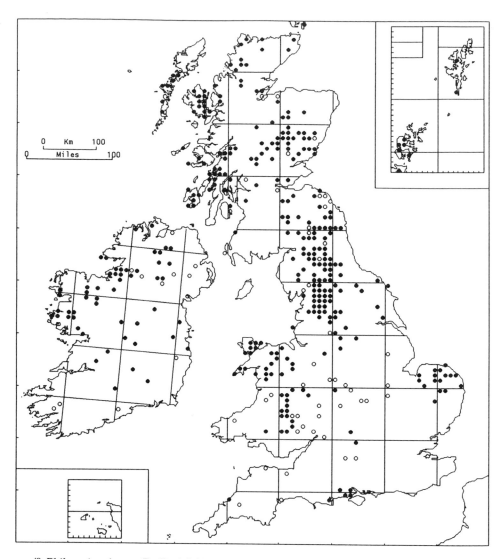

105/8. **Philonotis calcarea** (Br. Eur.) Schimp.

An aptly named plant, found in open or lightly shaded sites kept continually moist by a flow of calcareous water. Habitats include calcareous springs, marshes and dune-slacks, wet cliffs (especially of limestone), dripping rocks in ravines, tufa and, rarely, lock pounds and gates on canals. Typical associates in calcareous upland flushes are *Carex panicea*, *C. pulicaris*, *Campylium stellatum*, *Cratoneuron commutatum*, *Ctenidium molluscum* and *Drepanocladus revolvens* sensu lato; in lowland sites *Calliergon cuspidatum* and *Cratoneuron filicinum* are perhaps more frequent. 0–640 m (Glen Markie). GB 282+48*, IR 52+12*.

Dioecious; capsules rare. There is normally no specialized method of vegetative dispersal, but deciduous branchlets have been found on a Warwickshire plant, which continued to produce them when subsequently cultivated in a greenhouse (Field, 1988).

Widespread in Europe north to N. Scandinavia. Macaronesia, N. Africa, Turkey, Caucasus, Iran.

M. O. HILL

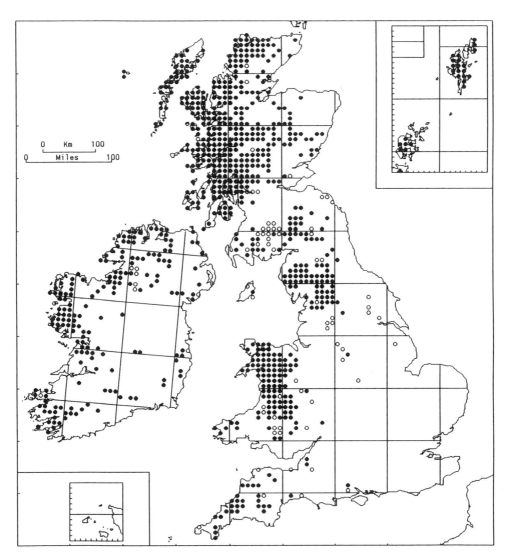

106/1. **Breutelia chrysocoma** (Hedw.) Lindb.

A hygrophilous species forming conspicuous tufts on unshaded or lightly shaded, wet or flushed surfaces in the upland zone. It is common in a wide range of habitats including damp heath, wet moorland, grassy slopes, streamsides, banks and cliff-ledges. In Scotland it is recorded from birch and birch-hazel woodland. Its most characteristic habitat is acidic grassland flushed by slightly basic water, with associates such as *Carex panicea*, *Aneura pinguis*, *Campylium stellatum* and *Drepanocladus revolvens*; only in the very wettest regions does it extend to ombrotrophic mires. In the Yorkshire Dales the main habitat is moist limestone grassland. 0–750 m (Glen Duror). GB 614+66*, IR 172+9*.

Dioecious; sporophytes rare, apparently because of a shortage of males, ripe in autumn. Bedford (1940) found that fruiting plants always grew within 2.5 cm of male plants.

A mainly northern oceanic species, possibly a European endemic, locally frequent in Britain, Ireland, the Faeroes and S.W. Norway; scattered and very rare elsewhere, recorded from the Pyrenees, Brittany, Corsica, Switzerland, Ardennes (extinct) and Westphalia (probably extinct). Reported from C. America.

J. W. Bates

173

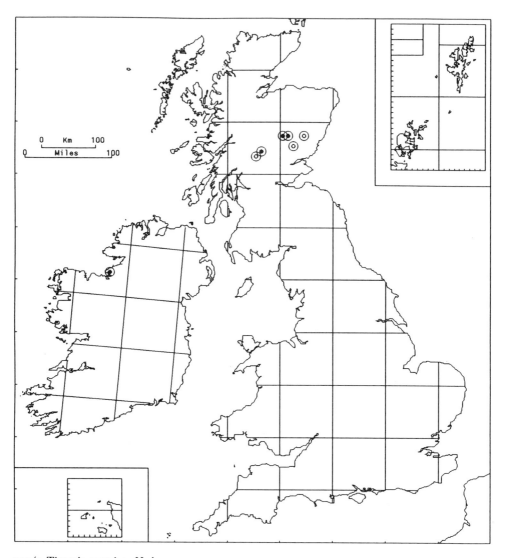

107/1. **Timmia austriaca** Hedw.

It usually occurs in very small quantities, often as one or two isolated stems on bare soil amidst short, open turf on ledges of dry or slightly irrigated calcareous schist or limestone cliffs. In most localities it grows with *T. norvegica* and where the two are found together, *T. austriaca* is invariably the less plentiful. Other associates include *Saxifraga oppositifolia*, *Bryoerythrophyllum ferruginascens*, *Ctenidium molluscum*, *Hypnum hamulosum* and *Tortella tortuosa*. Nowadays confined to higher altitudes from 500 m (W. Ireland) to 900 m (Breadalbane), but in the 19th century it was found at about 160 m in E. Scotland (near Airlie Castle). GB 3+3*, IR 1.

Dioecious; capsules unknown in the British Isles.

Circumpolar with an Arctic-montane distribution including all the main European mountain ranges; widespread in the High Arctic but absent from C. Asia, Himalaya and Japan.

It is often inconspicuous and its isolated shoots can easily be overlooked in the field because they superficially resemble a stunted *Polytrichum*. For a map of its world distribution see Brassard (1980).

H. J. B. BIRKS

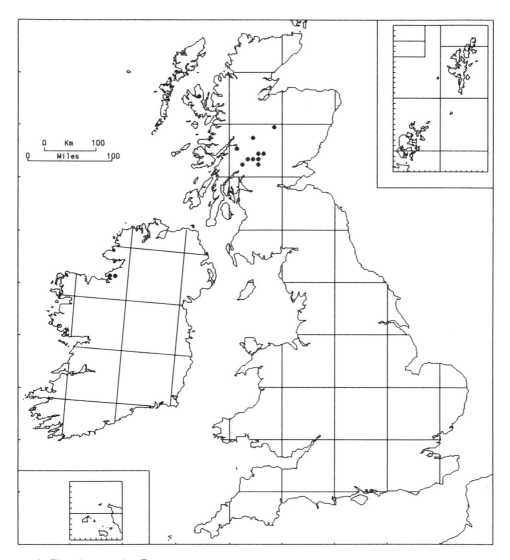

107/2. **Timmia norvegica** Zett.

Usually present in small quantities as scattered stems in moss cushions in crevices and amidst short, open turf on ledges of dry or slightly irrigated calcareous schist cliffs, or, more rarely, limestone or calcareous basalt cliffs. It favours, but is not restricted to, N. and E. aspects. Associates include *Saxifraga oppositifolia*, *Ctenidium molluscum*, *Encalypta alpina*, *Hypnum bambergeri*, *Lescuraea plicata* and *Tortella tortuosa*. Confined to higher altitudes, from just below 500 m in W. Ireland and 550 m in Skye to 1170 m (Ben Lawers). GB 11, IR 2.

Dioecious; capsules unknown in the British Isles.

Circumpolar, with an Arctic-montane distribution extending as far north as there is land; disjunct in the Southern Hemisphere. Iceland, N., W., C. and E. Europe. Turkey, Caucasus, N. and C. Asia, N. America, Greenland, New Zealand.

Its small populations are often inconspicuous and can easily be overlooked, especially as in the field it superficially resembles a *Dicranum*. It may be commoner in the C. Highlands than the map suggests.

H. J. B. BIRKS

175

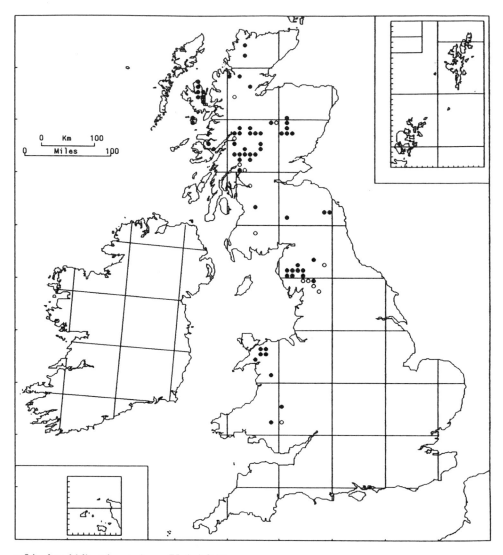

108/1. **Amphidium lapponicum** (Hedw.) Schimp.

This species occurs on moist base-rich montane cliffs and crags, both on exposed rocks and in gullies. Occasionally it descends to lower altitudes in ravines and on stream-banks. It usually forms small isolated tufts in rock crevices and is indifferent to rock type provided there is adequate base enrichment. 210 m (Skye) to 1070 m (Aonach Beag). GB 64+13*.

Autoecious; capsules common, maturing in summer.

Circumboreal arctic-alpine, occurring north to the High Arctic. Common in Scandinavia, extending south in Europe to the mountains of Spain and Bulgaria. It has recently been found in southern Africa (Drakensberg Mountains of Natal and Lesotho) and is reported from N. Africa and Hawaii (Rooy, 1991).

T. L. BLOCKEEL

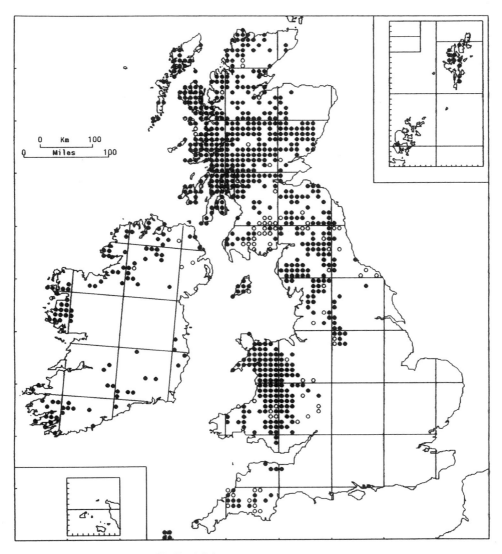

108/2. **Amphidium mougeotii** (Br. Eur.) Schimp.

On moist or intermittently irrigated siliceous rocks, most often on vertical surfaces and in crevices where it may form large swollen cushions. Habitats are varied and include woods, ravines, gorges, montane cliffs and gullies. In the west it occurs also on coastal rocks. Although it is able to survive intermittent desiccation it is not found in stations that are dry for long periods. It is most characteristic of acid rocks, often in company with *Blindia acuta*, but there is usually slight mineral enrichment and it may occur in weakly basic sites with *Gymnostomum* spp. and *Tortella tortuosa*. 0–1180 m (Ben Lawers). GB 672+77*, IR 100+7*.

Dioecious; capsules very rare.

Circumboreal. Widespread in Europe but confined to mountains in the south and rare or absent in the Mediterranean region.

T. L. Blockeel

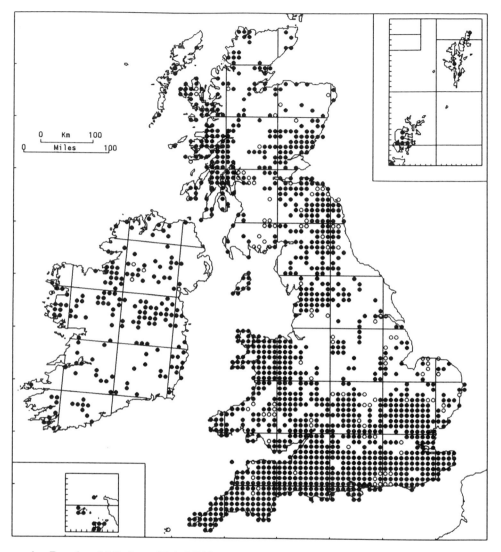

109/1a. **Zygodon viridissimus** (Dicks.) Brid. var. **viridissimus**

It forms small cushions or more extended patches on the bark of trees, on shady base-rich rock outcrops and walls, and on concrete. As an epiphyte it avoids the deepest shade, and trees with strongly acid bark. It is frequent on ash, sycamore, field maple, elder and sallow, but less common on beech and oak; often associated with *Frullania dilatata, Metzgeria furcata, Hypnum cupressiforme* var. *resupinatum, Orthotrichum affine* and *O. diaphanum.* 0–300 m (High Force). GB 1243+107*, IR 175+6*.

Dioecious; sporophytes occasional. Vegetatively propagated by gemmae which are almost always present on the stems and are also produced on the protonema in culture (Whitehouse, 1987).

Widespread in Europe north to Scandinavia and east to Poland and Romania (Transylvania). Macaronesia, Algeria, Japan, eastern N. America, C. America.

M. C. F. PROCTOR

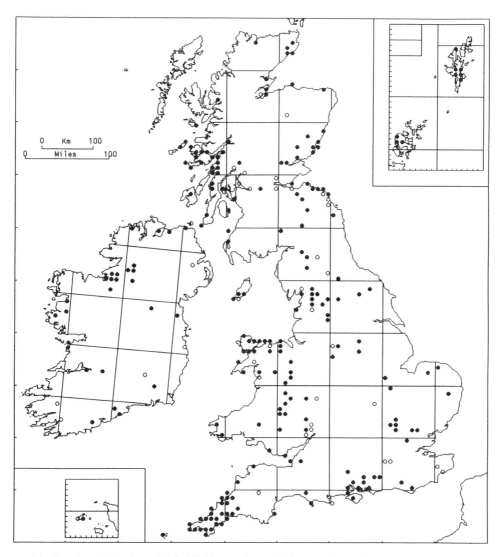

109/1b. **Zygodon viridissimus** (Dicks.) Brid. var. **stirtonii** (Schimp. ex Stirt.) Hagen

Forming cushions or patches on more-or-less shaded surfaces of base-rich rocks; sometimes on stone or brick walls (especially along the courses of mortar); rarely on trees and shrubs, mainly elder. Often associated with var. *viridissimus*. 0–300 m (Lammermuir Hills). GB 194+33*, IR 26+8*.

Dioecious; sporophytes rare. Ovoid stem gemmae usually present.

W. and N.W. Europe from N. France and W. Germany to Scandinavia.

This plant is variously treated by different authors, some regarding it as a species, others merely as a form of *Z. viridissimus*.

M. C. F. Proctor

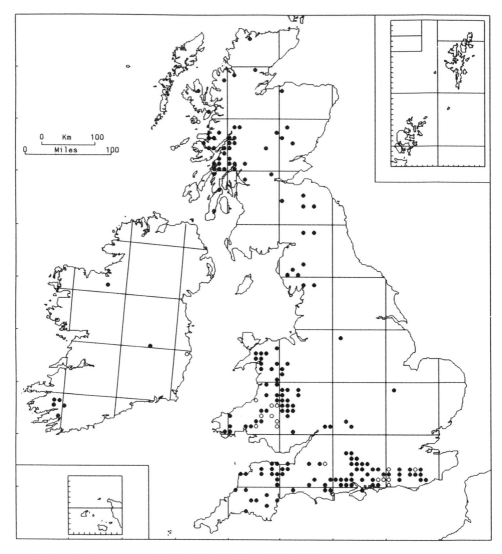

109/2. **Zygodon baumgartneri** Malta (*Z. viridissimus* (Dicks.) Brid. var. *rupestris* Lindb. ex C. Hartm., *Z. rupestris* Schimp. ex Lor.)

An occasional epiphyte on oak, beech, ash, sycamore and other trees, in similar habitats to *Z. viridissimus* and sometimes associated with that species. It has a distinct preference for old trees in long-established parks and at the edge of ancient woodland – conditions favoured by some other uncommon epiphytes, e.g. *Leucodon sciuroides*, *Pterogonium gracile* and the lichen *Lobaria pulmonaria*. Occasional on shady base-rich rocks. 0–500 m (Ben Hope). GB 175+17*, IR 7+1*.

Dioecious; sporophytes occasional. Vegetatively propagated by fusiform gemmae, which are almost always present on the stems and are produced on the protonema in culture (Whitehouse, 1987).

Europe north to Scandinavia and east to Austria and the Crimea; it is commoner than *Z. viridissimus* in the Mediterranean region. Macaronesia, Tunisia, Asia Minor, Caucasus, C. Asia, Japan, western and eastern N. America.

M. C. F. PROCTOR

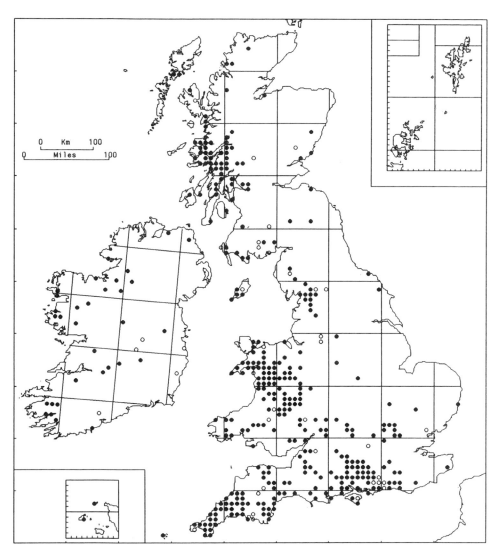

109/3. **Zygodon conoideus** (Dicks.) Hook. & Tayl.

An epiphyte which is commonest on elder, but also occurs on sallow, oak, beech, ash, apple and occasionally on birch and other trees. Associates include *Frullania dilatata*, *Metzgeria* spp., *Cryphaea heteromalla*, *Hypnum cupressiforme* var. *resupinatum*, *Neckera pumila*, *Orthotrichum* spp. and *Zygodon viridissimus*. Lowland. GB 328+33*, IR 32+6*.

Dioecious; sporophytes frequent. Vegetatively propagated by narrowly fusiform gemmae, which are almost always present on the stems and are also produced on the protonema in culture (Whitehouse, 1987).

W. Europe from Spain and Portugal east to S. Scandinavia and Switzerland. Azores, Madeira, Canary Islands, E. Canada (Newfoundland, Nova Scotia).

Z. conoideus is being recorded with increasing frequency in S.E. England. It may have been overlooked by earlier bryologists, but equally it may have genuinely become commoner.

M. C. F. PROCTOR

181

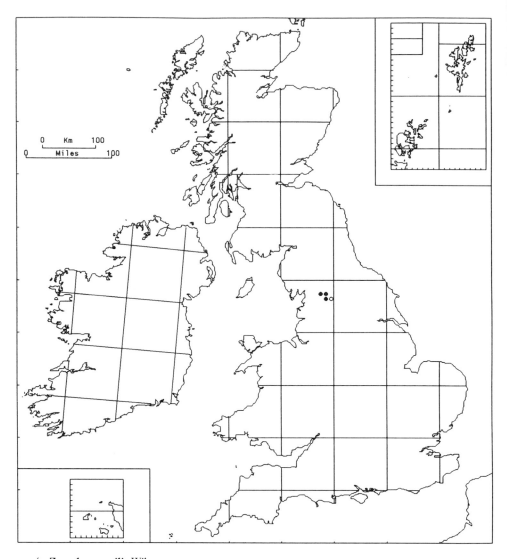

109/4. **Zygodon gracilis** Wils.

Very local and rare on Carboniferous Limestone rocks and drystone walls in the Craven Pennines. 270 m (near Clapham) to 425 m (Dale Head Farm); most of the known sites are at about 380–400 m. GB 3+1*.

Dioecious; only found once with fruit in Britain, by John Nowell, its discoverer, in 1866. Gemmae are produced on the protonema in culture (Whitehouse, 1987).

Known only from Europe, where its distribution is widely disjunct, with scattered localities on limestone in Switzerland, Italy, Bavaria, Austria and Poland.

This species was first described from Yorkshire. As Shaw (1962) points out, most of the known occurrences within its very circumscribed British area are on drystone walls, but there are also sites on natural rock outcrops.

M. C. F. Proctor

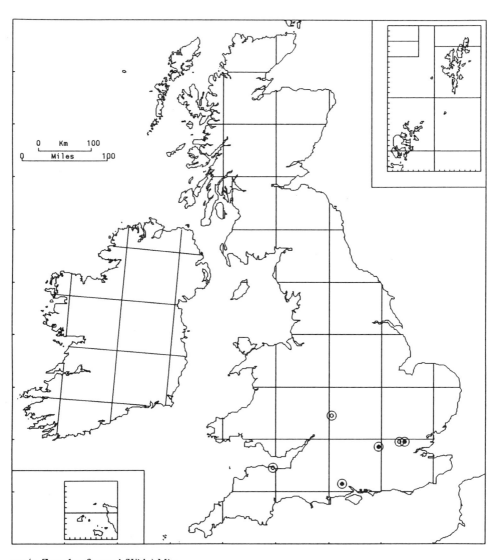

109/5. **Zygodon forsteri** (With.) Mitt.

An epiphytic species, occurring in a narrow range of microhabitats on the trunks and branches of trees, usually beech, typically where water streams down in wet weather below the scars left by shed branches and similar old bark wounds; less often in hollows and crevices on the exposed roots of old trees (Proctor, 1961). Lowland. GB 3+3*.

Autoecious; sporophytes regularly produced. Gemmae are produced on the protonema (Malta, 1926).

W. and S. Europe east to Germany, Switzerland, Croatia, Greece and Bulgaria. Algeria.

Although rare, *Z. forsteri* is remarkably persistent in some localities. It still survives in Epping Forest (whence came the felled tree on which it was originally discovered by Edward Forster in a Walthamstow timber yard, c. 1790) and at Burnham Beeches (where it was discovered by W. E. Nicholson in 1902). For a map of its European distribution see Adams (1984).

M. C. F. PROCTOR

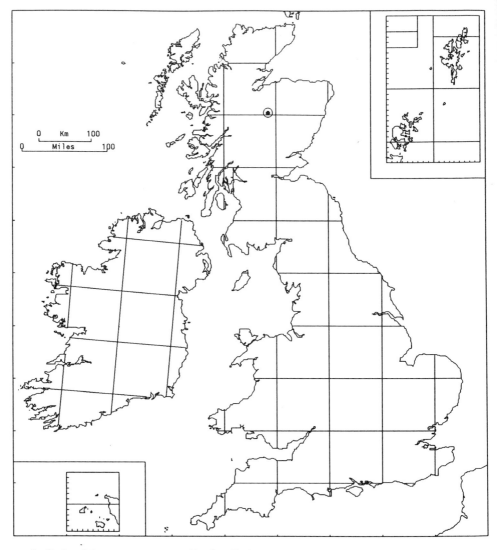

110/1. **Orthotrichum gymnostomum** Bruch ex Brid.

An epiphyte, found on the trunk of aspen (*Populus tremula*) in open pine-birch woodland on a N.-facing slope, associated with *Frullania dilatata, Ptilidium pulcherrimum, Radula complanata, Dicranoweisia cirrata, Hypnum cupressiforme* sensu lato, *Ulota crispa* var. *crispa, U. crispa* var. *norvegica* and *U. drummondii*. 240 m. GB 1.

Dioecious; sex organs and sporophytes unknown in Britain. Gemmae frequent.

C. and N. Europe. Turkey, Afghanistan, Newfoundland.

Known in Britain as a single tuft, collected by J. Dransfield in Rothiemurchus Forest in 1966. The habitat details are taken from the report of this discovery (Perry & Dransfield, 1967). The species has not been refound and may have been a casual occurrence, perhaps resulting from long-distance dispersal from the European mainland. In Europe, *O. gymnostomum* occurs most frequently on *Populus tremula*; in Newfoundland it favours *P. tremuloides* and *P. balsamifera*.

C. D. PRESTON

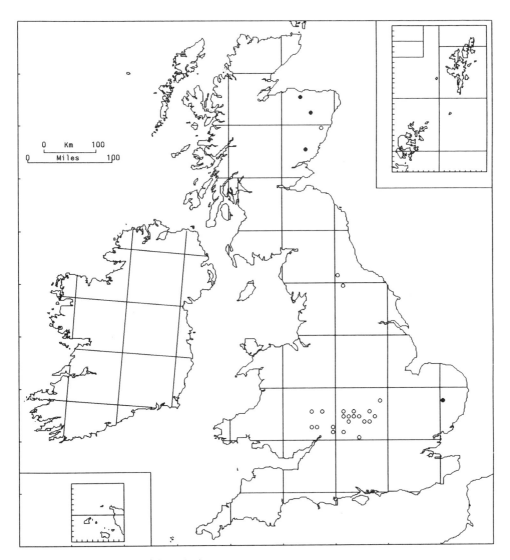

110/2. **Orthotrichum obtusifolium** Brid.

An epiphyte, growing as a pioneer on the exposed roots, trunks and twigs of trees and shrubs. It has been found on roadside, parkland and woodland trees, probably always in well-illuminated localities. Ash (14 sites) and elm (6) are the most frequent hosts, but it is also recorded from elder (2), oak (1) and willow (1). Associated species include *Frullania dilatata*, *Amblystegium serpens*, *Leucodon sciuroides*, *Orthotrichum affine*, *O. diaphanum* and *Zygodon viridissimus*. At one Scottish locality it is known on rotting logs as well as on a nearby elm tree. Lowland. GB 4+23*.

Dioecious; sporophytes unknown in Britain. Gemmae are produced abundantly on the leaves and develop on the protonema in culture (Goode *et al.*, 1993).

Circumboreal. C. and N. Europe.

Although *O. obtusifolium* was widespread, but not common, in the nineteenth century it has been recorded from only three localities in the last sixty years. At these recent sites it has been found on elm (Scotland) and on an elder twig (Norfolk). In northern Europe and N. America it has a distinct preference for *Populus* bark; in subarctic Europe it can also grow on calcareous rock.

C. D. PRESTON

185

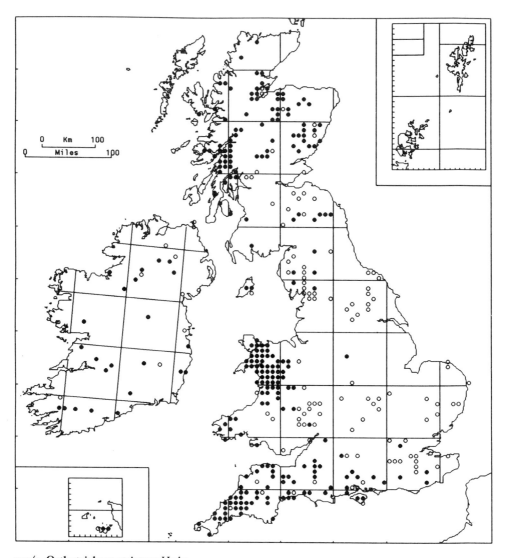

110/3. **Orthotrichum striatum** Hedw.

Orthotrichum striatum is an epiphyte which usually grows in small quantity on the base-rich bark of trees and shrubs such as ash, elder, elm, hazel, poplar, sycamore and willow but it also occurs on alder, beech and oak and is recorded from exotics such as buddleia. It is found on the trunks of trees in sheltered but open sites by roads, streams and rivers and on the trunks, branches and twigs of shrubs, often in sheltered valleys. *O. striatum* also grows occasionally on concrete in humid situations and on a serpentine rock outcrop on the Lizard Peninsula. Lowland. GB 232+107*, IR 23+10*.

Autoecious; sporophytes abundant, maturing from September to May.

Widespread in Europe north to Iceland and 70°N in Scandinavia. Macaronesia, N. Africa, Asia east to Kashmir, north-west N. America.

The marked decline of *O. striatum* in areas of high SO2 pollution is clearly shown by the map. The discovery in 1989 of a population on crack willow (*Salix fragilis*) on Hampstead Heath, London, is one of the most remarkable signs of falling SO2 levels (Adams & Preston, 1992).

C. D. PRESTON

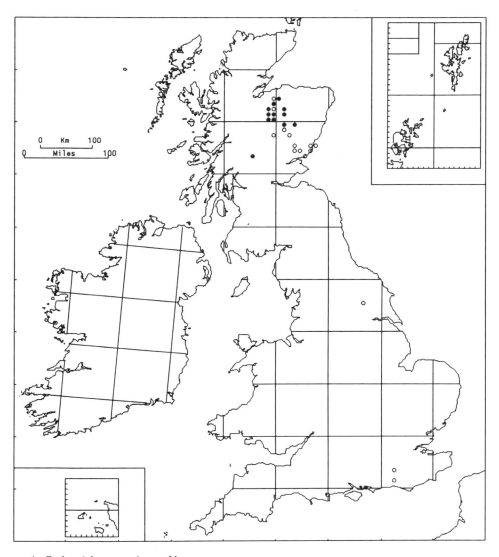

110/4. Orthotrichum speciosum Nees

An epiphyte, found on the branches of old hazel and juniper shrubs and on the trunks and branches of a range of trees (including ash, aspen, birch, poplar, rowan, sycamore and wych elm). Associated species include *Frullania dilatata, Radula complanata, Amblystegium serpens, Antitrichia curtipendula, Homalothecium sericeum* and *Hypnum cupressiforme*. The first British record was from 'trees and stones' and a specimen collected early this century was apparently growing on the stonework of a bridge. 0–350 m (Linn of Corriemulzie). GB 12+15*.

Autoecious; sporophytes common.

Circumboreal. Widespread in Europe, extending north to Spitsbergen; frequent on rocks and trees in some northern areas with a more continental climate than the British Isles (e.g. southern Scandinavia) and growing on rocks and dry earth in the tundra.

O. speciosum may have been eliminated from its more southerly British localities by air pollution. It was formerly widespread in the Netherlands but there too has suffered a similar decline (Touw & Rubers, 1989). It has recently been found as a colonist of woods in the Dutch polders and it might be expected to recolonize eastern England and the Scottish lowlands following the recent reduction in SO₂ levels.

C. D. Preston

187

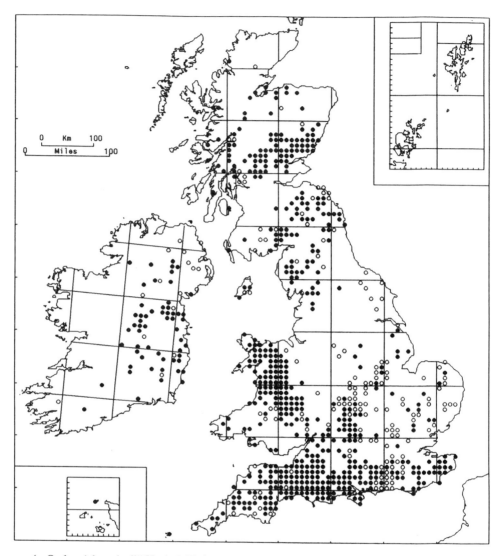

110/5. **Orthotrichum lyellii** Hook. & Tayl.

On trunks of trees in woods or on riversides and roadsides, often growing in small quantity amongst associates such as *Frullania dilatata*, *Metzgeria furcata*, *Neckera pumila*, *Orthotrichum striatum*, *Ulota crispa*, *Parmelia* spp. and (in drier places) *Leucodon sciuroides* and *Tortula laevipila*. Although its distribution is predominantly western it is absent from areas of highest rainfall where the tree-trunks are smothered in a carpet of *Hypnum mammillatum* (Proctor, 1962). It is found on a wide range of host-trees, most frequently on ash, elm, oak and sycamore. *O. lyellii* also grows on branches and twigs of elders and sallows in sheltered, humid thickets, and on the branches of hazel. It has occasionally been recorded on stonework. Lowland. GB 526+151*, IR 57+17*.

Dioecious; sporophytes scarce, maturing spring to late summer. Gemmae are always present on the leaves, sometimes in abundance, and develop on the protonema in culture (Whitehouse, 1987).

Europe north to southern Scandinavia. Macaronesia, N. Africa, S.W. Asia, western N. America.

O. lyellii has declined in areas with high SO_2 levels; plants still present in these areas are less robust and fruit less frequently than in the early 19th century.

C. D. Preston

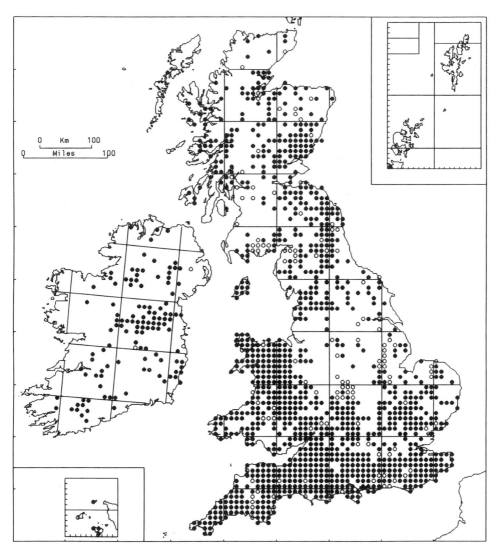

110/6. **Orthotrichum affine** Brid.

An epiphyte growing as a pioneer on the exposed roots, trunks and branches of trees and shrubs on roadsides and riversides and in scrub, fen carr and open woodland. It favours sheltered but not heavily shaded localities, being most luxuriant in rather humid microclimates and often occurring abundantly on twigs in the moist, relatively unpolluted atmosphere of the west. It is sometimes found on silty bark in the flood zone of streams and rivers. Although particularly frequent on elder, it also grows on a wide range of other hosts including alder, alder buckthorn, ash, beech, birch, buckthorn, elm, hazel, oak, poplar, sycamore and willow. Less frequently it grows on inorganic substrates such as rocks in natural outcrops and quarries, stonework of walls and bridges, concrete walls and culverts, and asbestos roofs. 0–530 m (Glen Clova). GB 1080+123*, IR 118+4*.

Autoecious; sporophytes frequent, often abundant, maturing in spring, summer and autumn. Gemmae occasional on the leaves.

Widespread in Europe. Macaronesia, N. Africa, Asia, western N. America.

O. affine extends further into polluted areas than the other epiphytic *Orthotrichum* species (except *O. diaphanum*).

C. D. Preston

189

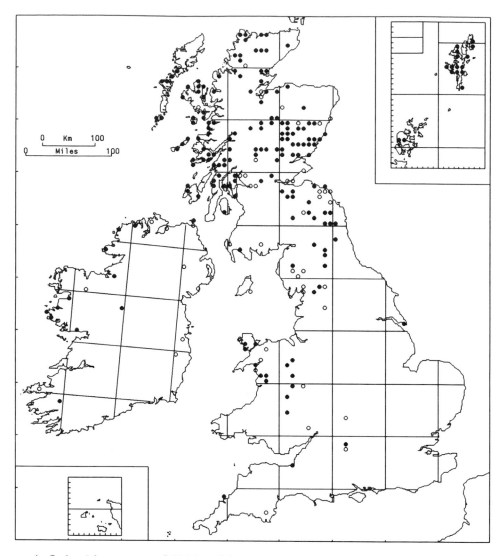

110/7. **Orthotrichum rupestre** Schleich. ex Schwaegr.

A plant of well-illuminated rocks and boulders, growing by lakes and rivers, on coastal outcrops, on boulders in scree, on sarsen stones and on walls and bridges. It is most frequent on basic igneous rocks including basalt, dolerite and tuff, but it also occurs on acidic rocks affected by calcareous sand and has been found on a concrete dam. In many areas it is absent from limestone, but in Skye it grows on limestone pavement and rock outcrops (Birks & Birks, 1974). *Grimmia trichophylla* var. *robusta*, *G. trichophylla* var. *trichophylla* and *Schistidium apocarpum* are recorded as associates. *O. rupestre* is occasionally found on exposed tree-roots by streams or on the trunks of trees such as ash and elm. 0–700 m (Whernside). GB 174+41*, IR 12+9*.

Autoecious; sporophytes abundant, maturing from June to August. Gemmae develop on the protonema in culture (Whitehouse, 1987).

One of the most widespread of *Orthotrichum* species, found throughout Europe north to Iceland and northern Scandinavia and in all continents including Antarctica.

C. D. Preston

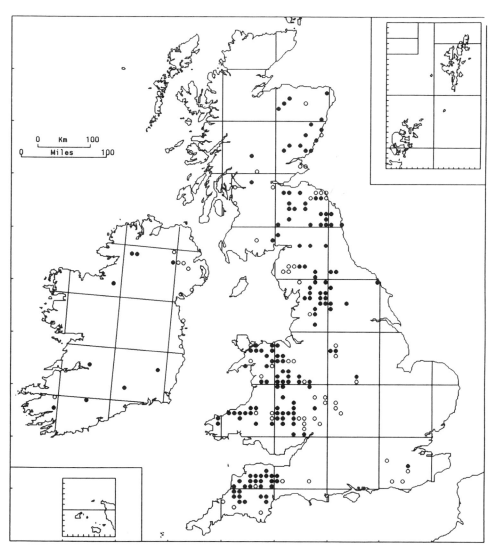

110/8. **Orthotrichum rivulare** Turn.

O. rivulare grows by large streams and rivers, above the normal water-level but in a zone liable to occasional flooding. The substrates on which it grows include the exposed roots, trunks and branches of trees and shrubs (including alder, ash, sycamore and willow), stones, rocks and occasionally the bases of walls. Associated species include *Cinclidotus fontinaloides*, *Racomitrium aciculare* and *Schistidium alpicola* by more rapidly flowing rivers, and *Homalia trichomanoides*, *Leskea polycarpa*, *Orthotrichum diaphanum*, *O. sprucei* and *Tortula latifolia* where the current is slower and more silt accumulates. Lowland. GB 157+55*, IR 9+10*.

Autoecious; sporophytes frequent, maturing in summer. Gemmae are not recorded on the leaves, but develop on the protonema in culture (Whitehouse, 1987).

Western Europe from Spain to the Netherlands, western Germany, Austria and Yugoslavia. Western N. America from central California to Washington, reaching its eastern limit in the Yellowstone National Park, Wyoming.

C. D. PRESTON

191

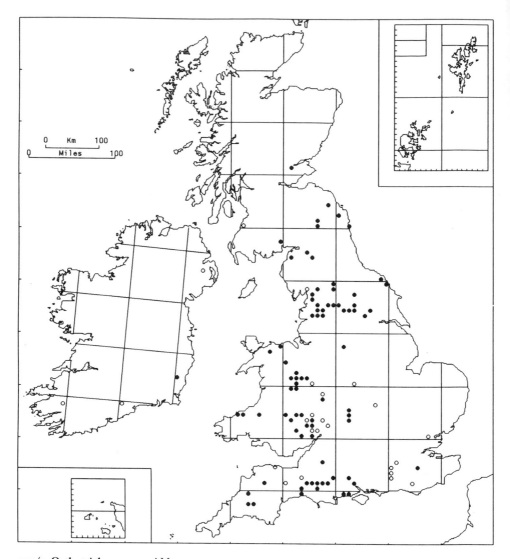

110/9. **Orthotrichum sprucei** Mont.

Confined to the exposed roots, trunks and branches of trees and shrubs by streams and rivers, growing above the normal water-level in a zone which is usually dry but liable to occasional flooding. The lower parts of the plant are often embedded in fine silt. The usual hosts are alder, ash and willow, probably because these are the commonest riverside trees; it is also recorded from wooden palings. Characteristic associates include *Bryum capillare*, *Leskea polycarpa*, *Orthotrichum diaphanum* and *Tortula latifolia*. Lowland. GB 81+22*, IR 1+3*.

Autoecious; sporophytes usually present, maturing in summer. Gemmae occasional on the leaves, and produced on the protonema in culture.

Endemic to western Europe. Outside the British Isles it is known only from France (rare, but possibly under-recorded), Belgium (rare) and the Netherlands (recorded in 1878).

The ecological requirements of *O. sprucei* overlap with those of the related *O. rivulare*, with which it often grows. However, *O. sprucei* rarely, if ever, grows on rock; it tends to be a plant of more mature, siltier rivers and although a rarer plant nationally it can outnumber or replace *O. rivulare* in such habitats.

C. D. Preston

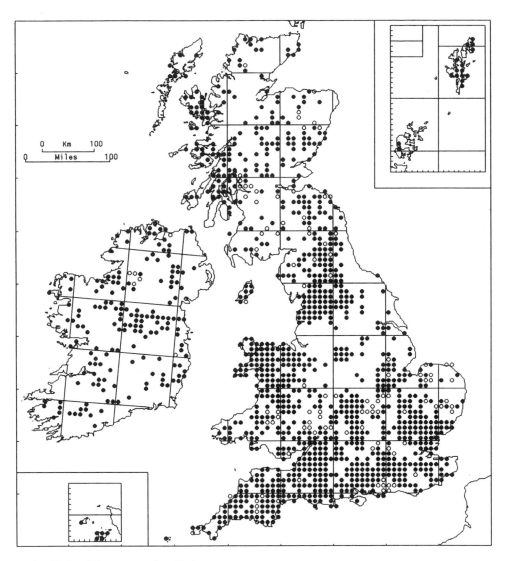

110/10. **Orthotrichum anomalum** Hedw.

A plant of base-rich rock, usually in open sites but sometimes in moderate shade. It is especially abundant in areas of Carboniferous Limestone, growing on rock outcrops, boulders, limestone pavement, drystone walls and in quarries. It is also frequent on other hard limestones and on basic igneous rocks such as basalt, but it is not found on chalk. Typical artificial substrates include concrete, stonework ledges and buttresses of bridges and churches, and mortar on otherwise non-calcareous walls; slate and asbestos roofs are colonized occasionally. 0–630 m (Gragareth). GB 1015+105*, IR 166+8*.

Autoecious; sporophytes abundant, maturing in May and June.

Throughout Europe north to Iceland and 67°N in Norway. Widespread in the Northern Hemisphere, from the boreal zone south to N. Africa, Turkey, Himalaya, Japan, Guatemala and Haiti.

Although absent from highly polluted cities during the main period of the mapping scheme it has recently begun to colonize limestone surfaces in London in response to falling SO_2 levels. The British record of *Orthotrichum urnigerum* Myr. has proved to be based on plants of *O. anomalum* growing in an unusual habitat (Blockeel, 1987).

C. D. Preston

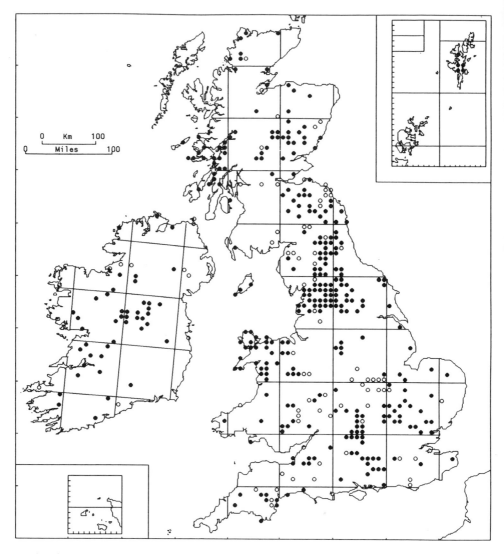

110/11. **Orthotrichum cupulatum** Brid.

A calcicole, found in a similar range of habitats to *O. anomalum* and usually accompanied by that species. It grows on flat limestone rocks, stone and mortared walls, parapets of bridges, flat tomb-tops, concrete walls and blocks and on roof tiles. It is also found on the silty branches and roots of trees by streams and rivers. In the drier southeast of England, *O. cupulatum* differs from *O. anomalum* in its tendency to prefer damper, shadier sites near water, although thriving populations can be found in other sites; further north and west the two species have very similar ecological requirements. 0–620 m (Gragareth). GB 301+73*, IR 44+10*.

Autoecious; sporophytes abundant, maturing in May. Gemmae develop on the protonema in culture (Whitehouse, 1987).

Widespread in Europe north to Iceland and northern Norway. Macaronesia, N. Africa, Asia east to the Himalaya, western and eastern N. America, S. America.

Var. *riparium* Hüb. is doubtfully distinct from var. *cupulatum*; both varieties are included in the map.

C. D. PRESTON

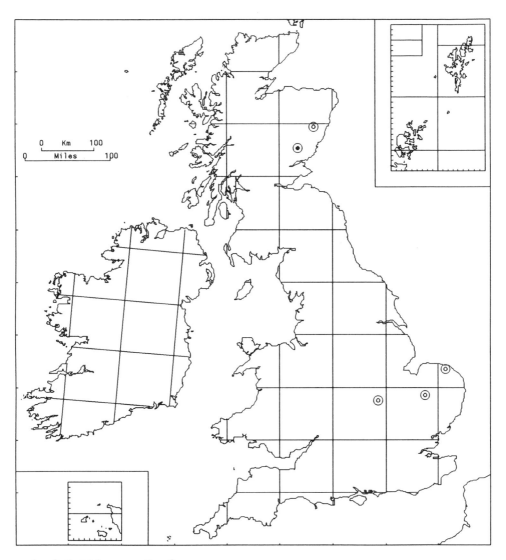

110/13. **Orthotrichum pumilum** Sw.

A tiny epiphyte, recorded from the bark of ash, sweet chestnut, elm and willow trees. Lowland. GB 1+4*.

Autoecious; sporophytes common. Gemmae are occasionally produced on the leaves (Vitt, 1973) and develop on the protonema in culture (Whitehouse, 1987), both reports being based on non-British material.

Widespread in Europe north to 64°N in Scandinavia. Macaronesia, N. Africa, Asia east to the Caucasus, N. America.

In Britain this plant has previously been called *Orthotrichum schimperi* Hammar, a species which is regarded as a synonym of *O. pumilum* by most recent authors (e.g. Vitt, 1973; Corley *et al.*, 1981). Many older British records have proved to be erroneous and the remainder require critical revision.

C. D. PRESTON

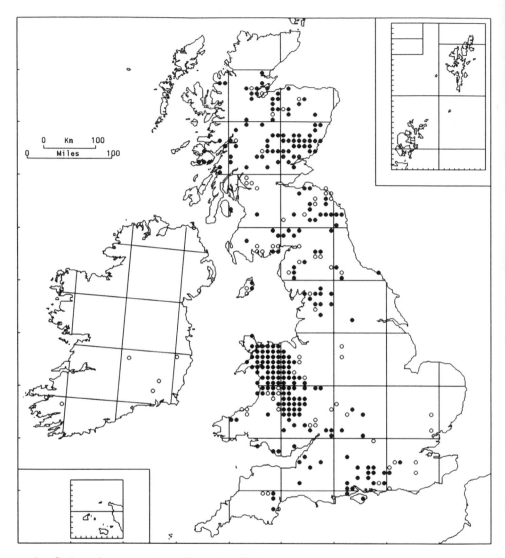

110/14. **Orthotrichum stramineum** Hornsch. ex Brid.

Usually found on vertical or sloping tree-trunks or the trunks and branches of shrubs, sometimes persisting on fallen trunks in the early stages of decay. It is particularly frequent in sheltered wooded valleys in upland areas. Although it shows some preference for the base-rich bark of ash, elder and sycamore it also occurs on a wide range of other species including alder, beech, birch, elm, field maple, hazel, oak and willow. *Amblystegium serpens*, *Bryum capillare*, *Hypnum cupressiforme*, *Orthotrichum affine*, *Ulota crispa* and *Zygodon conoideus* are characteristic associates. Confined to low altitudes. GB 266+68*, IR 8*.

Autoecious; sporophytes abundant, maturing in early summer.

Widespread in Europe north to Iceland and 66°N in Norway. N. Africa, S.W. Asia, Canada (Newfoundland – possibly introduced).

The absence of *O. stramineum* from well-worked regions such as Cornwall and Skye exemplifies the fact that several *Orthotrichum* species, in marked contrast to most *Ulota* species, decrease in frequency in highly oceanic areas.

C. D. PRESTON

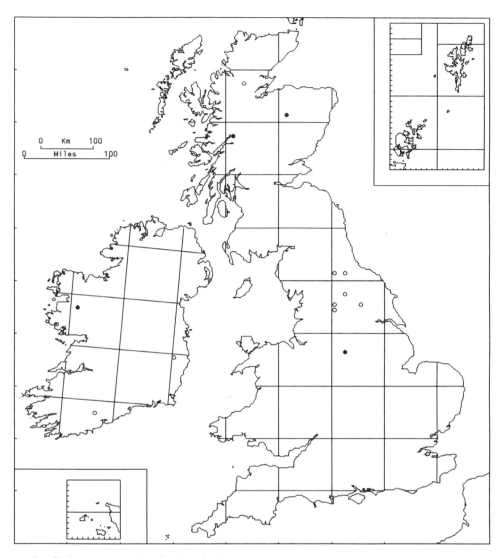

110/15. **Orthotrichum pallens** Bruch ex Brid.

It grows on the bark of ash, hazel, sycamore, willow and wych elm, often in rather open sites such as riversides. In Ireland it has been found on the trunk of a roadside sycamore at the edge of a village in unremarkable flat country with little shelter. In a Scottish locality it was also found on the trunk of a solitary tree, a wych elm by a river, but here with more shelter. Lowland. GB 3 + 8*, IR 1 + 3*.

Autoecious; sporophytes common. Gemmae are sometimes present on the leaves of American plants (Vitt, 1973) but have not been reported from Britain or Ireland.

Circumboreal. N. and C. Europe; rare in S. Europe and absent from most of the Mediterranean islands.

O. pallens is a somewhat characterless species, which may therefore be under-recorded. At most of its British and Irish sites it is known only from a single collection.

C. D. Preston

197

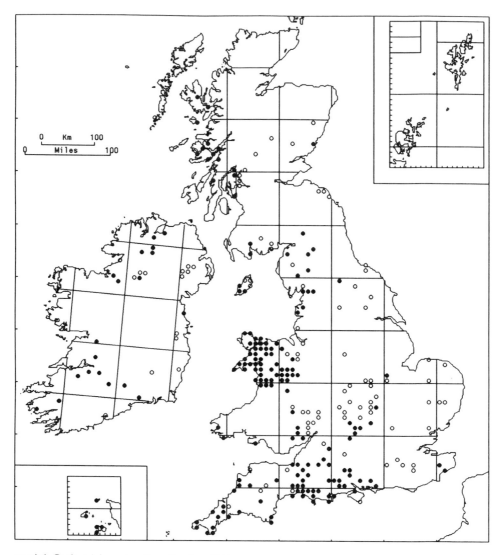

110/16. **Orthotrichum tenellum** Bruch ex Brid.

Orthotrichum tenellum usually grows in small quantity on the bark of trees and shrubs in open sites such as hedgerows, roadsides, streamsides and in pastures, occurring much less frequently in woods and carr. It is particularly frequent on old elders but is not uncommonly found on other hosts including alder, apple, ash, beech, elm, field maple, hazel, oak, pear, poplar, sallow, sycamore and willow. Less usual habitats include silty, riverside alders and damp concrete. *Orthotrichum affine*, *O. diaphanum* and *Tortula laevipila* are frequent associates. Lowland. GB 159+79*, IR 19+15*.

Autoecious; sporophytes common, maturing from March to July. Gemmae are sometimes found in abundance on the leaves (Appleyard, 1986) and are produced on the protonema in culture (Whitehouse, 1987).

Widespread in Europe north to southern Scandinavia, most frequent in the south. Macaronesia, N. Africa, Turkey, western N. America.

Like several of its congeners, *O. tenellum* appears to be sensitive to atmospheric pollution; it has declined in areas where SO2 levels are high.

C. D. PRESTON

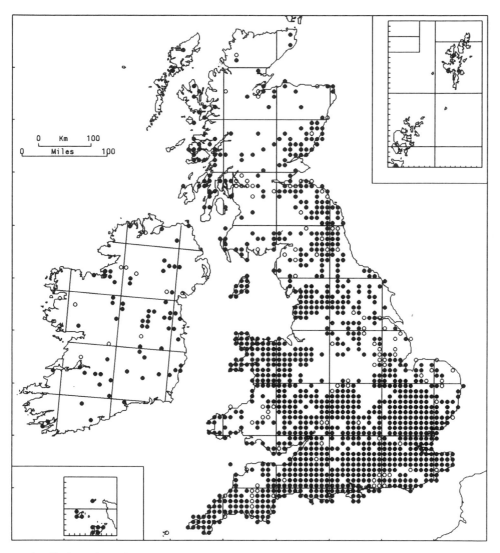

110/17. **Orthotrichum diaphanum** Brid.

Frequent both on inorganic substrates and as an epiphyte. On stonework, brickwork and concrete it grows with *Bryum argenteum, B. capillare, Grimmia pulvinata, Schistidium apocarpum* and *Tortula muralis*. It also occurs on rocks of natural outcrops and in quarries, and on artificial substrates such as asbestos and old tarmac. As an epiphyte it is often abundant on nutrient-enriched bark, e.g. of roadside trees in areas of intensive agriculture, but occurs in smaller quantity in less-polluted sites and on silty bark in the flood zone of streams and rivers. It is most frequent on elder; other hosts include alder, ash, elm, hazel, oak, sycamore, tamarisk and willow. It colonizes fence-posts and wooden gates. Lowland. GB 1148+105*, IR 65+12*.

Autoecious; capsules abundant, usually maturing in winter and spring but recorded throughout the year. Gemmae frequent on the leaves and, in culture, produced on the protonema.

Europe north to Shetland and southern Norway; a common epiphyte on southern European street trees. Macaronesia, Africa, S.W. Asia, Hawaii, N. and S. America.

It is the only member of the genus found in many urban areas, although it is absent in the most-polluted cities (Gilbert, 1968).

C. D. PRESTON

199

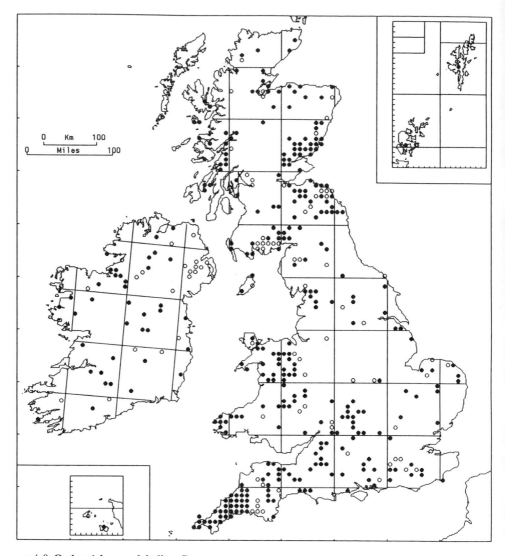

110/18. Orthotrichum pulchellum Brunton

An epiphyte which grows on the trunks and branches of trees and shrubs in sheltered but not heavily shaded situations, including open woodland and woodland rides, patches of scrub, fen carr, quarries and streamsides. It is much more frequent on elder and willows than on other hosts, but is occasionally found on alder, ash, beech, hawthorn, hazel, oak, rowan and sycamore. It only rarely colonizes wooden palings and rotting logs or inorganic substrates such as concrete and stone walls. Associated species include *Frullania dilatata*, *Metzgeria fruticulosa*, *M. furcata*, *Radula complanata*, *Amblystegium serpens*, *Cryphaea heteromalla*, *Hypnum cupressiforme*, *Orthotrichum affine*, *Rhynchostegium confertum*, *Ulota crispa* and *Zygodon viridissimus*. Lowland. GB 294+71*, IR 38+20*.

Autoecious; sporophytes abundant, maturing in late spring and early summer. Gemmae are not recorded.

N.W. Europe from France to Norway, Sweden and Poland; rare and scattered southwards in Spain, Sicily, Switzerland, Yugoslavia and Crete. Western N. America, in wet forests from Oregon to S.E. Alaska and Kodiak Island; disjunct in Idaho. Its British and European distribution resembles that of *Zygodon conoideus*.

O. pulchellum may be spreading in S.E. England. Since 1984 it has been found in five vice-counties from which it was not previously recorded.

C. D. PRESTON

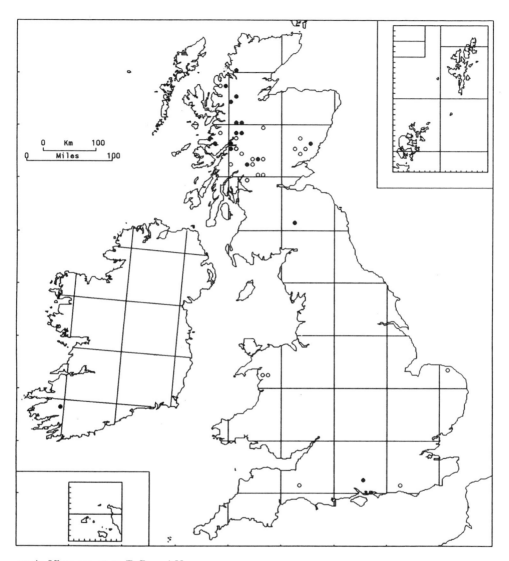

111/1. **Ulota coarctata** (P. Beauv.) Hammar

An epiphyte forming small yellowish or dark green tufts on birch, oak, sallow and other trees and bushes in sheltered humid places. In southern England it has been found on elder in calcareous woodland. Lowland. GB 17+22*, IR 1.

Autoecious; sporophytes regularly produced, autumn.

A mainly northern and montane species, rare in the lowlands of W. and C. Europe and in mountains of S. Europe, more frequent in the C. European mountains and W. Fennoscandia, north to about the Arctic Circle. Caucasus, eastern N. America.

M. C. F. Proctor

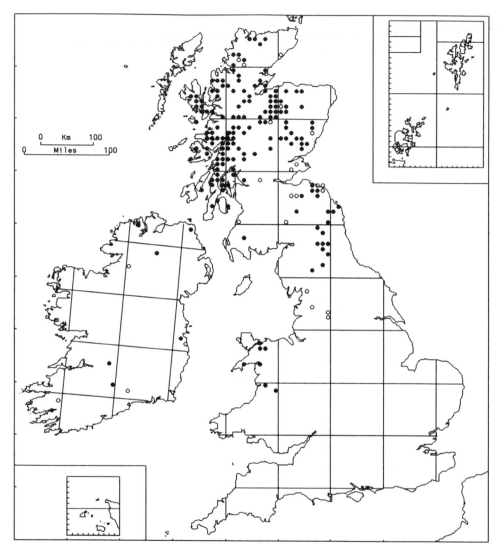

111/2. **Ulota drummondii** (Hook. & Grev.) Brid.

A locally frequent epiphyte forming tufts or spreading patches on branches of trees and shrubs, including birch, hazel, rowan, alder, juniper and sallow in wet upland areas. 0–620 m (Creag Meagaidh). GB 161 + 22*, IR 7 + 4*.
   Autoecious; sporophytes common, autumn.

   N. Europe south to the Tatra Mts and N. Germany, most frequent in W. Fennoscandia, where it reaches the Arctic; outlying stations in the Vosges and Carpathians. Japan, N. America (Pacific Northwest, Quebec (Bonaventure Island), Newfoundland).

M. C. F. Proctor

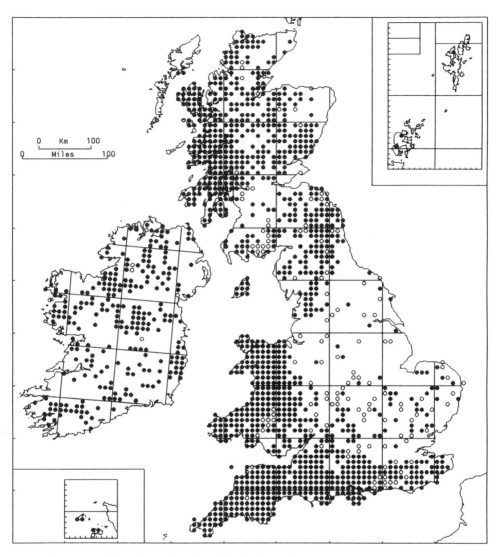

111/3. **Ulota crispa** (Hedw.) Brid. (including *U. bruchii* Hornsch.)

A characteristic epiphyte of the upper twigs and branches of a wide range of trees and shrubs including oak, birch, elder and hazel, where it is commonly associated with *Frullania dilatata*. Often prominent on branches of sallows in wet, sheltered places. It also occurs on trunks beneath the tree-canopy where it is not too deeply shaded and not subject to the competition of strongly-growing pleurocarps, especially on smooth-barked trees such as ash, and on coppice shoots of hazel and oak. Very rarely on acid rocks under trees. 0–410 m (Black Tor Copse). GB 1150+137*, IR 253+8*.

Autoecious; sporophytes abundant, ripe summer. Gemmae are produced on the protonema in culture (Whitehouse, 1987).

Throughout most of Europe. Macaronesia, W. Asia, Far East, eastern N. America.

Sensitive to atmospheric pollution and absent until recently from much of the Midlands and E. England. The status of var. *norvegica* (Grönvall) A. J. E. Smith & Hill is disputed; Rosman-Hartog & Touw (1987) argue for its retention as a separate species, *U. bruchii*. Var. *crispa* and var. *norvegica* have not been recorded with sufficient consistency to be mapped separately.

M. C. F. PROCTOR

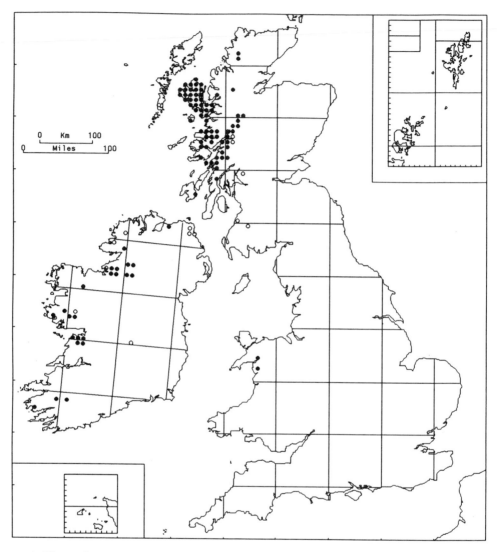

111/4. Ulota calvescens Wils.

An epiphyte forming small tufts on twigs and branches of trees and shrubs, especially hazel, birch and rowan, in woods, wooded ravines and wind-pruned coastal scrub. Associates include *Frullania teneriffae*, *Ulota crispa* and *U. phyllantha*. In Shetland it was recorded as an epiphyte on *Calluna*. 0–430 m (Ben Lui). GB 71+8*, IR 23+8*.

Autoecious; sporophytes common, dehiscing in autumn.

Portugal, Spain. Azores, Madeira, Canary Islands.

M. C. F. Proctor

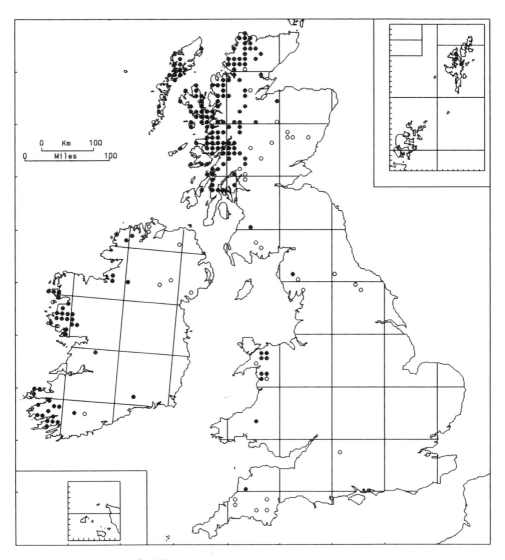

111/5. Ulota hutchinsiae (Sm.) Hammar

The most regularly saxicolous species of the genus in the British Isles, found on dry or intermittently flushed acid or basic igneous rocks and boulders, very rarely on trees. Habitats include coastal outcrops, streamsides, lake margins, stable block-scree, and sheltered mountain-corries. 0–490 m (Beinn Dorain). GB 145+29*, IR 43+8*.

Autoecious; sporophytes common, summer.

Widespread in the European mountains and southern Fennoscandia, to 68°N in Norway, extending east to the Tatra Mts and Carpathians. Turkey, western N. America (Alaska, Arizona), eastern N. America (N. Quebec south to Alabama and Georgia).

M. C. F. PROCTOR

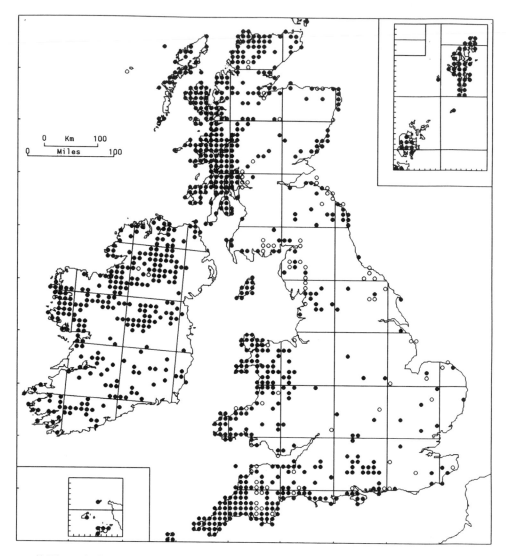

111/6. Ulota phyllantha Brid.

Saxicolous or epiphytic, forming small to large cushions on rocks, walls and the trunks or branches of a wide variety of trees and shrubs. Uncommon and almost exclusively epiphytic inland (most frequently in calcareous districts), but often abundant near western coasts where it may grow freely on maritime rocks and is usually the commonest epiphytic *Ulota* in exposed situations within 1–2 km of the sea. Frequent associates are *Schistidium maritimum* on rocks and *Cololejeunea minutissima*, *Frullania dilatata*, *Lejeunea ulicina*, *Metzgeria* spp. and *Ulota crispa* on trees and shrubs. Lowland. GB 670+86*, IR 282+4*.

Dioecious; sporophytes very rare. Conspicuous brown clusters of gemmae are almost always present on the tips of the upper leaves; gemmae also develop on the protonema in culture (Duckett & Ligrone, 1992).

Coasts of Iceland, Faeroes and N. Europe, from Kola Peninsula (Russia), Finnmark (Norway) and the Baltic coast of Fennoscandia south to the Gironde (France); in the southern part of its range it reaches about 120 km inland. Western and eastern coasts of N. America, Fuegian region of S. America.

Apparently spreading in S.E. England; it has been discovered in several inland counties since 1980.

M. C. F. PROCTOR

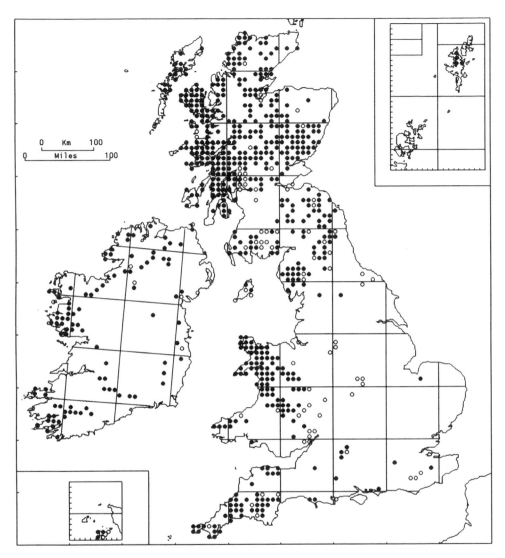

112/1. **Hedwigia ciliata** (Hedw.) P. Beauv.

Common and locally frequent on dry, sun-exposed, S.- or W.-facing rock-faces of low cliffs and tors, and on the tops of angular, detached boulders in moorland, in block-screes and below cliffs, growing with *Andreaea* spp., *Racomitrium* spp. and saxicolous lichens such as *Parmelia omphalodes* and *P. saxatilis*. It also occurs on the tops of large boulders by lakes and, more rarely, on acid drystone walls and roofing slates, and in old quarries. Restricted to hard acid or mildly basic rocks, absent from limestone and schist and commonest on crystalline siliceous rocks. Mainly at low and medium altitudes, ascending to 720 m (Ben Lawers). GB 531+78*, IR 95+7*.

Autoecious; capsules frequent, spring and summer.

Cosmopolitan. In the Northern Hemisphere its main distribution is in the boreal and cool-temperate forest zones, and in cool or cold continental interiors, extending north to the Low Arctic.

H. J. B. Birks

207

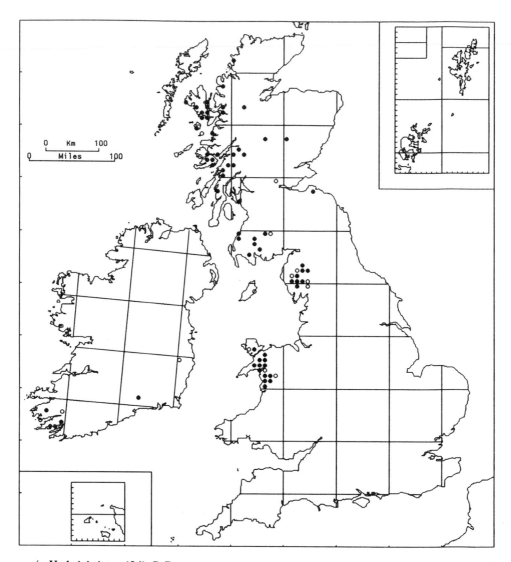

112/2. **Hedwigia integrifolia** P. Beauv.

Hedwigia integrifolia occurs in similar habitats to *H. ciliata* and is almost always associated with it, growing on dry, acid or mildly basic angular detached boulders, usually sun-exposed with a S. to W. aspect, in block-screes, on moorland and below cliffs. It is most frequent on basic igneous rocks of gabbro, basalt, dolerite, tuff and peridotite, and is absent from cliff-faces and man-made habitats except, rarely, drystone walls. Common associates include *Andreaea* spp., *Racomitrium* spp., *Parmelia* spp. and *Umbilicaria* spp. Mainly at low and moderate altitudes, ascending to 450 m (Moel Hebog). GB 60+11*, IR 5+6*.

Autoecious; capsules occasional, early summer.

W. Europe from N. Spain to S.W. Norway, extending east to C. and E. France (Massif Central, Vosges) and N. Italy. Cameroon, E. and S. Africa, Caucasus, Sri Lanka, Australia, Tasmania, New Zealand.

Normally less abundant than *H. ciliata*, it appears to be confined to warmer and slightly more basic habitats. For a map of its European distribution, see Schumacker (1985).

H. J. B. BIRKS

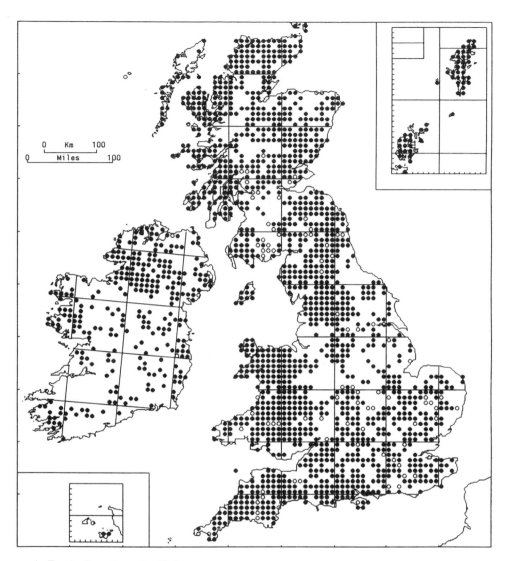

113/1. **Fontinalis antipyretica** Hedw.

An aquatic, found in ponds, ditches, swamps, canals, streams, rivers and even, occasionally, water-troughs. Although it often grows in oligotrophic water it is most frequent in neutral or basic, mesotrophic or eutrophic water and it can be particularly abundant in hitherto mesotrophic water which has recently received increased nutrient input. It tolerates periods of desiccation and insolation, and is frequent in the Irish turloughs where it forms a zone below that of *Cinclidotus fontinaloides* (Praeger, 1932). *F. antipyretica* descends to the lower limit of macrophyte growth, growing on mud at a depth of 12 m in the clear, calcareous water of Loch Baile a'Ghobhainn (Spence, 1967). 0–880 m (Coire an Lochain). GB 1449+92*, IR 278+4*.

Dioecious; sporophytes occasional, maturing in summer, only developing on emersed plants and most often found by ponds.

Europe from Iceland and N. Scandinavia to the Mediterranean islands. Widespread in the Northern Hemisphere, extending south to Macaronesia, Ethiopia, Japan and California; S. Africa.

Few Mapping Scheme recorders have distinguished the four varieties recognized in Britain and they are all mapped together.

C. D. PRESTON

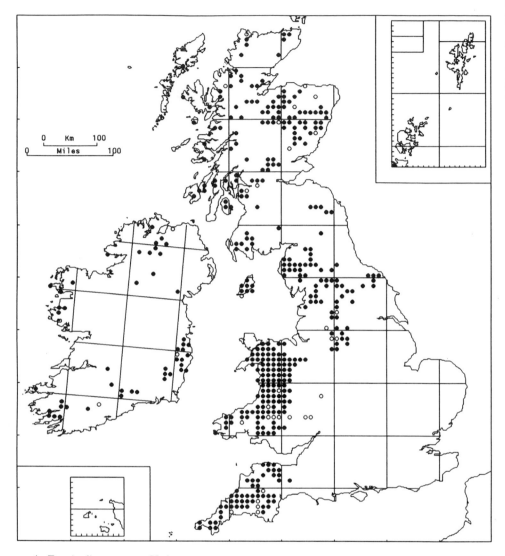

113/2. **Fontinalis squamosa** Hedw.

Fontinalis squamosa is much more restricted in its habitat range than *F. antipyretica*. It grows submerged in nutrient-poor, acidic water in fast-flowing streams and rivers, attached to rocks, boulders and tree-roots. Although less widespread than *F. antipyretica* it is commoner than that species in areas such as the Land's End peninsula and Bodmin Moor, where acidic rocks predominate (Paton, 1969). *F. antipyretica* is a frequent associate; others include the aquatic vascular plants *Callitriche hamulata*, *Myriophyllum alterniflorum* and *Potamogeton polygonifolius*. 0–830 m (Aonach Beag). GB 356+32*, IR 48+4*.

Dioecious; sporophytes occasional to rare, recorded from January to June and in September.

Widespread in Europe north to 68°N in Norway. Algeria.

The three varieties recognized in the British Isles grow in similar habitats. Var. *squamosa* is much the commonest. Var. *curnowii* Cardot is restricted to S.W. England, with an old record for Cheshire; elsewhere it is recorded from France and Norway. Var *dixonii* (Cardot) A. J. E. Smith occurs from N. Wales northwards and in Co. Fermanagh, Ireland; the only other records are from Portugal.

C. D. Preston

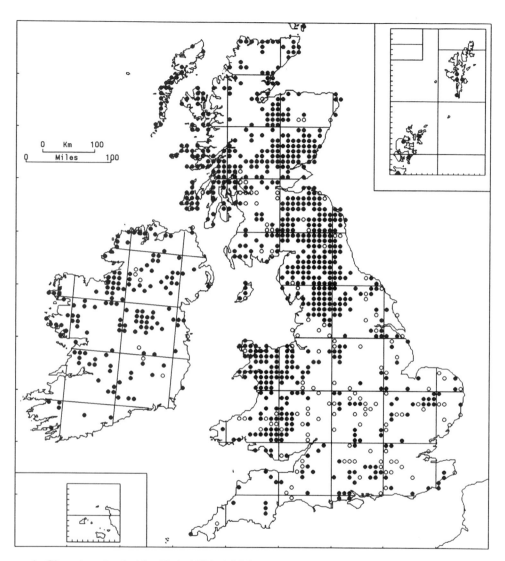

114/1. Climacium dendroides (Hedw.) Web. & Mohr

In damp, or, more rarely, dry habitats, which are at least mildly basic. It is found in damp grassland, especially in sites where the water-level fluctuates, such as the fringes of lakes, reservoirs and dune-slacks; wet meadows; mesotrophic fens; flushed slopes; damp willow scrub, fen carr and swampy alder woods; and in high-altitude, moss-dominated, snow-bed communities. Although normally in damp habitats, *Climacium* can grow in drier conditions in limestone grassland, on the fringes of limestone rock outcrops, at the foot of limestone walls, on stabilized sand-dunes and in short grazed turf inland, both on non-calcareous sandy ground and on calcareous Breckland soils. 0–900 m (Beinn Heasgarnich). GB 726+135*, IR 149+9*.

Dioecious; sex organs usually develop in permanently moist habitats but most colonies are unisexual and sporophytes are therefore rare. In N. England female clones outnumber males by 6:1, and even in mixed colonies female shoots are more frequent than males (Bedford, 1938). Capsules mature from September to March.

Circumpolar, mainly in the boreal zone and Arctic, and in mountains south to S. Europe, Turkey, C. Asia, Japan and south-west U.S.A. (New Mexico); in eastern N. America it reaches south only to Pennsylvania. Disjunct in New Zealand.

C. D. PRESTON

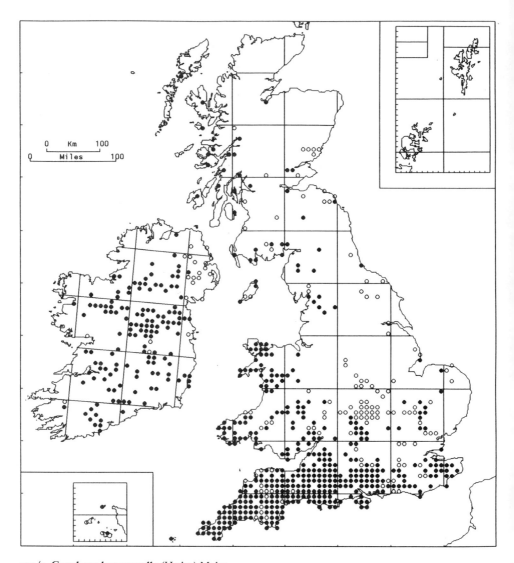

115/1. Cryphaea heteromalla (Hedw.) Mohr

An epiphyte, growing on bare bark or through a thin mat of pleurocarpous mosses on the trunks, branches and occasionally the exposed roots of trees and shrubs. It is most often found on the soft, water-retentive bark of mature elders but it also occurs on other hosts including apple, ash, beech, elm, field maple, hawthorn, ivy, oak, poplar, sycamore, willow and (very rarely) yew and conifers. It favours sheltered, often humid, situations such as the interior of scrub and woodland, and shrubs by streams, in fen carr and in chalkpits. Occasionally it grows on shaded rocks, concrete, masonry and gravestones. Lowland. GB 413+128*, IR 139+21*.

Autoecious; capsules usually present, maturing in winter. Gemmae are not recorded in the wild, but are produced on the protonema in culture (Whitehouse, 1987).

W. and C. Europe, north to Scotland and S. Sweden; rare in S. Europe and absent from most of the Mediterranean islands. Macaronesia, N. Africa, Israel, Turkey, Caucasus.

A pollution-sensitive species, less frequent than formerly in areas where SO2 levels are high.

C. D. Preston

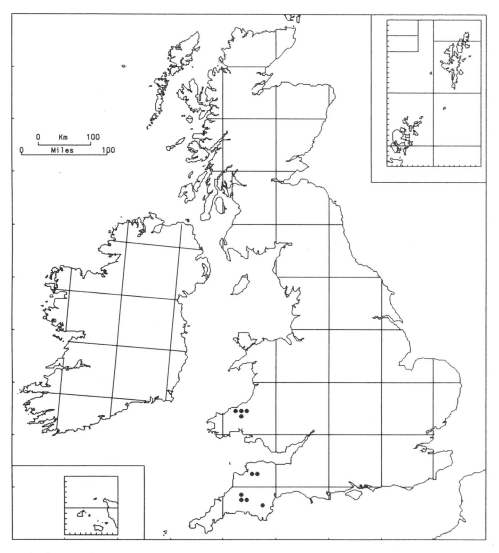

115/2. Cryphaea lamyana (Mont.) C. Müll.

On tree-trunks and rocks by major rivers, growing in a zone which is just above the normal water-level but is frequently flooded. As an epiphyte it apparently favours ash, elm and sycamore but is also recorded on alder, hazel and oak. Associated species include *Metzgeria furcata*, *Radula complanata*, *Cratoneuron filicinum*, *Homalia trichomanoides*, *Isothecium myosuroides*, *Leskea polycarpa*, *Neckera pumila*, *Orthotrichum affine*, *O. rivulare* and *Tortula latifolia*. Lowland. GB 10.

Autoecious; sporophytes frequent, maturing in autumn and spring. Gemmae are not recorded in the wild, but are produced in abundance on the protonema in culture (Whitehouse, 1987).

W. Europe and the W. Mediterranean region: Portugal, Spain, France, Switzerland and Italy. N. Africa.

Discovered in England by J. S. Tozer (who was drowned in 1836), beside the R. Dart above Hood Bridge, S. Devon. Although it was at one time thought to be extinct at this site (Crundwell, 1951), it was rediscovered there in 1980, on the stumps of recently felled sycamores.

C. D. PRESTON

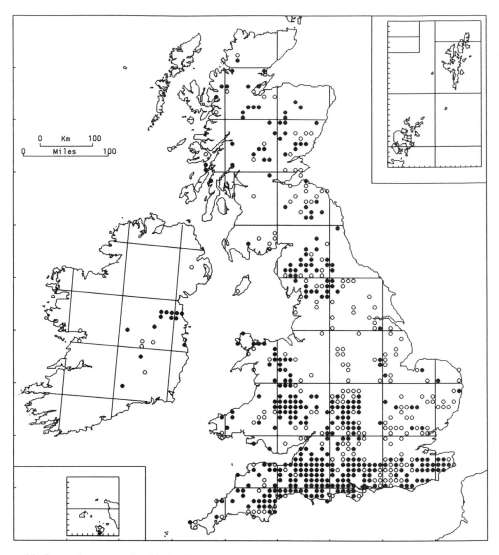

116/1. **Leucodon sciuroides** (Hedw.) Schwaegr.

Leucodon sciuroides is primarily an epiphyte, usually growing on base-rich tree-boles in open sites such as roadsides, hedges, parks and the sides of streams. It is most frequent on ash; other hosts include apple, beech, elder, elm, field maple, oak, poplar, sycamore and willow. It also grows on base-rich rock in natural habitats and on walls and tombstones. *Frullania dilatata, Hypnum cupressiforme, Orthotrichum lyellii* and *Tortula laevipila* are characteristic associates. 0–730 m (Creag na Caillich). GB 364+223*, IR 13+9*.

Dioecious; sporophytes rare. Vegetative propagation by deciduous branchlets.

Widespread in the Palaearctic. Europe north to Iceland and N. Scandinavia; a dominant epiphyte in Mediterranean countries above the coastal plain. Macaronesia, N. Africa, Asia.

Variable in size. Robust plants qualifying as var. *morensis* (Limpr.) De Not. are rarer than var. *sciuroides* but fruit more frequently. *Leucodon* has declined in lowland England because of SO_2 pollution coupled with eutrophication of bark. In some counties it is extinct as an epiphyte but persists on church walls and churchyard monuments made of oolitic limestone.

C. D. PRESTON

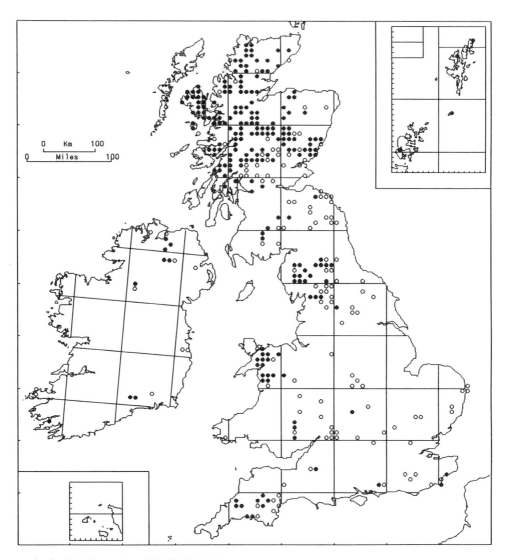

117/1. **Antitrichia curtipendula** (Hedw.) Brid.

Antitrichia is found on rocks and boulders on hillsides, in scree and by lakes, and on cliffs and walls. It also grows as an epiphyte on a range of trees and shrubs, and may be locally abundant on stunted trees in open habitats. On the dwarf oaks of Wistman's Wood in Dartmoor, it colonized 4–5 cm diameter twigs after *Frullania dilatata* and *Ulota crispa* but before a dense mass of *Hypnum mammillatum* developed (Proctor, 1962). It is also recorded, albeit rarely, in upland grassland, on sand-dunes and shingle, amongst *Thymus* over serpentine in Shetland and in N.-facing chalk grassland at Heyshott Down. 0–715 m (Ben Lawers). GB 212+128*, IR 9+10*.

Dioecious; sporophytes very scarce, perhaps more frequent formerly, maturing in spring.

Europe north to Iceland and Norway. Macaronesia, N. and C. Africa, S.W. Asia, N. America.

Antitrichia was widespread and abundant in the Late Glacial; it is also known from several archaeological sites in lowland England (Dickson, 1973, 1981; Stevenson, 1986). It disappeared from most of its south-eastern sites before 1900, almost certainly eliminated by SO2 pollution. It suffered a catastrophic decline at Wistman's Wood, possibly because of the increased growth of the host trees this century (Proctor, Spooner & Spooner, 1980).

C. D. PRESTON

215

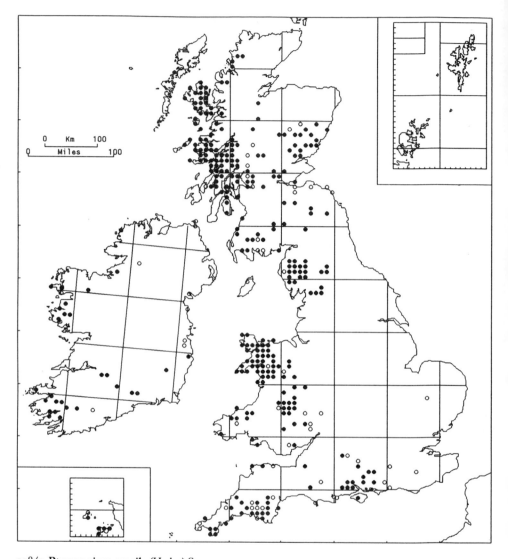

118/1. **Pterogonium gracile** (Hedw.) Sm.

Pterogonium grows on the sloping or vertical sides of rocks and boulders in sheltered, well-insolated sites or in shaded localities, sometimes by water on dry rocks in humid microclimates. It is most frequent on hard, basic igneous or metamorphic rocks such as basalt, dolerite, gabbro or calcareous tuff and is locally abundant on shaded serpentine on the Lizard Peninsula. It also grows on acidic rocks by the sea, but only rarely on calcareous sedimentary rocks or in turf over calcareous sand. *Pterogonium* is also found as an epiphyte on the bases or trunks of trees, usually on the base-rich bark of species such as ash, elder, elm, poplar and sycamore but also on oak. In the New Forest it is most frequent on ancient beeches, where it is sometimes abundant on individual trees (Paton, 1961). 0–500 m (Skye). GB 274+47*, IR 28+9*.

Dioecious; sporophytes very scarce, maturing in February.

S. and W. Europe, extending north to S.W. Norway; common in the Mediterranean countries at moderate altitudes. Macaronesia, Africa from the Mediterranean to the Cape, S.W. Asia, western N. America.

C. D. Preston

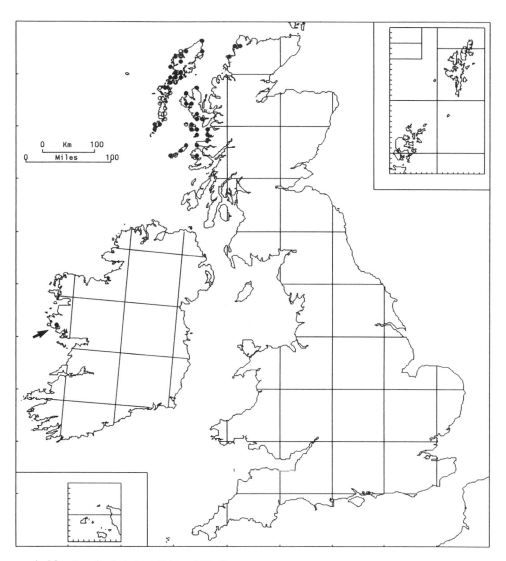

119/1. **Myurium hochstetteri** (Schimp.) Kindb.

This extremely western plant occurs most frequently as dense tufts in damp crevices of sunny or shaded wave-scoured rock platforms at the base of low sea-cliffs, often growing with *Frullania teneriffae*, *Bryum alpinum* and *Ulota phyllantha*. It also occurs in crevices of sheltered coastal rocks, on humus amongst heather on steep N.- to E.-facing block-strewn slopes below sea-cliffs, growing with *Hymenophyllum wilsonii*, *Bazzania tricrenata*, *Herbertus aduncus* and *Scapania gracilis*, and, more rarely, in open irrigated turf at the edge of exposed sea-cliffs. 0–110 m. GB 37+14*, IR 1.

Presumably dioecious; antheridia and archegonia are unknown in Britain, but capsules have been found in the Azores. Some Scottish plants produce shoots with deciduous leaves (Crundwell, 1981).

M. hochstetteri has a Macaronesian distribution, occuring in the Azores, Madeira and Canary Islands.

More frequent in the Outer Hebrides than elsewhere. It is inexplicably absent from many suitable-looking habitats between Ardnamurchan and the Point of Stoer, on Skye and Mull, and in W. Ireland.

H. J. B. BIRKS

217

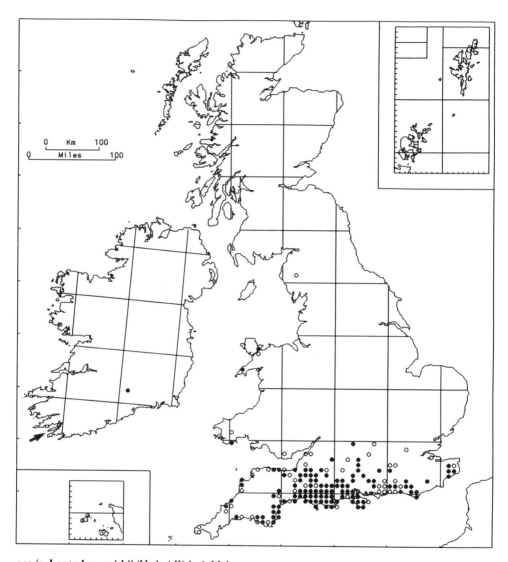

120/1. **Leptodon smithii** (Hedw.) Web. & Mohr

An epiphyte of tree-boles in woods, parks, churchyards, on roadsides and in hedgerows, also occurring on branches of shrubs in laid hedges. It is most frequent on hosts with basic bark, commonest (at least until recently) on elm, often found on ash and sycamore and also recorded from beech, elder, field maple, hawthorn, lime, oak, poplar, rhododendron, sallow and spindle. It is notably tolerant of dry shade, and thus able to grow on the underside of leaning trees. It is found rarely on walls (including concrete and flint) and calcareous rocks. Lowland. GB 109+53*, IR 1+1*.

Dioecious; sex organs frequent, sporophytes occasional, maturing in spring. Greene (1958) suggested that production of fruit was limited by spatial separation of the sexes.

Frequent in the Mediterranean countries, both on rocks and on trees, and extending north along the Atlantic coast to the British Isles; outlying localities in southern C. Europe and on the Black Sea coast. Macaronesia, N. Africa, Tanzania, S. Africa, S.W. Asia, southern S. America, S.E. Australia and New Zealand. Its European and world distributions are mapped by Pócs (1960).

C. D. Preston

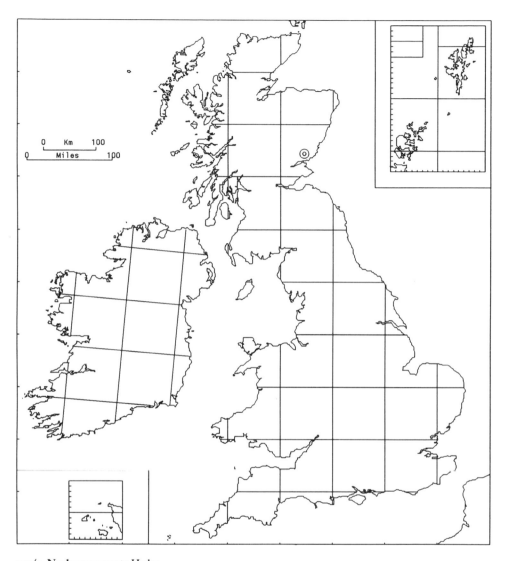

121/1. Neckera pennata Hedw.

An epiphyte, found only once in Britain, on the lower trunk of a beech-tree. Lowland. GB 1*.

Autoecious; sporophytes present. Vegetative propagation by flagelliform branches is frequent in Scandinavia.

C. and N. Europe. Widespread in the Northern Hemisphere; Australia, New Zealand, S. America (one old record).

The single British collection was made by T. Drummond at Fothringham in 1823. Although subsequently searched for, this distinctive plant has never been refound. It may have been a chance colonist rather than a long-term resident in Scotland. It is the only British species of *Neckera* that is not represented in the fossil record. In Scandinavia and N. America it usually grows on the trunks of deciduous or coniferous trees, but sometimes occurs on rock and rarely on logs and stumps. According to Duell (1992) it is now one of the most endangered mosses in C. Europe, although formerly widespread there.

C. D. Preston

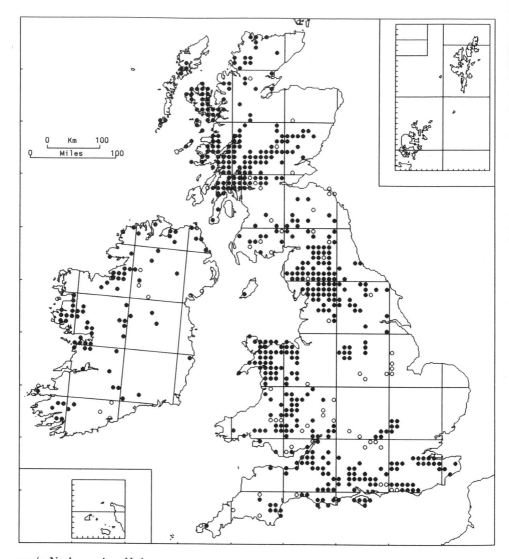

121/2. **Neckera crispa** Hedw.

One of the most reliable indicators of base-rich habitats. It grows on shaded or sheltered, dry or periodically irrigated rock-faces, including the sides of grikes in limestone pavement; on shaded walls, boulders and cliffs; in thin, species-rich calcareous turf; in lowland *Dryas* heath in northern Scotland; and in moist, tall-herb communities on montane cliff-ledges. It is particularly frequent on limestone, but also grows over chalk, calcareous schists, shales, sandstones and basic igneous rocks. Although shade-tolerant it is often replaced by *Thamnobryum alopecurum* on deeply shaded rocks. As an epiphyte, it occurs on the trunks and bases of trees, almost exclusively in habitats where it also occurs on rocks or the ground. Some of the most characteristic of its numerous associates are *Scapania aspera*, *Ctenidium molluscum*, *Fissidens cristatus*, *Homalothecium sericeum*, *Neckera complanata* and *Tortella tortuosa*. 0–700 m (Beinn Dorain). GB 503+62*, IR 80+11*.

Dioecious; sporophytes usually absent in southern England, occasional in the north and west. Gemmae not recorded in the British Isles.

Europe north to Iceland and Norway, becoming restricted to mountains in the south. Macaronesia, N. Africa, S.W. Asia.

C. D. PRESTON

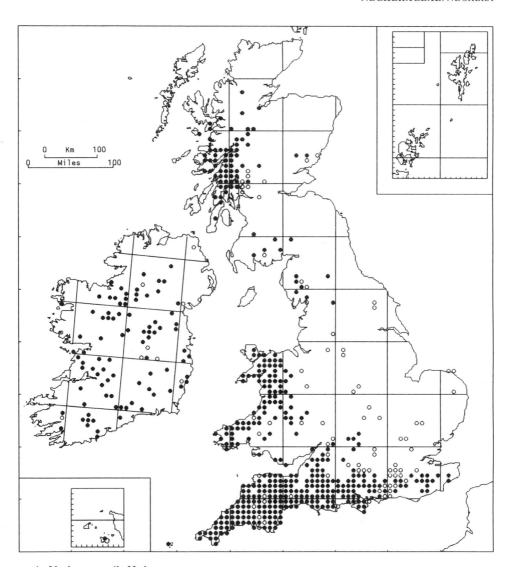

121/3. **Neckera pumila** Hedw.

An epiphyte, growing on trees and shrubs in woods and plantations or in sheltered sites such as valleys, sunken lanes and old quarries and marl-pits. It grows most luxuriantly on the smooth bark of trees such as ash, beech and sycamore but it is also found on a wide range of other hosts including alder, apple, birch, elder, elm, hawthorn, hazel, ivy, oak, poplar, rhododendron, sweet chestnut and willow. Associated species include *Frullania dilatata, F. tamarisci, Lejeunea ulicina, Metzgeria furcata, Cryphaea heteromalla, Hypnum cupressiforme, Leucodon sciuroides, Orthotrichum lyellii, Ulota crispa, U. phyllantha* and *Zygodon viridissimus*. Only very rarely is it recorded on inorganic substrates, e.g. on an old wall in Angus. Lowland. GB 396+78*, IR 87+9*.

Dioecious; sporophytes rare in S.E. England (although perhaps more frequent formerly), occasional elsewhere. Filamentous axillary branches are frequent.

W. and C. Europe north to northern Sweden; rare in S. Europe where it is restricted to areas of high rainfall. Macaronesia, Turkey.

N. pumila, like many epiphytes, has declined in areas of high atmospheric pollution.

C. D. Preston

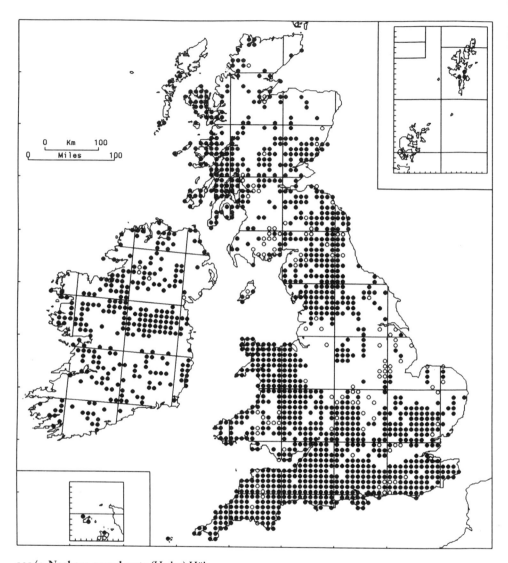

121/4. Neckera complanata (Hedw.) Hüb.

On sheltered, stable, usually shaded substrates. In woods, hedges and on streamsides it grows on the trunks, bases, exposed roots and gnarled coppice-stools of trees and shrubs, extending to branches of shrubs in the west. Inorganic substrates include steep or vertical, dry, calcareous rocks and boulders, including the grikes of limestone pavement; stone or mortared walls and bridges; Cornish hedges; turf over limestone; and ancient, grazed chalk grassland, especially on banks and steep slopes. It is particularly frequent in areas of calcareous soils and is rare (even as an epiphyte) over acidic clays, peats and sands. 0–600 m (Pen-y-Ghent). GB 1219+118*, IR 284+5*.

Dioecious; sporophytes occasional or rare, maturing from late winter to summer. Flagelliform branches are often produced in the leaf axils and act as a means of vegetative propagation.

Circumboreal. Widespread in Europe north to Iceland and northern Scandinavia.

Similar morphologically and ecologically to *Homalia trichomanoides*, which which it sometimes grows. *N. complanata* is less moisture-loving and is not normally found in the flood zone of streams and rivers.

C. D. PRESTON

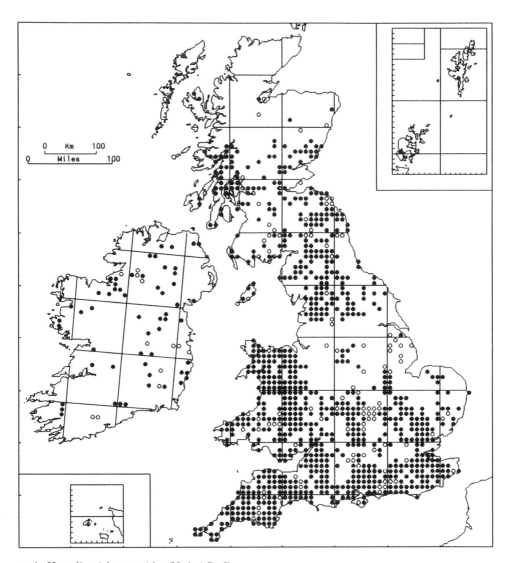

122/1. **Homalia trichomanoides** (Hedw.) Br. Eur.

On shaded, stable substrates. It grows on the exposed roots, bases, gnarled coppice-stumps and, rarely, branches of trees and shrubs, especially species with basic bark such as ash and maple. It is particularly frequent on trees by rivers, streams and ditches, growing above the water-level but in a zone liable to occasional flooding, but it also occurs away from water in damp woodland, wooded ravines, hedgerows and shaded lanes. *Homalia* is also found on a range of moist, shaded rocks, including limestone, basic igneous rock and non-calcareous sandstone, on streamside boulders, earthy banks, walls and stonework near the ground. Lowland. GB 820+113*, IR 56+13*.

Autoecious; sporophytes frequent, maturing in winter. Gemmae are produced on the protonema in culture (Whitehouse, 1987).

Widespread in the Northern Hemisphere, mainly in the broad-leaved forest zone. W., C. and N. Europe, as far north as N. Scandinavia.

Homalia is apparently decreasing in some areas of S.E. England because of hedgerow clearance, drainage and perhaps air pollution (Adams, 1974).

C. D. PRESTON

223

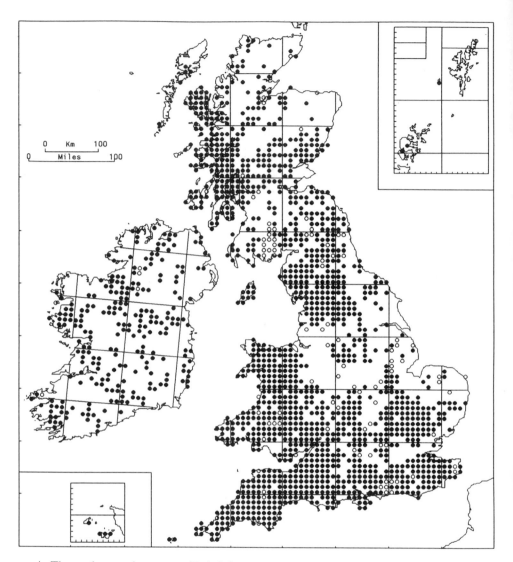

123/1. **Thamnobryum alopecurum** (Hedw.) Gang.

A shade-tolerant species which occurs in several distinct habitats. It grows on the ground, exposed tree-roots and tree-bases in woodland and on the banks of ditches and sheltered lanes, occurring on mildly acidic, neutral or basic soils but in particular abundance in woods over chalk, limestone and calcareous boulder clay. It is also found on dry or damp sheltered walls and rock-faces, the sides of grikes in limestone pavement and of sandstone sea-caves, the interstices of stabilized block-scree and the floors of quarries and chalkpits. In the uplands it is a characteristic plant of rocks, boulders and masonry in or alongside torrents, streams and rivers, where it often forms a dense band just above the water-level, and of wet rock-faces by waterfalls. In these waterside habitats it avoids the most acid sites, but does not behave as a calcicole. 0–500 m (Skye). GB 1366+100*, IR 242+6*.

Dioecious; sporophytes rare in the south-east, occasional in the north and west, maturing from September to May. Balls of *Thamnobryum*, detached from disturbed carpets of the moss on woodland floors, continue to grow as they are kicked and blown around and may play a minor role in dispersal.

Widespread in Europe north to Iceland and the coast of N. Norway. Macaronesia, N. Africa, Asia.

One of the most phenotypically plastic of British mosses.

C. D. Preston

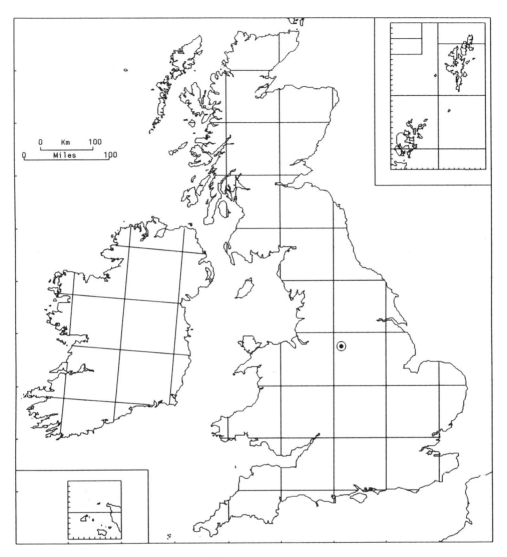

123/2. **Thamnobryum angustifolium** (Holt) Nieuwl.

On a shaded limestone rock in a ravine. *T. angustifolium* grows by a calcareous spring, where it is inundated by water in winter but dry in summer. It is surrounded by a dense sward of *T. alopecurum*. Lowland. GB 1.

Dioecious; only male plants known.

Endemic to Derbyshire. A specimen supposedly from Ireland is of doubtful provenance and a record from Madeira is erroneous.

T. angustifolium was discovered by G. A. Holt in 1883. A single colony, perhaps one clone, still occurs in his locality but the plants are apparently smaller than those collected last century. Furness & Gilbert (1980) showed that it maintains its characters in cultivation. Its taxonomy was reassessed by Hodgetts & Blockeel (1992), who concluded that it is more closely related to *T. cataractarum* and to the Madeiran *T. fernandesii* Sérgio than to *T. alopecurum*.

C. D. Preston

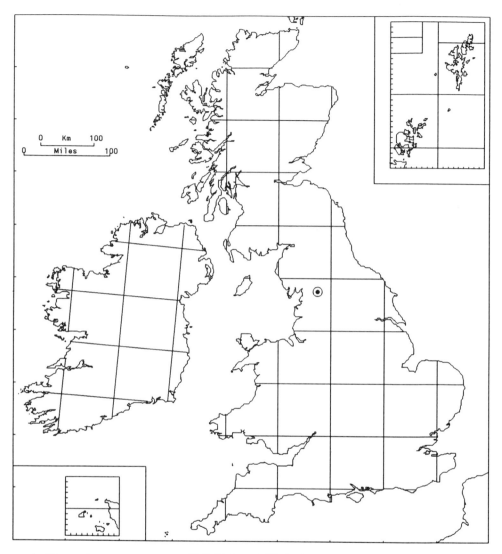

123/3. **Thamnobryum cataractarum** N.G. Hodgetts & Blockeel

A shade-tolerant species, found on vertical or steeply sloping streamside rocks in a narrow, deeply incised ravine. Here a calcareous stream flows over base-poor rocks of the Ingletonian strata. Pure swards of *T. cataractarum* grow in the swiftly flowing stream, submersed 10–25cm below the summer water-level. At shallower depths, plants extend into a zone where *Cinclidotus fontinaloides* and *Rhynchostegium riparioides* predominate and *Cratoneuron commutatum* var. *virescens*, *Fissidens crassipes* and *F. rufulus* occur as associates. Isolated stems can also be found amongst *T. alopecurum* just above the water-level. In waterfalls, *T. cataractarum* grows with *Rhynchostegium riparioides* just below the water surface. Lowland. GB 1.

Sterile; gametangia and sporophytes unknown.

Endemic to Yorkshire.

T. cataractarum was recently described from Twisleton Glen, a locality well known for oceanic bryophytes. It is closely related to the Madeiran *T. fernandesii* Sérgio. These notes are based on the detailed account by Hodgetts & Blockeel (1992).

C. D. PRESTON

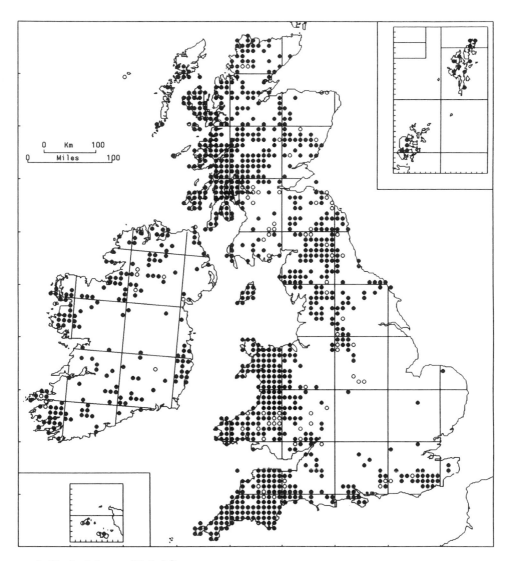

124/1. **Hookeria lucens** (Hedw.) Sm.

A locally frequent plant of damp shaded non-calcareous soils. Characteristic habitats include low-lying acid or mildly basic but fertile steep earthy banks by wooded streams, ditches, rivers and ravines with *Chrysosplenium oppositifolium*, *Pellia epiphylla*, *Fissidens adianthoides* and *Rhizomnium punctatum*; shaded soily ledges in ravines and gullies; humid, often deeply-shaded recesses in cliffs; crevices between boulders in stable block-screes on steep montane slopes; sheltered caves and crevices in sea-cliffs; and, in S.W. England, sheltered banks of wooded lanes. It also occurs in old quarries, in marshes, in wet alder and willow woods, on damp shaded rocks in woods on steep slopes, and in wet soaks and flushes within deciduous woods. In its less sheltered habitats it is often confined to N.- to E.-facing slopes. 0–1000 m (Aonach Mor). GB 850+77*, IR 171+13*.

Autoecious; capsules frequent, autumn and winter.

W. Europe from S. Spain to the Faeroes and W. Norway, extending east through C. Europe to the Carpathians and S. Sweden; absent from much of S. Europe. Macaronesia, Tunisia, Turkey, Caucasus, western N. America.

H. J. B. BIRKS

227

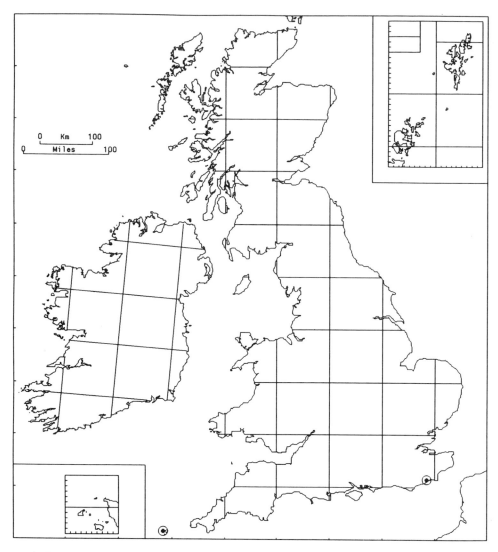

125/1. **Eriopus apiculatus** (Hook. f. & Wils.) Mitt. (*Calyptrochaeta apiculata* (Hook. f. & Wils.) Vitt)

On shaded substrata close to the coast. On Tresco, Isles of Scilly, it occurs on shaded earthy or peaty banks growing with *Lunularia cruciata, Atrichum undulatum, Ceratodon purpureus, Eurhynchium praelongum* and *Fissidens bryoides*. At Fairlight, East Sussex, it occurs on sheltered sandstone boulders on an undercliff, growing mixed with *Eurhynchium praelongum* and *Scorpiurium circinatum* under scrub of blackthorn, bramble and sallow. Lowland. GB 2.

Dioecious; only female plants known in Britain.

South Africa, Chile, Argentina (Tierra del Fuego), Australia, Tasmania, New Zealand.

In both localities it was first discovered in 1967 (Paton, 1968; Wallace, 1971). It is almost certainly an introduction from the Southern Hemisphere with horticultural plants (Paton, 1968) or from shipping or smuggling on the south coast of England (Stern, 1991).

H. J. B. BIRKS

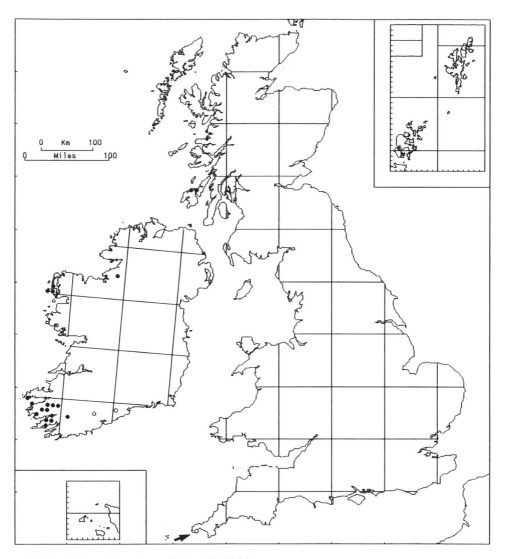

126/1. **Cyclodictyon laetevirens** (Hook. & Tayl.) Mitt.

It occurs, often as pure patches, on acid or mildly basic, deeply-shaded, wet rocks in dripping caves and recesses by waterfalls and cascades, and in holes in the sides of ravines. More rarely, it grows on stipes and fronds of senescent *Trichomanes speciosum* in deeply-shaded, dripping caves and recesses, and in deep caves between blocks on steep slopes where there is percolating water. On Islay and Jura it is found in wet, dripping caves along the raised-beach platform. Associates include *Jubula hutchinsiae*, *Lejeunea holtii*, *Radula holtii* and *Riccardia chamedryfolia*. 0–330 m (S.W. Ireland). GB 2+1*, IR 17+2*.

Synoecious, autoecious or dioecious; capsules rare, late summer and autumn.

A southern species confined in Europe to Portugal, Spain and the British Isles. Azores, Madeira, tropical Africa.

Probably commoner in W. Ireland than the map suggests. Extinct in its Cornish stations, apparently as a result of over-collecting.

H. J. B. Birks

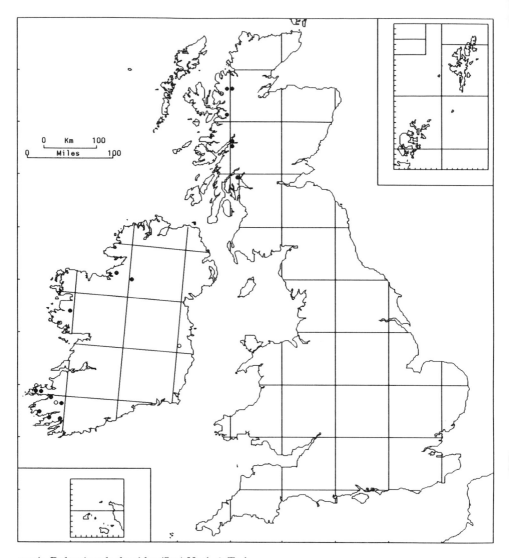

127/1. **Daltonia splachnoides** (Sm.) Hook. & Tayl.

It occurs in a range of habitats, including gently sloping flushed, often mildly basic, blocks in and by shaded wooded streams, flushed N-. to E.-facing rock-faces growing with species such as *Blindia acuta* and *Heterocladium heteropterum*, shaded rotten logs overhanging streams and rivulets, and tree-roots, tree-trunks and humus by or near streams or waterfalls in sheltered situations. From near sea-level to 460 m (Kerry), but not ascending above 360 m in W. Scotland. GB 6+1*, IR 10+3*.

Autoecious and dioecious; capsules common, summer.

Unknown on the European mainland. Azores, Madeira, W. Africa (Bioko, formerly known as Fernando Po), western N. America, Mexico, Antilles, New Zealand.

This elusive plant may possibly be commoner than the map suggests, but it is undoubtedly very rare and is absent from countless suitable-looking locations. It may be rather fugitive; in two sites it has been found on exposed roots of young ash-trees aged between ten and fifteen years. Here, it would probably be ousted by robust species such as *Isothecium myosuroides* and *Hypnum cupressiforme* after about ten years.

H. J. B. BIRKS

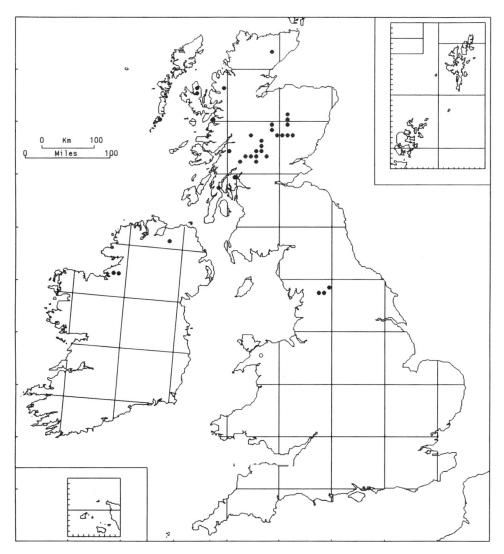

128/1. **Myurella julacea** (Schwaegr.) Br. Eur.

Usually occurs as scattered shoots growing amongst other basiphilous bryophytes such as *Amphidium lapponicum*, *Anoectangium aestivum*, *Encalypta ciliata*, *E. rhaptocarpa* and *Tortella tortuosa* on dry or slightly irrigated soil in cracks or on small ledges of lightly shaded limestone, calcareous mica-schist, or basalt cliffs, usually of north to east aspect. It can also occur as isolated shoots in shaded crevices on basic cliffs growing with *Distichium capillaceum*, *Mnium thomsonii*, *Plagiopus oederi* and *Pohlia cruda*, and as small but pure patches on dry limestone or mica-schist cliffs, often facing south or west, growing near *Pseudoleskeella catenulata* and *Schistidium* spp. 450 m (Skye) to 1200 m (Ben Lawers). GB 32+1*, IR 3.

Dioecious; capsules very rare, ripe in late summer.

Circumpolar, with a subarctic-subalpine distribution and a Southern Hemisphere disjunction. Iceland, Arctic Europe, mountains of W., C., N. and E. Europe south to the Crimea. Turkey, Caucasus, C., N. and E. Asia, N. America, Greenland (to 83°N), Antarctica.

Two varieties are recognized in the British Isles, var. *julacea* and var. *scabrifolia* Lindb. ex Limpr. The latter is a taxon of uncertain status and usually associated with var. *julacea*.

H. J. B. BIRKS

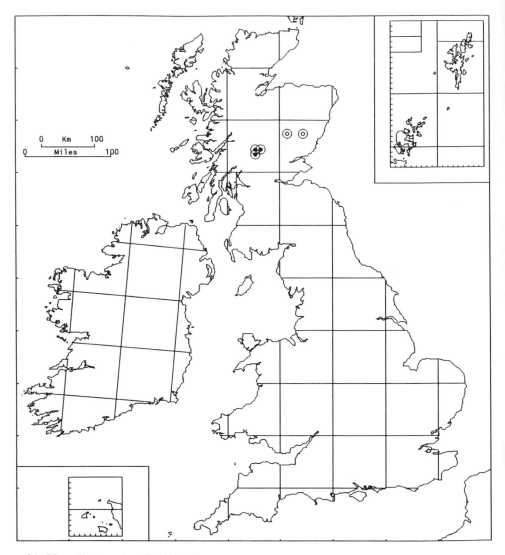

128/2. **Myurella tenerrima** (Brid.) Lindb.

On dry soil in crevices and on bare, often loose rock on calcareous mica-schist cliffs with a southern aspect. It usually occurs in small quantity as isolated stems growing among other basiphilous bryophytes such as *Encalypta ciliata*, *E. rhaptocarpa*, *Myurella julacea* and *Tortella tortuosa*. Confined to high altitudes, 700–1180 m (Ben Lawers). GB 3+2*.

Dioecious; capsules not known from Britain.

Circumpolar with a subarctic-subalpine distribution. Iceland, W., C. and N. Europe, Svalbard; montane except in the north. N. Africa, Turkey, Caucasus, C., E. and N. Asia, N. America (to 83°N on Ellesmere Island), Greenland.

It is invariably but inexplicably rarer and considerably more local than *M. julacea*. In view of the availability of seemingly suitable habitats in the C. Highlands it is probably commoner than the map suggests.

H. J. B. BIRKS

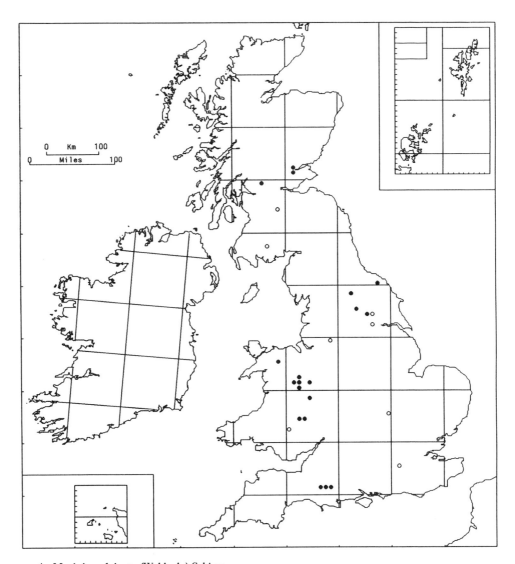

129/1. **Myrinia pulvinata** (Wahlenb.) Schimp.

Confined to tree-boles and roots, often of alder, on banks in the middle and lower reaches of streams and rivers where there is a well defined flood zone. It is usually embedded in silt or sand and associated with *Leskea polycarpa* and *Tortula latifolia*. However, it is much more restricted in occurrence than either of these two species, being absent from sites where flooding is shallow. Lowland. GB 19+8*.

Autoecious; capsules frequent, maturing in summer.

N. and N.W. Europe. N.W. Siberia, western and eastern Canada.

T. L. BLOCKEEL

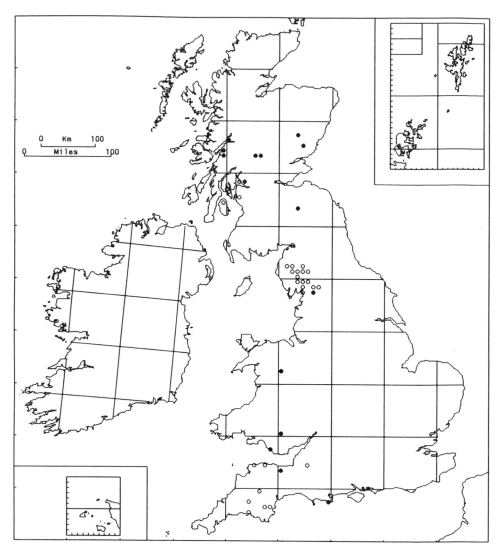

130/1. **Habrodon perpusillus** (De Not.) Lindb.

A slender epiphyte, found in tightly appressed patches or occasionally creeping through other bryophytes on well-illuminated trunks or branches of trees and shrubs. It usually grows on basic bark: sycamore is the most frequent host (perhaps surprisingly, as it is an introduction) and there are several records from ash and elm. It has also been recorded on aspen, birch, blackthorn, elder, hawthorn, lime, oak and willow. Associated bryophytes include *Frullania dilatata, Metzgeria furcata, Orthotrichum affine, O. lyellii, O. stramineum, O. tenellum, Tortula papillosa, Zygodon baumgartneri* and *Z. viridissimus*. Lowland. GB 15+29*.

Dioecious; sporophytes very rare. Effective reproduction is by gemmae, which are frequent on the stems; similar gemmae are produced on the protonema in culture (Whitehouse, 1987).

A Mediterranean-Atlantic species, reaching its northern limit in S.W. Norway. Macaronesia, N. Africa, S.W. Asia.

Habrodon has decreased in frequency in the last 100 years, perhaps because of air pollution. At least one colony has disappeared recently as the elm on which it grew was killed by disease. It is, however, an inconspicuous species and it may be somewhat more frequent in Scotland than the map suggests.

C. D. PRESTON

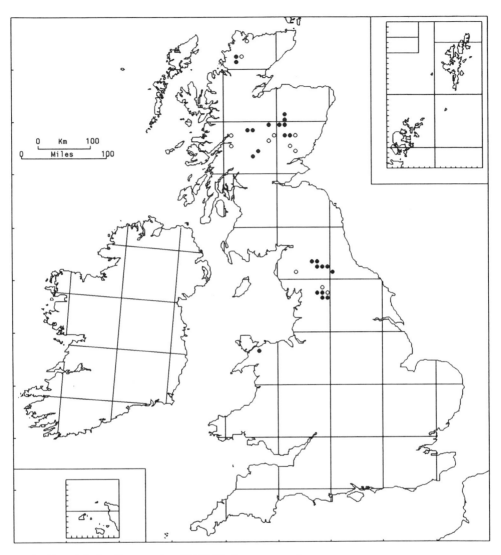

131/1a. **Pseudoleskeella catenulata** (Brid.) Kindb. var. **catenulata**

It forms dense dark- or olive-green patches on calcareous rocks in the hills, occurring on limestones of Carboniferous and Cambrian age in N. England and N.W. Scotland and on metamorphosed limestones in the central Scottish Highlands. It prefers dry, open sites, often S.-facing, where there is little competition. Its associates may include *Ctenidium molluscum*, *Schistidium* spp., *Tortella tortuosa* and, more rarely, *Campylium halleri* and *Lescuraea incurvata*. 250 m (Inchnadamph) to 1100 m (Ben Lawers). GB 24+12*.

Dioecious; sporophytes unknown in Britain.

Mountains of W., C. and N. Europe.

The genus *Pseudoleskeella* has recently been revised (Wilson & Norris, 1989). Records of *P. catenulata* var. *catenulata* from outside Europe apparently all refer to other taxa.

G. P. ROTHERO

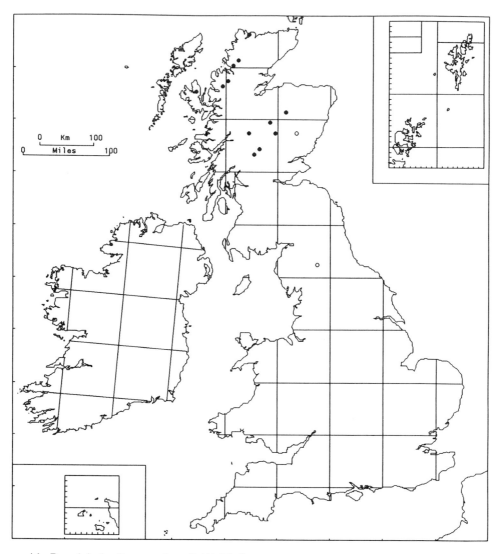

131/1b. **Pseudoleskeella catenulata** (Brid.) Kindb. var. **acuminata** (Culm.) Amann (*P. sibirica* (Arnell) P. Wilson & Norris)

On dry, calcareous, montane rocks in similar situations to var. *acuminata*, often with the same associates and sometimes mixed with it, but seemingly more tolerant of both shelter and competition. From near sea-level (Smoo Cave) to 1100 m (Ben Alder Forest). GB 12+2*.

Dioecious; sporophytes unknown. Lacking specialized means of vegetative propagation.

Widespread in European mountains, including the Pyrenees and Alps, and in Scandinavia. Siberia (Yenisey), N. America, Greenland.

P. catenulata var. *acuminata* has a very different appearance to var. *catenulata*, being more reminiscent of a *Lescuraea*. It was misunderstood in Britain until Crundwell (1953) found the two taxa growing intermixed. Following a revision of *Pseudoleskeella* by Wilson & Norris (1989), var. *acuminata* is now generally regarded as a distinct species, *P. sibirica* (Anderson *et al.*, 1990; Corley & Crundwell, 1991).

G. P. ROTHERO

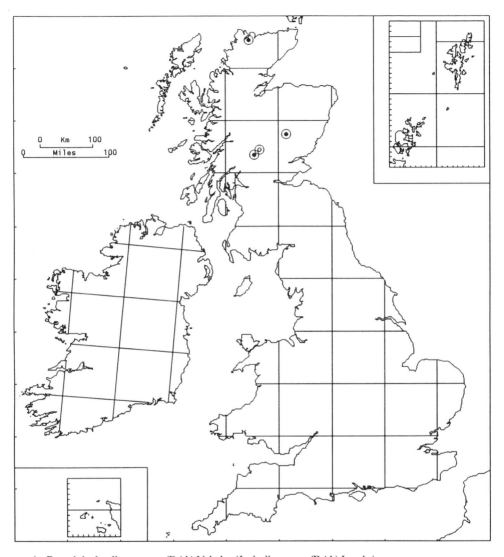

131/2. **Pseudoleskeella nervosa** (Brid.) Nyholm (*Leskeella nervosa* (Brid.) Loeske)

In recent years this species has been found only on open calcareous rocks in the mountains, but there is an old record from near Ben Lawers on tree-trunks, a common habitat in calcareous areas of the European continent and in N. America. 500 m (Ben Hope) and 820 m (Meall nan Tarmachan). GB 3+1*.

Dioecious; capsules unknown in Britain. Dwarf axillary branches frequently crowd the stem apex and presumably act as vegetative propagules.

Circumboreal, ranging from the Arctic south in mountains to Spain, Greece, Iran, C. Asia, Japan and southern U.S.A. (Arizona, N. Carolina). For a map of the world distribution, refer to Lewinsky (1974).

G. P. ROTHERO

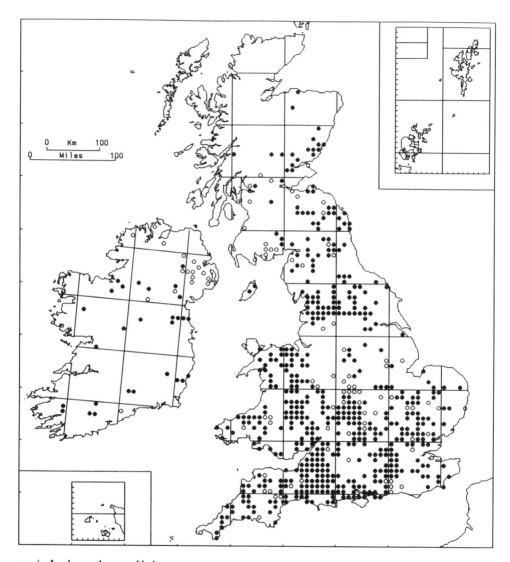

132/1. Leskea polycarpa Hedw.

Virtually confined to the borders of rivers, streams, ditches, lakes and pools in the eutrophic lowlands, growing in dry sites above the normal water-level but in a zone liable to flooding. It favours substrates where some silt has accumulated, being most frequent on the exposed roots and lower trunks of riverside trees but also found on the woodwork, brickwork and stonework of walls and bridges and, very rarely, on earthy waterside banks. Characteristic associates are *Amblystegium serpens*, *Barbula sinuosa*, *Bryum capillare*, *Cinclidotus fontinaloides*, *Cratoneuron filicinum*, *Homalia trichomanoides*, *Orthotrichum affine*, *O. diaphanum* and *Tortula latifolia*. *L. polycarpa* is very rarely found away from water but there are isolated records from such habitats, e.g. abundant plants on beech roots at a site on the chalk escarpment in Kent (Rose, 1951). Lowland. GB 511+78*, IR 28+18*.

Autoecious; sporophytes very common, maturing in summer.

Widespread in the North temperate zone, north to Iceland, central Scandinavia, Siberia and Canada (British Columbia, Newfoundland), and south to Macaronesia, N. Africa, Turkey, C. Asia, Japan and southern U.S.A. (Oregon, Louisiana).

C. D. Preston

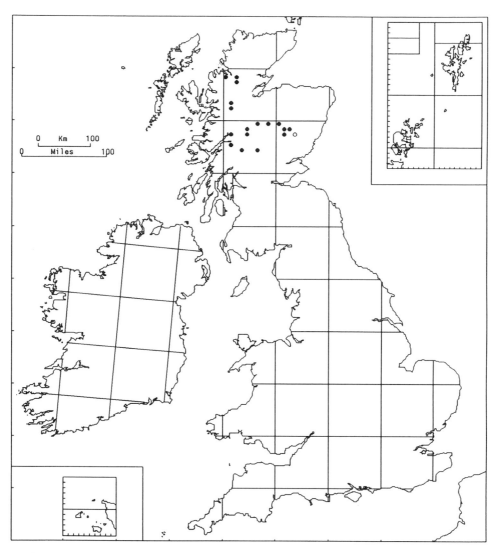

133/1. **Lescuraea patens** (Lindb.) Arn. & C. Jens. (*Pseudoleskea patens* (Lindb.) Kindb.)

Typically this species occurs on mica-schist or limestone rock surfaces or on soil among basic rocks in rather dry, sheltered situations, in gullies, on cliffs or between boulders. *L. patens* is, however, tolerant of a wide range of conditions, and can occur in exposed situations, in damp places by flushes and seepages, and on rocks beside streams and small lakes, often with *Pterigynandrum filiforme*. In several sites it grows in some abundance on the surface of blocks in scree subject to very late snow-lie. Here it seems indifferent to rock type, occurring on granite, acid granulite and schist, with an associated flora that varies from moderately calcicolous to definitely calcifuge and often includes *Brachythecium glaciale*, *Hylocomium umbratum* and *Plagiothecium* spp. 600–1200 m (Ben Nevis). GB 17+1*.

Dioecious; capsules occasional in snow-beds, very rare elsewhere, spring or summer.

Widespread in Europe, especially in the north, becoming montane further south. Turkey, E. Asia, N. America.

It has often been confused with *L. incurvata*. Specimens in herbaria have not been systematically revised; the map may include some errors.

N. F. Stewart

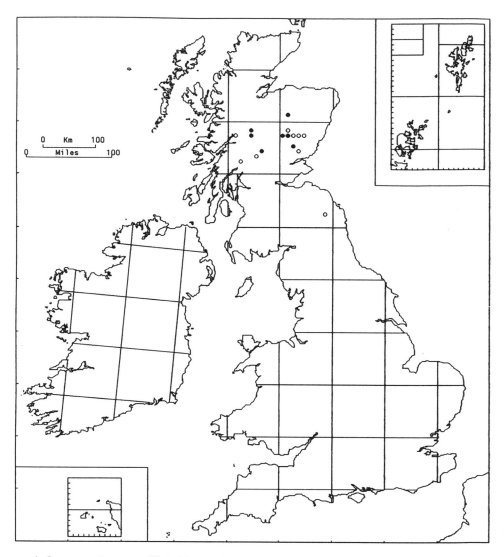

133/2. Lescuraea incurvata (Hedw.) Lawton (*Pseudoleskea incurvata* (Hedw.) Loeske)

A species of calcareous and basic siliceous rocks including mica-schist, limestone, calcareous sandstone and andesite. Most localities are in the mountains, where it grows in dry, sheltered situations, e.g. at the bases of boulders in block-scree and in gullies. It is also found occasionally on rocky banks in lowland river ravines, and, at one site, on a wall. Associated species include *Blepharostoma trichophyllum, Distichium capillaceum, Pseudoleskeella catenulata* and *Tortella tortuosa.* Usually in the altitudinal range 600–1100 m (Ben Lawers), but descending to 80 m (near Town Yetholm). GB 7+9*.

Dioecious; capsules rare, spring.

Boreal-montane, widespread in the Northern Hemisphere except for eastern N. America.

N. F. STEWART

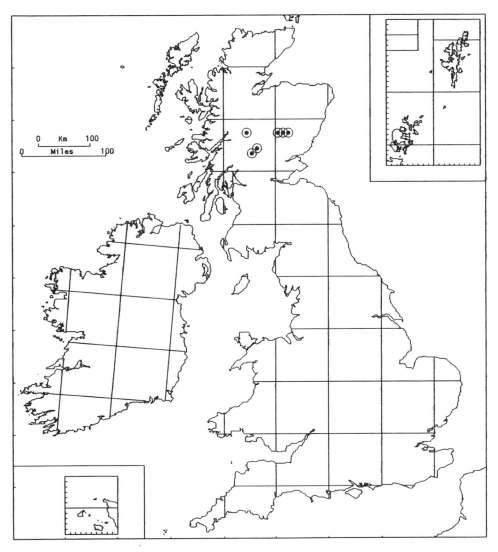

133/3. **Lescuraea plicata** (Schleich. ex Web. & Mohr) Lawton (*Ptychodium plicatum* (Web. & Mohr) Schimp.)

A calcicole of wet or dry rocks and rocky turf, on limestone or mica-schist. It often occurs in somewhat sheltered situations such as gullies and crevices among boulders. A typical habitat is in turf at the base of crags or at the base of boulders in block-scree, with associates including *Hypnum callichroum* and *Lescuraea incurvata*. 650–1200 m (Ben Lawers). GB 6.

Dioecious; capsules very rare in Britain and N. Europe, but occasional in the southern mountains, mature in autumn.

An arctic-alpine, occurring widely in Europe from the far north to the mountains of the south. Outside Europe known only from the Caucasus.

N. F. STEWART

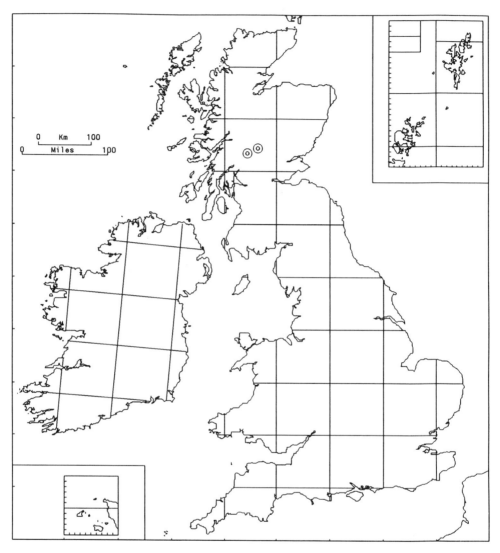

133/4. Lescuraea saxicola (Br. Eur.) Milde

It occurred on stones and broken pieces of rock, presumably on base-rich substrata. 700-950 m. GB 2*.
Dioecious; capsules unknown in Britain and very rare elsewhere.

Circumboreal, from the Arctic south to the mountains of Turkey, Caucasus, C. Asia and S. Canada; very rare in coterminous U.S.A. (Utah, Vermont).

First found by W. West on Ben Lawers in 1880. It was always very rare there and cannot have been helped by West himself, who distributed it to about three dozen people. It was found on a few subsequent occasions and was last seen in 1911. It has otherwise been found only on Meall na Samhna in 1905.

N. F. Stewart

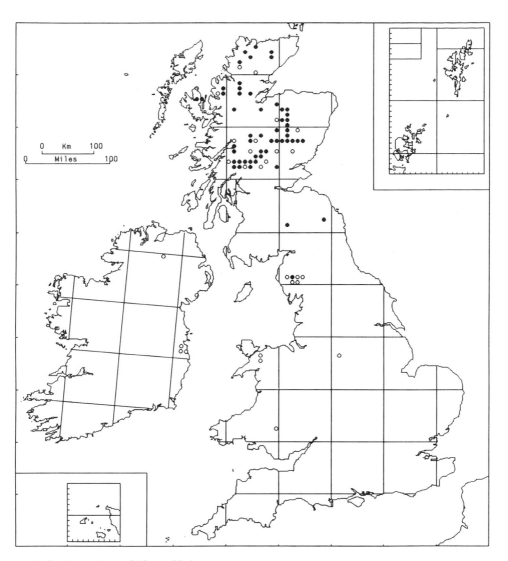

134/1. **Pterigynandrum filiforme** Hedw.

A plant of open sites in the mountains, where it forms neat yellow-green, green or brown mats on basic rocks and, rarely, branches and roots of trees. Frequent associates include *Andreaea* spp., *Isothecium myosuroides* and *Racomitrium* spp. or, on more strongly base-rich substrata, *Lescuraea incurvata*, *L. patens* and *Schistidium* spp. It often occurs on the larger boulders on the margins of montane lochs, a habitat it occasionally shares with *Lescuraea patens*. 10 m (Loch Maree) to 1100 m (Ben Lawers). GB 54+25*, IR 5*.

Dioecious; capsules rare. Gemmae often present on the stems.

Circumboreal and, towards the south, montane, ranging from the Arctic south to Madeira, Canary Islands, Algeria, Turkey, Kashmir, India, Japan and Mexico.

Although quite widespread in Britain and Ireland, it is uncommon except in the Breadalbane mountains, where it is locally frequent. Var. *majus* (De Not.) De Not., a poorly-marked variety found mainly on loch margins, is not mapped separately. It is probably only a habitat modification.

G. P. ROTHERO

243

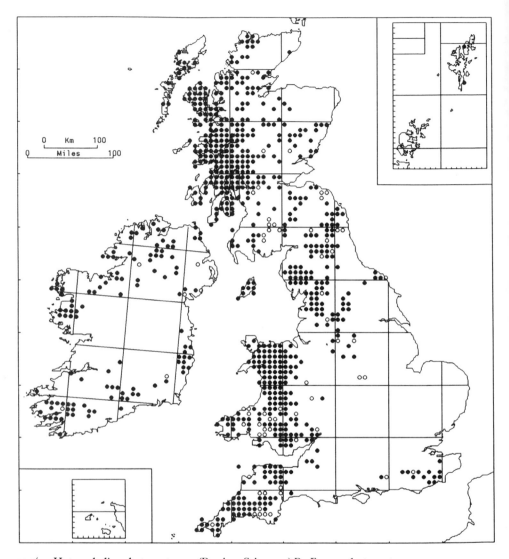

135/1a. **Heterocladium heteropterum** (Bruch ex Schwaegr.) Br. Eur. var. **heteropterum**

A plant occurring typically on damp, acidic to mildly basic rocks in deeply shaded and humid sites, normally in woodlands or ravines, often close to streams and rivers. It is commonly found on steep or vertical rock-faces, in fissures or on boulders, sometimes in periodically flooded sites. It is often associated with a rich and varied bryophyte flora. On moist rock-faces or boulders this may include such species as *Cephalozia bicuspidata*, *Diplophyllum albicans*, *Metzgeria temperata*, *Grimmia hartmanii*, *Isopterygium elegans* and *Rhabdoweisia crispata*, and on wetter rocks *Hyocomium armoricum* and *Racomitrium aciculare*. It has also been recorded in sheltered sea-caves, on cliffs, in quarries, on rocky lane-banks, and on granite tors and walls. It occurs less commonly on soil, and rarely as an epiphyte, for instance on tree-roots. Most frequent in lowland to low-montane areas, but ascends to 1150 m (Coire an Lochain). GB 631+61*, IR 115+6*.

Dioecious; sporophytes rare, spring.

Widespread in W. and C. Europe, extending east to Sweden and the Carpathians and north to 68°N in western Norway. Macaronesia, Turkey, Caucasus.

M. J. WIGGINTON

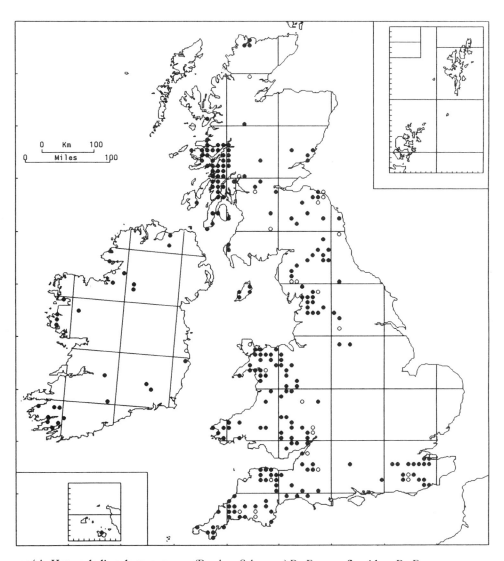

135/1b. **Heterocladium heteropterum** (Bruch ex Schwaegr.) Br. Eur. var. **flaccidum** Br. Eur.

A plant of densely shaded rocks in woodland and ravines, often near streams and rivers. It appears to favour a rather drier, more basic substrate than does var. *heteropterum*, occurring on limestone, calcareous sandstone and other rocks which are at least mildly basic. It has also been recorded growing on flint in chalky districts and on ragstone in S.E. England. On rock-faces, boulders or small stones, its associates include such species as *Metzgeria furcata*, *Ctenidium molluscum*, *Fissidens pusillus*, *Homalia trichomanoides*, *Hygrohypnum luridum* and *Taxiphyllum wissgrillii*. It occurs infrequently on soil, and rarely epiphytically on tree-roots. Lowland. GB 223+27*, IR 26+3*.

Dioecious; sporophytes unknown.

W. and C. Europe, north to southern Scandinavia.

M. J. WIGGINTON

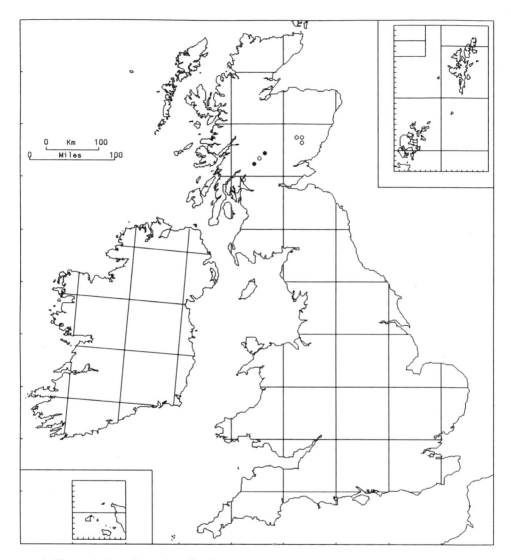

135/2. **Heterocladium dimorphum** Br. Eur.

On shaded montane rocks. On Ben Lawers it occurs on sheltered, and usually dry, ledges on crags and isolated boulders and also at the base of boulders in vegetated scree. *Diplophyllum taxifolium*, *Lophozia sudetica*, *Dryptodon patens* and *Encalypta alpina* have been noted as associates. A montane plant, ascending to 1100 m (Ben Lawers). GB 2+5*.

Dioecious; sporophytes unknown in Britain.

Arctic and N. Europe, including Iceland and Svalbard, and in the major mountain ranges of C. and S. Europe. Turkey, Caucasus, Siberia, eastern and western N. America, Greenland.

In N. America it is normally found on boulders or in the crevices of cliffs, also occasionally on soil or humus on shaded stream-banks, or epiphytically on tree-roots.

M. J. WIGGINTON

246

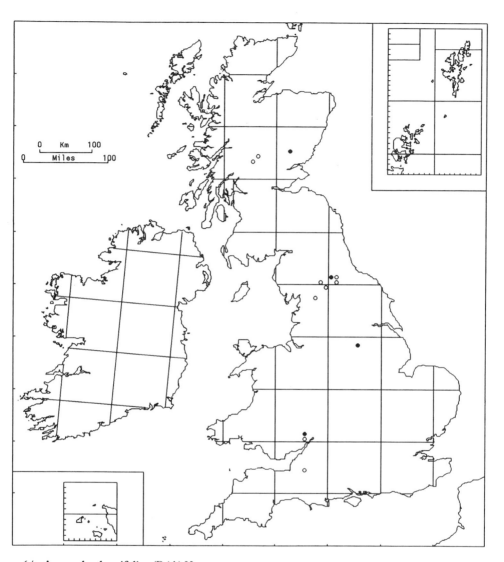

136/1. Anomodon longifolius (Brid.) Hartm.

On shaded, steep or vertical limestone or basic sandstone rocks, often in wooded valleys and ravines, and on limestone rock-ledges. Associated species include *Scapania aspera*, *Anomodon viticulosus*, *Isothecium striatulum*, *Neckera complanata* and *Thamnobryum alopecurum*. Since 1950 it has not been found above 300 m, but it formerly occurred on Ingleborough and Ben Lawers, presumably at higher elevations. GB 4+9*.

Dioecious; sporophytes not known in Britain. The species lacks special means of vegetative propagation.

Widespread in Europe north to N.W. Norway; absent from the south and the Mediterranean islands. Caucasus, Siberia, Russian Far East (Sakhalin), Japan.

The apparent decline of *A. longifolius* is clearly shown on the map. However, it has recently been rediscovered in Yorkshire (1986) and Co. Durham (1988), and it may be overlooked elsewhere.

C. D. Preston

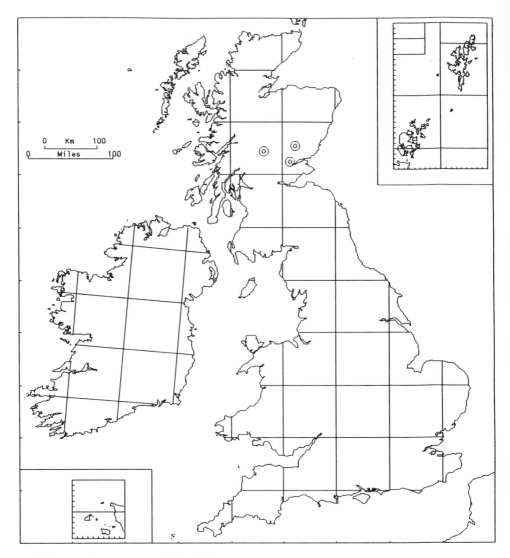

136/2. **Anomodon attenuatus** (Hedw.) Hüb.

In mats on rotting tree-trunks and damp, shaded, probably basic, sandstone rocks in the Den of Airlie and as an epiphyte on alders at Elcho. It is also recorded, without habitat details, from Ben Lawers (Braithwaite, 1887-1905). In Europe and N. America it is found in neutral or basic, often moist or heavily shaded habitats, growing on bark (especially at the base of trees), stumps, rotting logs, rock and sometimes soil. Lowland. GB 3*.

Dioecious; sporophytes not known in Britain. The species lacks specalized means of vegetative propagation.

C. and N. Europe north to Iceland and S. Scandinavia. W. and C. Asia, N. and C. America, West Indies.

Apparently extinct in Scotland. It was recorded at the Den of Airlie between 1868 and 1871 and at Elcho from 1900 until 1911. At Airlie it occurred 'very sparingly' (Braithwaite, 1870); nevertheless, J. Fergusson, who discovered it there, collected at least 350 cm² in 1868 (specimens at BM, CGE and E). It is tempting to conclude that his over-collecting contributed to its demise, but the fact that several species formerly known from the Den of Airlie have not been found there this century (e.g. *Encalypta brevicollis, Lescuraea incurvata, Pseudoleskeella catenulata, Timmia austriaca*) suggests that other factors may be involved.

C. D. Preston

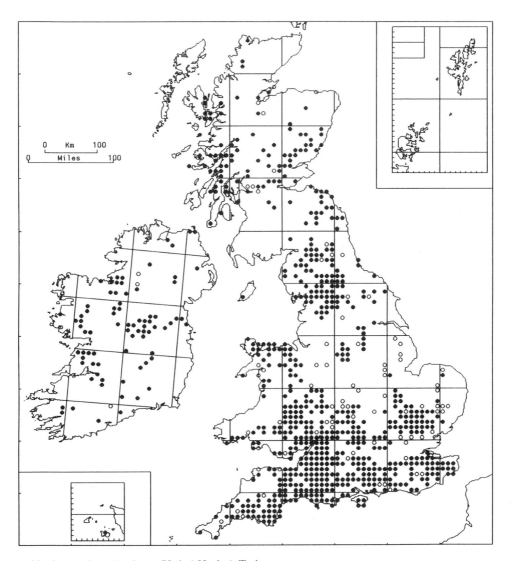

136/3. **Anomodon viticulosus** (Hedw.) Hook. & Tayl.

Frequent and locally abundant in well-drained calcareous habitats, especially where shaded. On limestone rock outcrops and drystone walls it often grows in profusion; other characteristic habitats include basic sandstone rocks, steep chalk hedgebanks, thin closely-grazed chalk grassland (at one site as an epiphyte on *Helianthemum* stems) and the exposed roots and bases of trees, especially ash and elm, in woods and hedges. It also occurs on silt-covered tree-roots, bridges, brick walls and concrete by streams and rivers, and on shaded brickwork and concrete away from water. It is occasionally found in smaller quantity in non-calcareous habitats, including sandstone rocks, siliceous boulders and Cornish hedges, but it is absent from areas of highly acidic or peaty soil. Characteristic associates include *Porella platyphylla*, *Ctenidium molluscum*, *Fissidens cristatus*, *Homalothecium sericeum* and *Neckera complanata*. Mainly lowland, no localized record above 300 m. GB 642+80*, IR 78+4*.

Dioecious; sporophytes very scarce, maturing in winter and spring.

Circumboreal. Widespread in Europe, although rare in the Mediterranean lowlands.

A. viticulosus appears to have declined since 1950 in some parts of S. England, for reasons which are far from obvious (Adams & Preston, 1992)

C. D. PRESTON

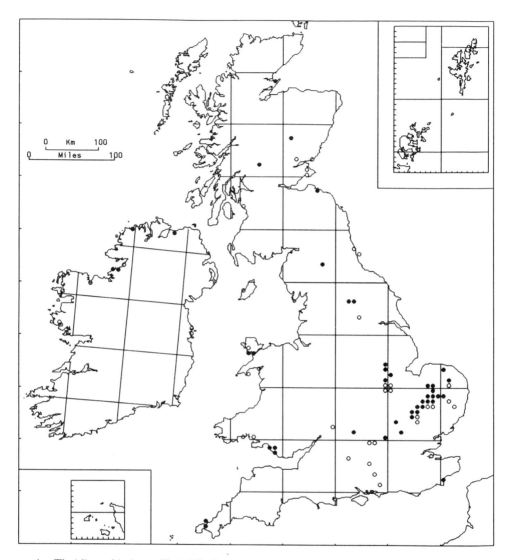

137/1a. **Thuidium abietinum** (Hedw.) Br. Eur. ssp. **abietinum** (*Abietinella abietina* (Hedw.) Fleisch.)

In East Anglia, it grows on thin soils in open, short-grazed chalk grassland, normally where the turf is unimproved and undisturbed. It also grows on the calcareous sandy soils of Breckland. Associated species include *Leiocolea turbinata*, *Campylium chrysophyllum*, *Ditrichum flexicaule*, *Rhytidium rugosum* and *Trichostomum crispulum*. Elsewhere in lowland Britain it occurs in old calcareous dune grassland in a few areas, and is occasionally found on banks in quarries and in thin turf on Magnesian limestone. In upland areas, where it is very rare, it has been recorded on montane rock-ledges, in *Dryas* heath and on basic talus slopes. 0–850 m (Caenlochan Glen). GB 40+32*, IR 4+6*.

Dioecious; sporophytes very rare.

Arctic, northern and montane regions of the Northern Hemisphere, south to southern Europe, S.W. Asia, Himalaya and southern U.S.A. (Colorado, Virginia). In the Southern Hemisphere only in Lesotho.

M. J. WIGGINTON

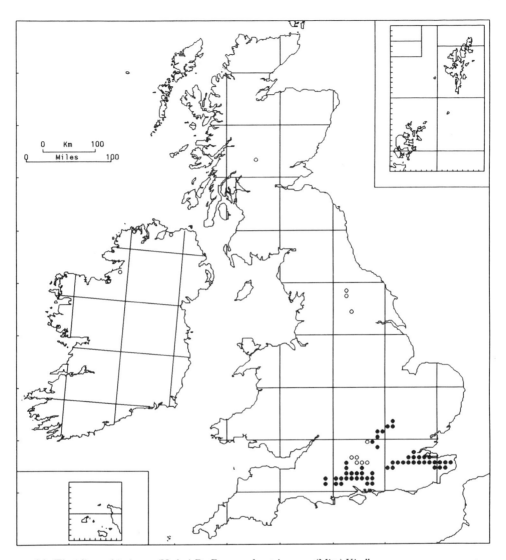

137/1b. **Thuidium abietinum** (Hedw.) Br. Eur. ssp. **hystricosum** (Mitt.) Kindb.

On the chalk downlands of southern England its preferred habitat is ancient, unimproved grassland which has been traditionally managed. It grows on thin soil in open, short-grazed turf, on banks and in quarries. It is often associated with *Leiocolea turbinata*, *Ditrichum flexicaule*, *Entodon concinnus*, *Weissia* spp. and a range of acrocarpous mosses characteristic of open chalk soils. In Ireland and northern Britain it also occurs in calcareous turf. Mainly lowland. GB 53+10*, IR 4*.

Dioecious; sporophytes unknown in Britain.

Spain, France, C. Europe. Japan, China.

Intermediates between ssp. *abietinum* and ssp. *hystricosum* are rare in Britain and Ireland, but occur widely through the range of ssp. *abietinum* outside the range of ssp. *hystricosum* (Duell-Hermanns, 1981).

M. J. WIGGINTON

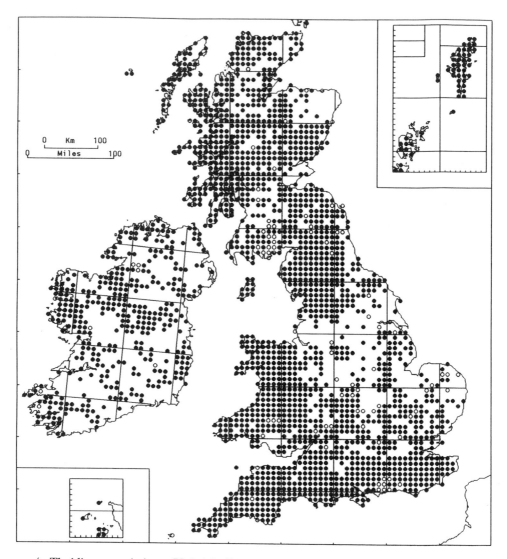

137/2. **Thuidium tamariscinum** (Hedw.) Br. Eur.

This species occurs in a wide range of moist lowland and montane habitats, preferring a mesic substrate and avoiding very wet, acid, exposed or nutrient-rich conditions. It is common on banks and flatter ground in woods, where it often accompanies other robust bryophytes such as *Plagiochila asplenioides*, *Cirriphyllum piliferum*, *Eurhynchium striatum*, *Hylocomium splendens* and *Rhytidiadelphus triquetrus*. It also grows on shaded woodland boulders, on rotting logs, and in a wide range of lowland and montane grasslands. Other habitats include dunes, limestone pavements, lowland and montane dwarf-shrub heaths, wall-tops, and montane cliffs and ledges. 0–880 m (Ben Lawers). GB 1796+85*, IR 373+10*.

Dioecious; sporophytes occasional in the north and west, very rare elsewhere.

Europe north to Iceland and southern Scandinavia. Macaronesia, Turkey, Caucasus, Russian Far East (Sakhalin), Japan, Newfoundland (probably introduced), Tanzania, Réunion (Indian Ocean).

M. J. Wigginton

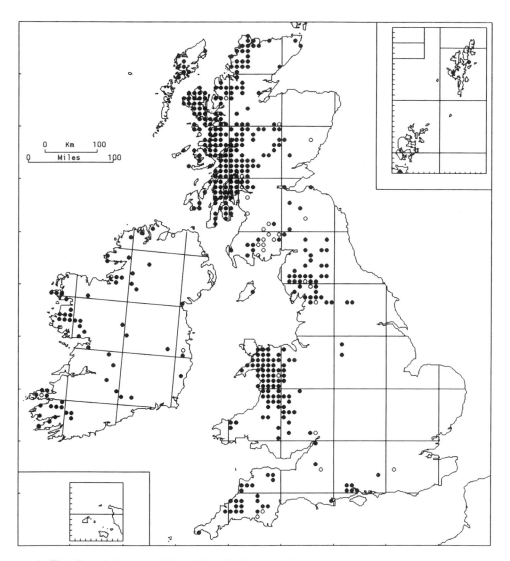

137/3. Thuidium delicatulum (Hedw.) Mitt. (*T. elatum* Duby)

The most frequent habitats are damp grassland, seepage slopes and flushes. It also occurs on damp montane cliff-ledges, in *Calluna-Vaccinium* heaths and in woods, where it grows on shaded rocks, on banks or on the woodland floor. Less frequently it is found in quarries, in moist woodland rides, on roadside banks, on peaty soil on limestone pavement, and, very rarely, on tree-stumps. In flushes and seepages it may accompany such species as *Saxifraga aizoides*, *Bryum pallens*, *Fissidens adianthoides* and *Philonotis fontana*. On shaded boulders and woodland banks, associates may include *Plagiochila porelloides*, *Antitrichia curtipendula*, *Ctenidium molluscum* and *Isothecium myurum*. Lowland to montane, ascending to 800 m (Seana Bhraigh). GB 376+28*, IR 59+6*.

Dioecious; sporophytes occasional, winter.

Europe including Iceland. Caucasus, Siberia, C. Asia, China, Japan, N., C. and S. America.

The distinction between *Thuidium delicatulum* and *T. philibertii* is often difficult, and apparent intermediates occur, especially in the north and west (Tallis, 1961). The map must contain some errors.

M. J. WIGGINTON

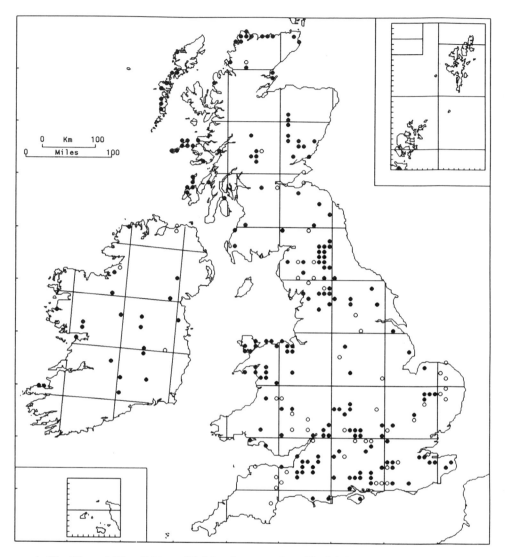

137/4. **Thuidium philibertii** Limpr. (*T. delicatulum* var. *radicans* (Kindb.) Crum, Steere & Anders.)

This species occurs principally in calcareous grasslands, especially in short-grazed turf which has not been enriched with fertilizer or recently disturbed. It grows on chalk and limestone soils, on calcareous clays and on basic sand-dunes. It also grows on damp basic rocks in woods, and in a variety of montane habitats including ravines and damp rock-ledges. It is a frequent machair species in parts of western Scotland, where it is accompanied by such mosses as *Amblystegium serpens* var. *salinum*, *Entodon concinnus*, *Pottia heimii* and *Rhodobryum roseum*. 0–800 m (Caenlochan Glen). GB 194+50*, IR 21+4*.

Dioecious; sporophytes not known in Britain.

Europe including Iceland. Caucasus, Asia, N., C. and S. America.

This species is difficult to separate from *T. delicatulum* (Tallis, 1961); the map must contain some errors. Some authorities (e.g. Crum & Anderson, 1981) treat *T. philibertii* as a variety of *T. delicatulum*, and many workers have noted their inconsistent and intergrading character.

M. J. WIGGINTON

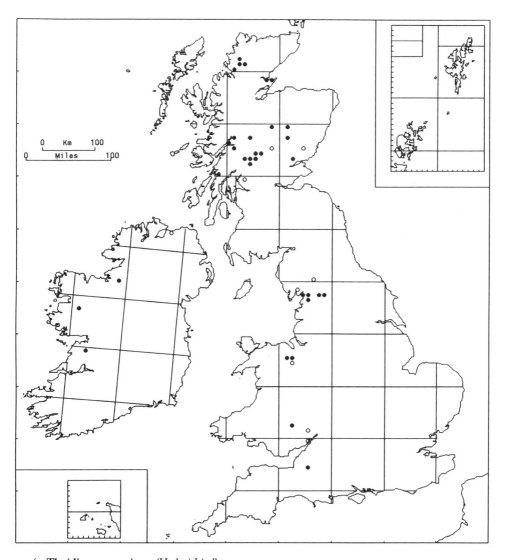

137/5. **Thuidium recognitum** (Hedw.) Lindb.

A calcicolous species of Cambrian and Carboniferous Limestone areas, and other basic formations in the
Scottish Highlands. It is normally a plant of woodland and shaded places, where it grows over boulders or on
banks. However, it also occurs in open habitats. It is locally frequent on limestone pavements in N.W. England,
where associated species include *Barbula reflexa, Ctenidium molluscum, Pleurochaete squarrosa* and *Tortella tortuosa.*
Other habitats include limestone grassland, basic scree, montane rock-ledges, *Dryas* heaths, limestone quarries
and, very rarely, sand-dunes. 0–950 m (Coire Cheap). GB 30+8*, IR 3+1*.
    Dioecious; sporophytes not known in Britain.
    Europe including Iceland. Algeria, Caucasus, N. and C. Asia, N. America.
    Early records of *T. recognitum* were often erroneous, but following a revision by Tallis (1961) the species has
generally been well understood in Britain and Ireland. The map should contain very few errors.

M. J. WIGGINTON

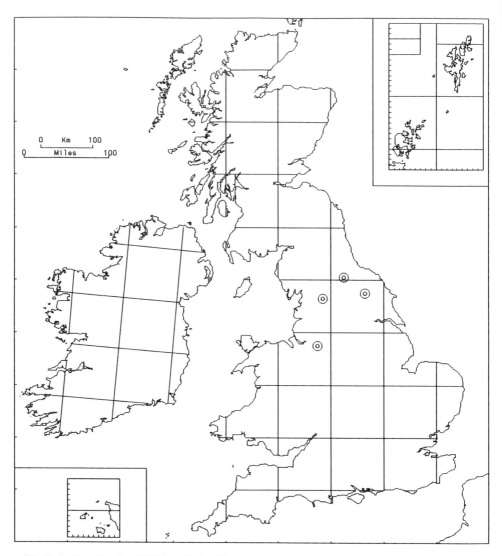

138/1. **Helodium blandowii** (Web. & Mohr) Warnst.

At Terrington North Carr, a site described as a small swampy hayfield, it grew with *Calliergon cuspidatum*, *Climacium dendroides*, *Homalothecium nitens* and *Plagiomnium elatum*. In Cheshire it grew among *Carex rostrata* with *Calliergon cuspidatum* and *Sphagnum* cf. *teres*. In Europe it occurs in intermediate to rather poor fens and spring bogs, especially those with light scrub cover, sometimes in wet birch forest at fen margins. *Homalothecium nitens* and *Sphagnum warnstorfii* are frequent associates. It frequently shares sites with *Paludella squarrosa* as it did in England. Lowland to about 400 m (Malham Tarn Moss). GB 4*.

Autoecious, fruiting from April to June. Unlike many relict species it produced sporophytes fairly frequently until its demise.

Widespread in boreal and sub-Arctic Eurasia and N. America but absent from the High Arctic; scattered southwards to the Alps, Carpathians, Turkey, Mongolia and central N. America (Colorado, New England).

A relict species greatly reduced following contraction about 5,000 B.P. By the 19th century it was restricted to four localities. In Cheshire it persisted until at least 1885. In 1858 it was found at Halnaby Carr, persisting until 1894, since when it has not been seen in England.

F. J. RUMSEY

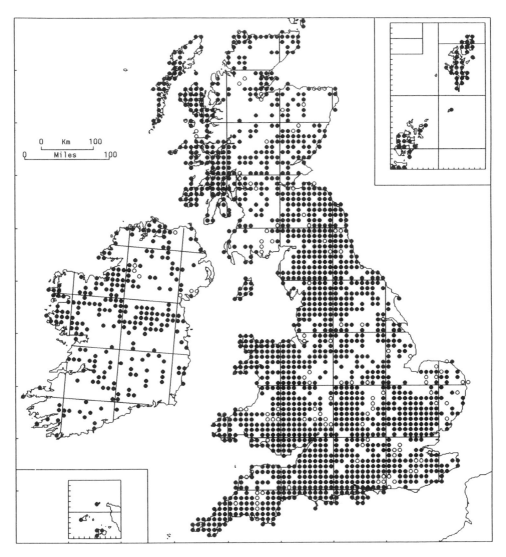

139/1. **Cratoneuron filicinum** (Hedw.) Spruce

It grows best around calcareous springs and rills, where it favours rather more fertile sites than *C. commutatum*. It is also widespread (often in attenuated or reduced forms) in a great variety of other damp, more-or-less base-rich situations, e.g. in moist open turf, on damp stonework or brickwork and on trampled paths. Common associates include *Brachythecium rivulare, Calliergon cuspidatum. Campylium stellatum* and *Cratoneuron commutatum*. 0–850 m (Caenlochan Glen). GB 1625+116*, IR 254+9*.

Dioecious; sporophytes uncommon, late spring.

Widespread in the Northern Hemisphere from the High Arctic south to N. Africa, Himalaya and C. America; reported from New Zealand.

An extremely variable plant, nicely characteristic in appearance in its well-grown forms, but easily passed over as a poorly-grown *Amblystegium* when depauperate or immature (as it often is).

M. C. F. PROCTOR

257

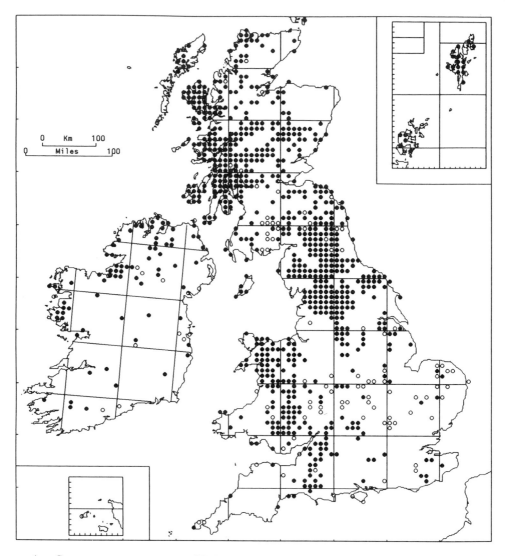

139/2a. **Cratoneuron commutatum** (Hedw.) Roth var. **commutatum** (*Palustriella commutata* (Hedw.) Ochyra)

A characteristic plant of calcareous springs and seepages, locally common in districts where limestones or other calcium-rich rocks outcrop, particularly in the mountains. Often in almost pure masses, which may be encrusted with calcium carbonate, forming banks or covering the rock or soil surface around and below the issue of water. Typical associates are *Cardamine pratensis*, *Cratoneuron filicinum*, *Philonotis* spp. and, from Upper Teesdale and the Lake District northwards, *Saxifraga aizoides*. 0–900 m (Ben Lawers). GB 735+93*, IR 73+13*.

Dioecious; sporophytes rare, mature in summer.

Widespread in the temperate Northern Hemisphere, from Greenland and Arctic Europe south to Madeira, N. Africa, Turkey, Himalaya and Colorado (U.S.A.).

In its typical form, the neatly and regularly pinnate shoots make this a distinctive plant, but apparent intermediates between var. *commutatum* and var. *falcatum* occur occasionally. Var. *virescens* is probably a habitat form induced by growing submerged in running water.

M. C. F. PROCTOR

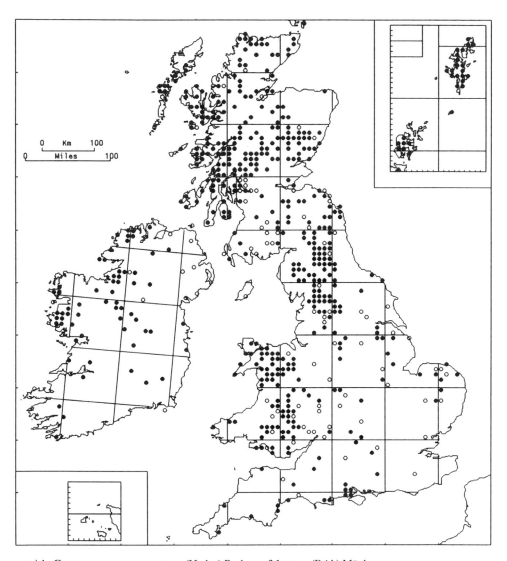

139/2b. **Cratoneuron commutatum** (Hedw.) Roth var. **falcatum** (Brid.) Mönk.

A locally common plant of calcium-rich flushes, seepages and open small-sedge fens with lateral movement of highly calcareous (but probably generally phosphate-poor) water. Typical associates are *Aneura pinguis*, *Bryum pseudotriquetrum*, *Campylium stellatum*, *Drepanocladus revolvens* and *Scorpidium scorpioides*. It is conspicuous in some grazed upland calcareous mires on Carboniferous Limestone in the Craven Pennines and Upper Teesdale. 0–1070 m (Aonach Beag). GB 433+81*, IR 61+12*.

Dioecious; sporophytes uncommon, maturing in summer.

Widespread in N. Europe, including Iceland and the Faeroes, and in mountains further south. Algeria, Caucasus, C. Asia, Himalaya, Sichuan (China), N. America.

Typically occupies a different ecological niche from var. *commutatum*. It is uncertain how far the morphological differences between the varieties are genetically determined, and how far they are a consequence of the different habitats.

M. C. F. Proctor

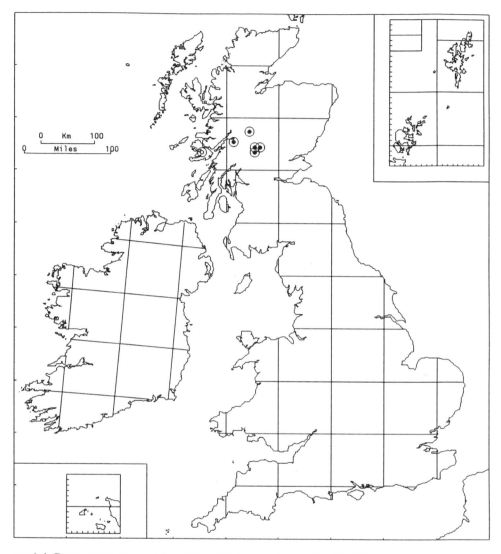

139/2d. **Cratoneuron commutatum** (Hedw.) Roth var. **sulcatum** (Lindb.) Mönk.

A very local plant of moist (or intermittently moist) calcareous mica-schist or limestone mountain rocks. Ascends to 1175 m (Ben Lawers). GB 5+1*.

Dioecious; capsules unknown.

Mountains of N., C. and E. Europe. Alaska, Greenland.

A distinctive slender small-leaved plant, locally abundant in the calcareous Alps, where it forms dense golden-brown patches tightly appressed to limestone rock surfaces.

M. C. F. PROCTOR

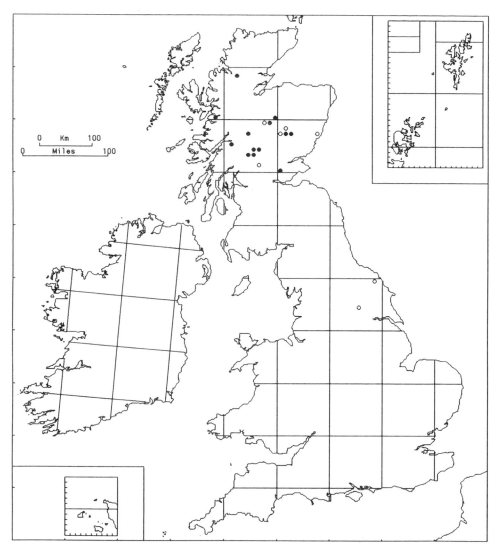

139/3. **Cratoneuron decipiens** (De Not.) Loeske (*Palustriella decipiens* (De Not.) Ochyra)

Springs and flushes on calcareous mica-schists. In the eastern Highlands, *Philonotis seriata* has been noted as an associate. Mostly above 550 m, ascending to 980 m (Ben Lawers). GB 13+7*.

Dioecious; sporophytes rare, mature in summer.

Iceland, Fennoscandia north to Finnmark (Norway), N. Russia, mountains of C. and S. Europe. Caucasus, C. Asia (Sayan Mts), Japan, Alaska. In Norway, *Saxifraga aizoides* has been noted as an associate (Dahl, 1956).

M. C. F. Proctor

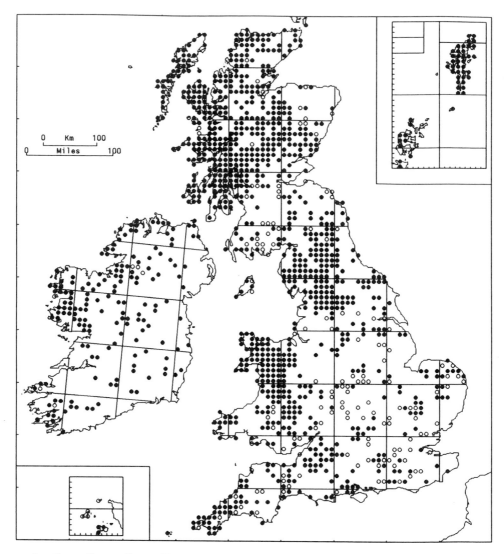

140/1a. **Campylium stellatum** (Hedw.) J. Lange & C. Jens. var. **stellatum**

It is found on wet, open base-rich ground, in a range of habitats including moorland soaks, flushed grassland, dripping calcareous rocks, dune-slacks and fens. Where it occurs on otherwise acid moorland it is a sure sign of basic seepage, typically growing with small sedges such as *Carex demissa, C. flacca* and *C. panicea* and with other bryophytes of basic conditions such as *Calliergon cuspidatum, Cratoneuron commutatum* and *Drepanocladus revolvens* sensu lato. 0–1070 m (Aonach Beag). GB 983+144*, IR 163+6*.

Dioecious; capsules rare, ripe in early summer.

Circumboreal; widespread in the Northern Hemisphere from the Arctic south to the mountains of S. Europe, W. and C. Asia, southern U.S.A., Mexico and Guatemala.

In much of C. and S. England, var. *stellatum* is less frequent than var. *protensum*. This does not show up very clearly on the maps, partly because var. *stellatum* is a more attractive plant and was therefore more often collected in the past and partly because of difficulties in distinguishing the two varieties. When recorders are in doubt, they tend to opt for var. *stellatum*. In the moorland country of the north and west, var. *stellatum* is very common, certainly much commoner than var. *protensum*, which there behaves as a scarce montane calcicole.

M. O. HILL

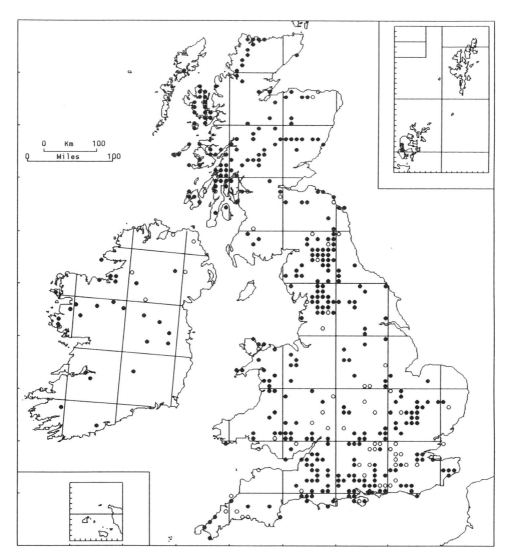

140/1b. **Campylium stellatum** (Hedw.) J. Lange & C. Jens. var. **protensum** (Brid.) Bryhn

This plant of neutral and calcareous soils occurs on tracks, in quarries, on clay banks and in grassland, often in seepages where the ground is intermittently moist but not permanently wet. It is also found in dry grassland and chalkpits, where it may be mixed with *C. chrysophyllum*. It sometimes colonizes temporary habitats such as newly-constructed banks and newly-sown pastures. Although frequent in woodland rides, it is uncommon on the ground and stumps under the tree-canopy. In the mountains, it is recorded from moist tall-herb and saxifrage communities, often with *Ctenidium molluscum*. 0–850 m (Creag Meagaidh). GB 353+60*, IR 26+6*.

Dioecious; capsules are rare in Britain and Ireland but have been recorded in Berkshire (Jones, 1953) and Essex (Adams, 1974).

Circumboreal, widely distributed in cold and cool regions of the Northern Hemisphere, south to S. Europe, Turkey, Himalaya and northern U.S.A.

Probably under-recorded because of confusion with var. *stellatum* and *C. chrysophyllum*. Many authors treat it as a species, *C. protensum* (Brid.) Kindb., but plants seemingly intermediate to var. *stellatum* occur in Britain. According to Fremstad (1978), it commonly produces capsules in Norway.

M. O. HILL

263

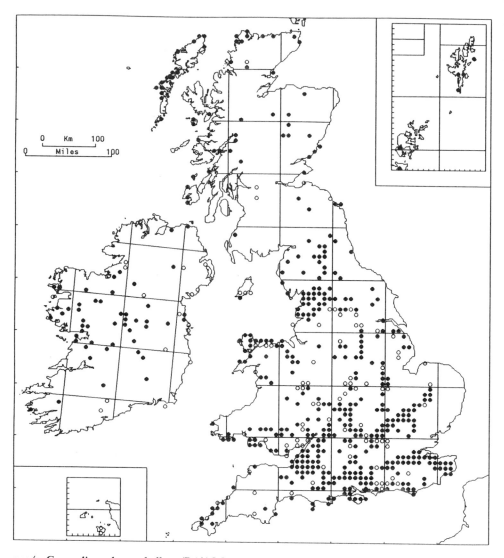

140/2. **Campylium chrysophyllum** (Brid.) J. Lange

A calcicole of dry, unshaded or lightly-shaded short turf and rocks. On chalk and limestone it grows in grassland, in old quarries, on scree, and, to a small extent on walls and bridges. Like many calcicoles of open ground it is frequent also on dunes. On other substrata it is generally uncommon, but occurs locally where suitably calcareous conditions exist. The communities in which it grows are often species-rich; typical associates include *Ctenidium molluscum*, *Fissidens cristatus* and *Weissia* spp. Almost exclusively lowland but ascending to 630 m in E. Scotland (Loch Brandy). GB 416+86*, IR 49+18*.

Dioecious; capsules rare, spring.

Circumboreal; widespread in the Northern Hemisphere, from the Arctic south to N. Africa, Turkey, C. Asia, Japan and Mexico, reaching the tropics in Guatemala and Colombia.

The scarcity of records of this widespread boreal species from higher altitudes in Britain and Ireland is rather surprising and may be due in part to poor recording. Duncan (1966) indicates that *C. chrysophyllum* is fairly frequent in the highland part of Angus, but gives no localized records.

M. O. HILL

264

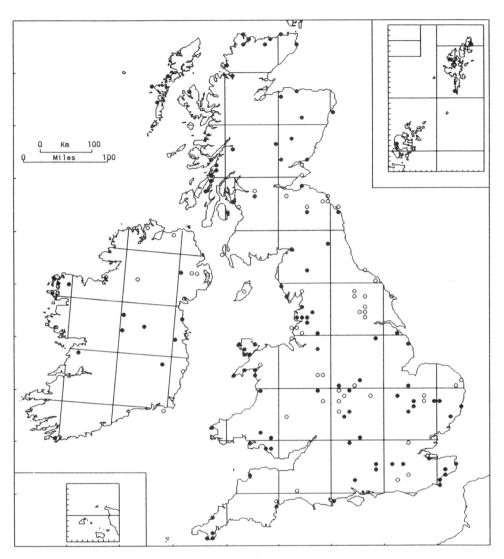

140/3. **Campylium polygamum** (Br. Eur.) J. Lange & C. Jens.

It grows in unshaded, moist grassland and marshy turf where the ground-water is either calcareous or salty. In western and northern districts it is almost exclusively coastal, occurring in dune-slacks, spray-washed grassland and flushes on sea-cliffs. In Scotland, where it may be one of the most abundant pleurocarpous mosses in salt-marshes, it is found down to levels flooded by a hundred tides a year; however, it is rare in English salt-marshes (Adam, 1976). Inland, it is recorded from fens, basic flushes, gravel-pits and moist pastures. Almost exclusively a lowland plant, it has been found at 850 m on serpentine near Meikle Kilrannoch. GB 105+51*, IR 13+11*.

Autoecious or synoecious; capsules frequent, ripe early summer.

An exceptionally widespread species, found almost throughout the boreal and temperate zones of the Northern Hemisphere and extending north to the Arctic, where it is rare; widespread in cool regions of the Southern Hemisphere, including Antarctica; scattered but very rare on tropical mountains.

In the field it closely resembles *C. stellatum* and it may sometimes therefore be overlooked, especially when growing inland.

M. O. HILL

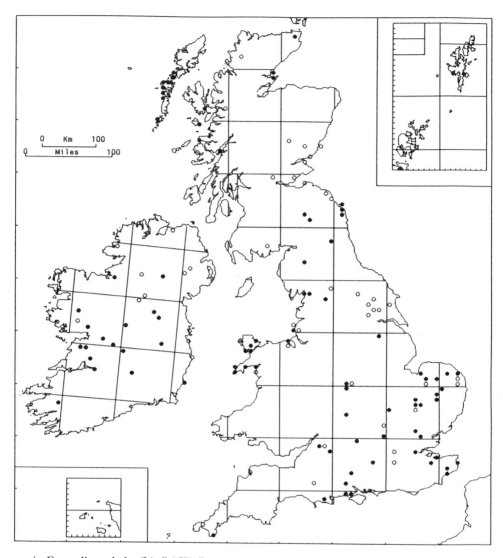

140/4. **Campylium elodes** (Lindb.) Kindb.

This calcicole of wet or swampy ground grows in turf, in moss-carpets and on dead vegetation or, rarely, wood, in well-illuminated situations or light shade. Many of its occurrences are in dune-slacks or moist sandy grassland; one Scottish locality is in a salt-marsh. Inland, it is found where there is seepage of calcareous water, mainly in fens but also in wet heaths and swamp woodland, sometimes on tree-bases. It is often associated with other Amblystegiaceae, including *Calliergon cuspidatum*, *Campylium polygamum*, *C. stellatum*, *Drepanocladus aduncus*, *D. revolvens* sensu lato, and *Scorpidium scorpioides*. Lowland. GB 72+39*, IR 18+11*.

Dioecious; capsules rare, early summer.

Widespread in temperate Europe, becoming rare towards the north and confined to higher elevations in the south. Himalaya, Japan, Russian Far East.

M. O. HILL

266

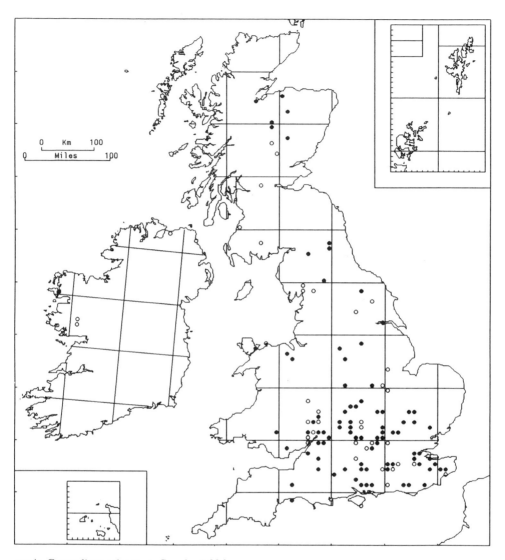

140/5. **Campylium calcareum** Crundw. & Nyh.

This moss is a strong calcicole, growing on chalk and limestone rocks, on flints, on roots and stumps, and on hard basic earth in well-drained situations, often in deep shade. Rarely, it is found on walls. It is commonest in beech woods on chalk, but occurs in a wide range of other habitats, including quarries that are reverting to woodland, limestone ravines, wooded undercliffs and shady lane-banks. Lowland. GB 85+35*, IR 1+4*.

Autoecious; capsules common, ripe early summer.

W. and C. Europe north to S. Fennoscandia. Turkey.

Campylium calcareum belongs to the species-complex of *C. hispidulum* (Brid.) Mitt. Crundwell & Nyholm (1962) recognize two members of the group in Europe, namely *C. calcareum* and *C. sommerfeltii* (Myr.) J. Lange, which is circumboreal and is found in N. Europe but not Britain. *C. hispidulum* sensu stricto is confined to N. America. However, American authors, e.g. Anderson *et al*. (1990), do not accept the subdivision of *C. hispidulum* into distinct taxa. This results in continuing nomenclatural confusion, from which *C. calcareum*, although both literally and figuratively a small species, is fortunately exempt.

M. O. HILL

267

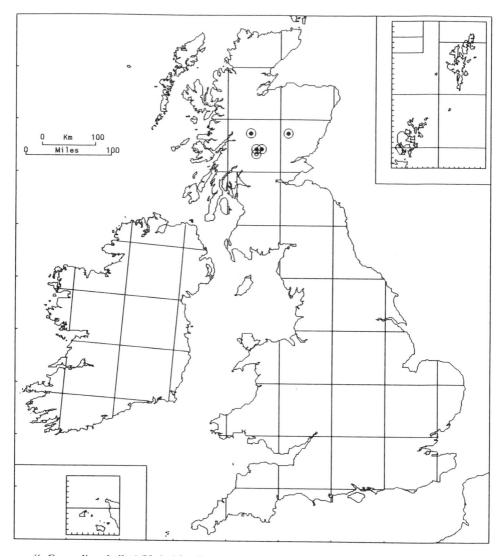

140/6. **Campylium halleri** (Hedw.) Lindb.

A montane calcicole which is found on well-drained mica-schist or limestone rock. Recorded habitats include large blocks of schist, a limestone boulder and a low vertical rock-face. 850 m (Caenlochan Glen) to 1000 m (Ben Lawers). GB 4+1*.

Autoecious; capsules frequent, apparently produced abundantly in some years but not others (Long, 1982b).

Boreal-montane, widespread in mountain ranges of the Northern Hemisphere, occurring both below and above the tree-line and reaching the Arctic in Fennoscandia, European Russia and Alaska.

M. O. HILL

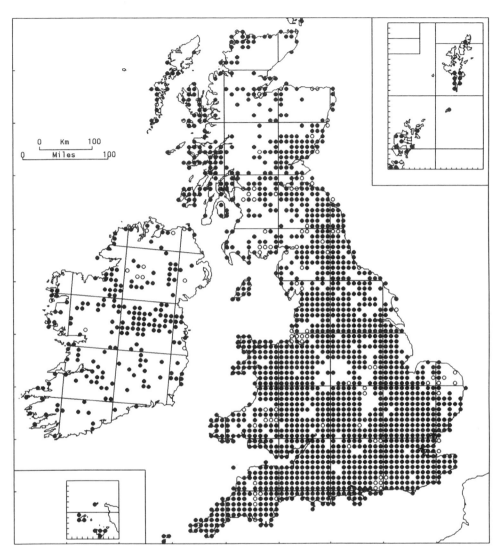

141/1. **Amblystegium serpens** (Hedw.) Br. Eur.

Forming extensive, flat mats on a wide range of moisture-retentive substrates. It grows on soil, both in the open and in deep shade, on stones, rocks or masonry, as well as on dead wood and as an epiphyte, particularly on willows and elder. Inland, it is common in damp woodland and carr, but its habitats include stable sand-dunes, brackish pools and upper salt-marsh turf in coastal regions. Occurring on both calcareous and non-calcareous substrata, it grows on chalk, limestone, concrete, flints, sandstone, flagstone and brick, but tends to favour nutrient-rich sites. Mainly lowland, ascending to 380 m (Malham Tarn). GB 1720+92*, IR 200+13*.

Autoecious and almost invariably fertile; sporophytes common, ripe in summer and autumn.

Arctic and temperate regions of the Northern Hemisphere south to N. Africa, Himalaya and Mexico. Ecuador, Peru, Tasmania, New Zealand.

Within this species there is wide cytological variation based on polyploidy (Fritsch, 1991). Var. *salinum* Carringt. is a poorly-recorded segregate, which is not mapped separately. Protonemal gemmae have been induced in subantarctic material (Selkirk, 1981), but this may be a distinct taxon.

M. E. NEWTON

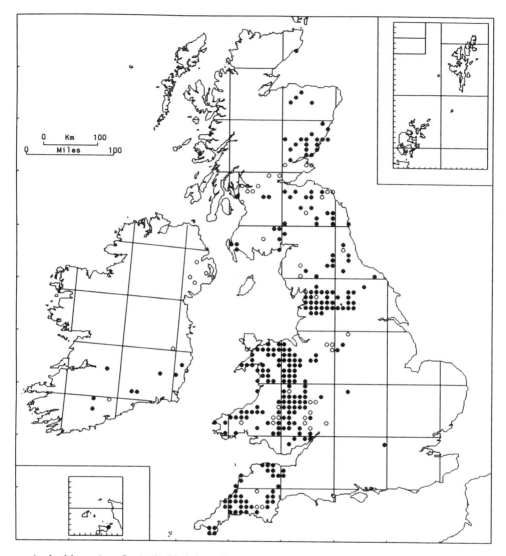

141/2. **Amblystegium fluviatile** (Hedw.) Br. Eur. (*Hygroamblystegium fluviatile* (Hedw.) Loeske)

Attached to rocks, stones and wood in fast-flowing streams and rivers where permanently or frequently inundated. It tends to favour sandstones, shales and, less commonly, limestones, and may be locally abundant in non-calcareous rivers with *Fontinalis squamosa* and *Rhynchostegium riparioides*. Lowland. GB 254+42*, IR 8+8*.

Autoecious, occasionally fertile; sporophytes maturing in late summer. Specialized means of asexual reproduction absent.

In much of Europe except the extreme north. Turkey, temperate N. America south to Arizona and Texas. Reported from Guatemala and Peru.

According to Crum & Anderson (1981), it is more strictly calcicolous in N. America than in Britain.

M. E. Newton

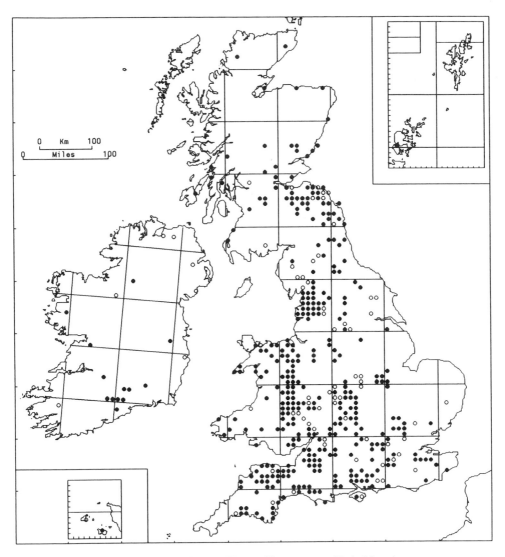

141/3. **Amblystegium tenax** (Hedw.) C. Jens. (*Hygroamblystegium tenax* (Hedw.) Jenn.)

Predominantly a species of basic water, growing submerged or in situations subject to frequent flooding on stones, rocks, masonry and exposed tree-roots in streams and rivers and, where there is significant flow of water, in canals and ditches. It occurs on chalk, limestone and concrete, and also on siliceous substrates where the water is basic. *Fissidens crassipes* is a common associate. 0–440 m (Crooke Gill). GB 341+65*, IR 13+9*.

Autoecious, frequently fertile; sporophytes maturing in late summer. No special means of asexual reproduction.

Throughout most of Europe. Macaronesia, N. Africa, W., C. and E. Asia, N. America.

It can be confused with *A. varium* and *Cratoneuron filicinum*. The overall distribution pattern should be correct, but the map is likely to contain some errors.

M. E. NEWTON

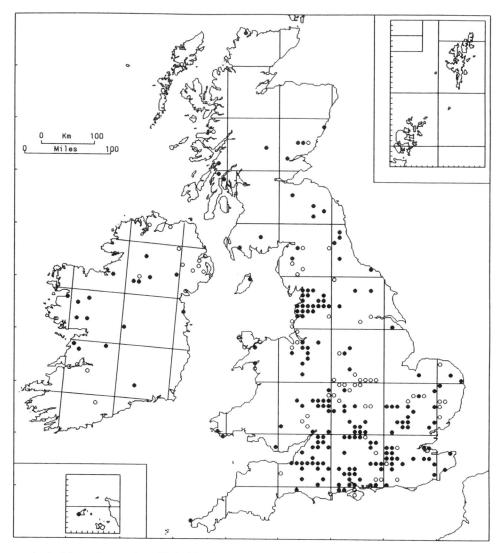

141/4. **Amblystegium varium** (Hedw.) Lindb.

On wood, stones, soil or clay in wet places, particularly by streams and pools. Common habitats are carr, where it grows on stumps and bases of alder and willows, and marshy meadows, where it occurs amongst rushes and grasses. It is found occasionally on stony ground in old chalk and limestone workings. Lowland. GB 184+60*, IR 17+15*.

Autoecious; occasionally fertile, sporophytes maturing in late summer. No special means of asexual reproduction.

Widespread in Europe except the south. Macaronesia, W., C. and E. Asia, N. and C. America.

Notoriously difficult to distinguish in some of its forms from closely related species, *A. varium* deserves taxonomic revision in the context of the extensive polyploidy for which it is remarkable (Fritsch, 1991).

M. E. NEWTON

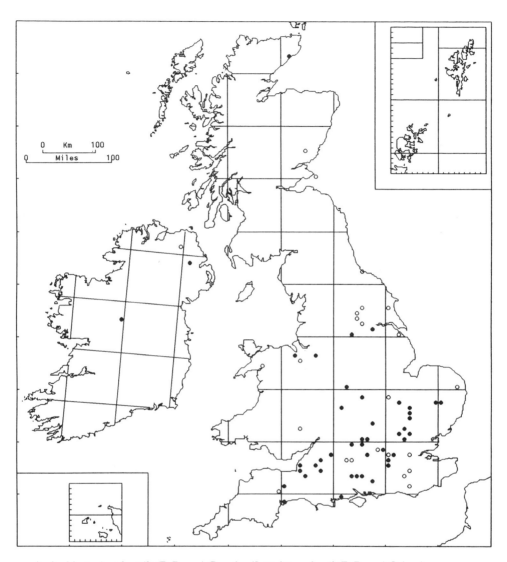

141/5. **Amblystegium humile** (P. Beauv.) Crundw. (*Leptodictyum humile* (P. Beauv.) Ochyra)

In small quantity among vascular plants near water, particularly in damp pasture, in carr, on open marshy ground and on tree-boles by streams and ponds. Lowland. GB 42 + 23*, IR 2 + 1*.

Autoecious; frequently fertile, sporophytes maturing in summer. No special means of asexual reproduction.

Widely distributed but generally uncommon through the temperate Northern Hemisphere south to the Himalaya and southern U.S.A. Disjunct in New Guinea.

Almost certainly under-recorded, since it bears a close resemblance to other species, notably *A. riparium*, *A. saxatile*, *A. tenax* and *A. varium* and, at least in part, occupies a similar ecological niche. Gardiner (1981) presented evidence of its persistence by the R. Thames near Kew for at least a century. Its occurrence in America is based on the probable synonymy, accepted by most recent authors (e.g. Anderson *et al.*, 1990, Ochyra *et al.*, 1991), of *A. humile* and *A. trichopodium* (Schultz) Hartm.

M. E. NEWTON

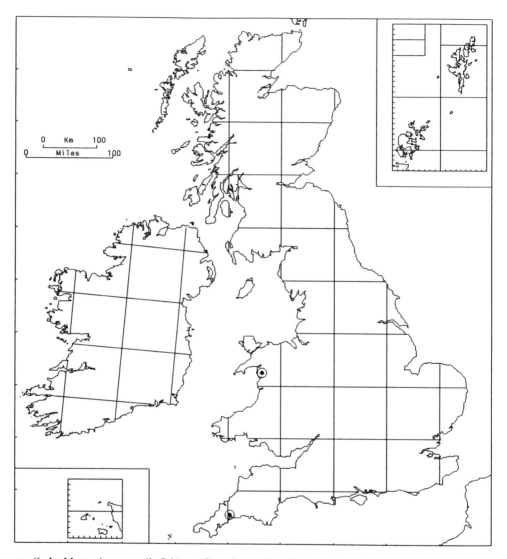

141/6. **Amblystegium saxatile** Schimp. (*Campylium radicale* (P. Beauv.) Grout)

In damp, stagnant and acid conditions on decaying vegetation. At one site it grew on dead *Juncus* in a marshy hollow with willows in old china-clay workings; at the other it was on recently burnt tussocks of *Molinia caerulea* in a shallow peat-cutting. The only associated bryophyte common to both sites is *Calliergon cordifolium*. Lowland. GB 2.

Autoecious; frequently fertile, sporophytes mature in late June.

Scattered through W., C. and N. Europe south to the former Yugoslavia and the Crimea. Japan, N. America, Ecuador.

This recent addition to the British flora (Crundwell & Nyholm, 1964b) occurs in unattractive habitats and is likely to have been overlooked. Details of its habitat in Wales are given by Yeo & Blackstock (1988).

M. E. NEWTON

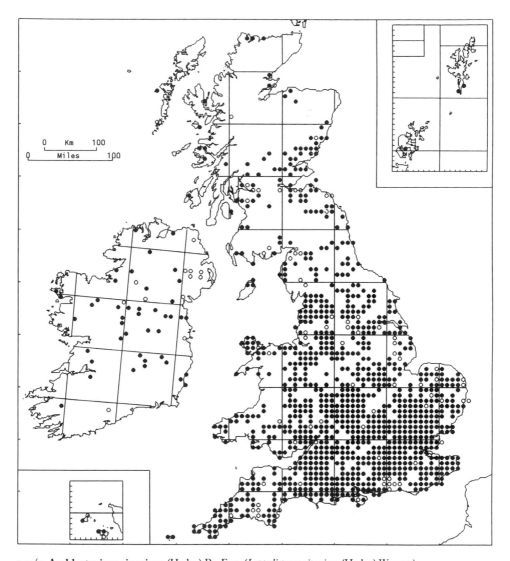

141/7. **Amblystegium riparium** (Hedw.) Br. Eur. (*Leptodictyum riparium* (Hedw.) Warnst.)

In loose, often extensive mats in eutrophic conditions on wet substrates, particularly wood, but also stone, masonry and soil. It is especially common in and by ditches, ponds, canals and slow-moving streams and rivers, at and just above the water-level. The plant is conspicuous in filter-beds of sewage treatment works, and proliferates in brewery effluent channels (Kelly & Huntley, 1987). It is tolerant of urban pollution (Gardiner, 1981). Lowland. GB 971+78*, IR 43+13*.

Autoecious; usually fertile, sporophytes reaching maturity over a period of several months from summer to early winter.

Widespread in the Northern Hemisphere. S. Africa, Australia, New Zealand, Kerguelen Island.

M. E. NEWTON

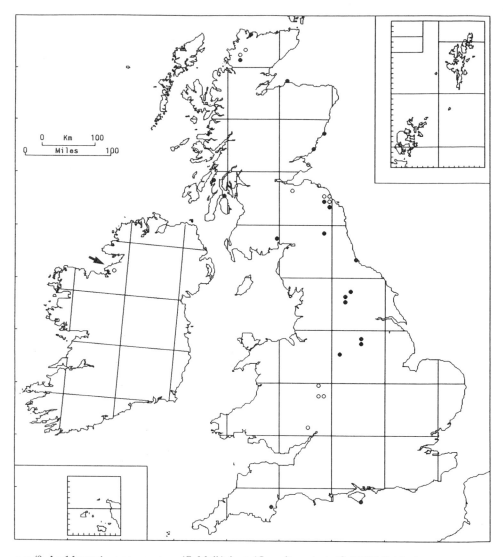

141/8. **Amblystegium compactum** (C. Müll.) Aust. (*Conardia compacta* (C. Müll.) Robins.)

In extensive, dense mats on calcareous substrata in deep shade, including damp rocks and soil below overhanging crags. Mostly found on limestone cliffs, it also occurs in sandstone sea-caves. Lowland. GB 20+12*, IR 1*.

Dioecious; sporophytes unknown in Europe. Copious multicellular, filamentous gemmae arise from the abaxial surface of the nerve towards the tip of the leaf.

Widespread but rather rare in Europe north to Iceland and Arctic Norway (Finnmark). C. Asia, N. and C. America, Greenland.

In N. Europe it can grow on more open ground and occurs on damp brackish sand by the Gulf of Bothnia in full sunshine.

M. E. Newton

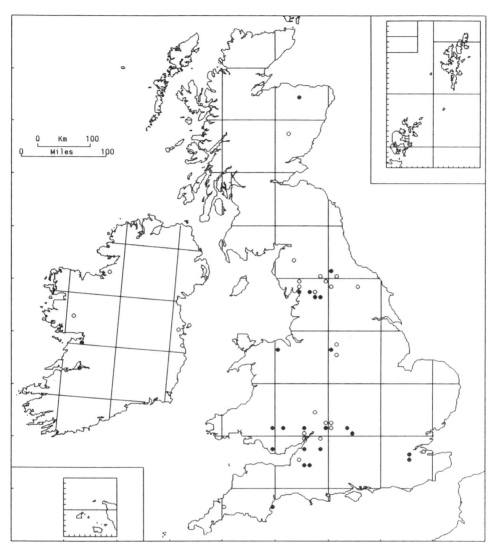

142/1. **Platydictya confervoides** (Brid.) Crum (*Amblystegium confervoides* (Brid.) B., S. & G.)

A calcicolous species typically occurring on limestone rock in shaded sites, usually in woodland. It is found in deep clefts in rock-faces, or on vertical faces of rock-scars or boulders, often amongst or shaded by robust bryophytes such as *Anomodon viticulosus*, *Brachythecium populeum*, *Ctenidium molluscum*, *Isothecium myurum*, *Neckera complanata* or *Thamnobryum alopecurum*, and sometimes associated with *Lejeunea cavifolia* and *Metzgeria furcata*. It has also been reported growing on lumps of chalk and limestone on the ground in woodland, and on damp basic montane rocks. Usually, perhaps always, below 300 m elevation. GB 22+21*, IR 2+5*.

Autoecious; sporophytes frequent, summer.

Widespread in Europe north to northern Norway. Madeira, Caucasus, N. America.

Because of its small size this species is easily overlooked. It may be more frequent than the map suggests, particularly in the limestone regions of the Pennines.

M. J. Wigginton

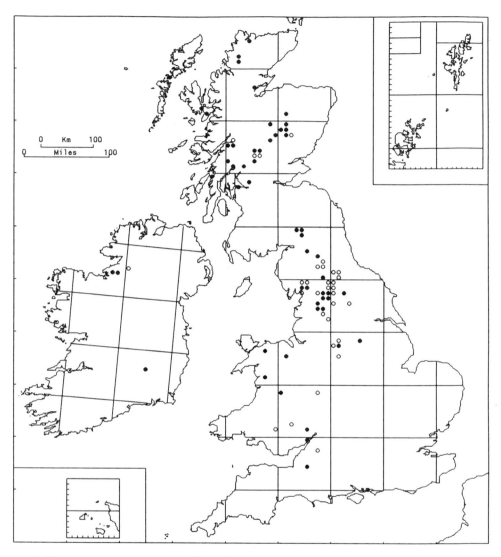

142/2. **Platydictya jungermannioides** (Brid.) Crum (*Amblystegium jungermannioides* (Brid.) A. J. E. Smith)

A plant of damp basic rocks and rock crevices in shaded locations, often in moist woodlands or wooded ravines, but also on montane rocks. It is found on limestone, calcareous sandstone and schist, and occasionally on tufa and other basic rocks, growing either directly on the rock surface or intermixed with other bryophytes. Associated species may include *Lejeunea cavifolia*, *Brachythecium populeum*, *Cirriphyllum crassinervium*, *Ctenidium molluscum*, *Orthothecium rufescens* and *Rhynchostegiella tenella*. It also grows on basic soil on rock surfaces or in crevices. From low altitudes to 1175 m (Ben Lawers). GB 51+30*, IR 3+1*.

Dioecious; sporophytes very rare, summer. Axillary gemmae are sometimes produced.

Widespread in Europe north to Iceland and the Arctic. Caucasus, N. and C. Asia, China, Japan, N. America, Greenland.

In N. America it has also been recorded on peaty soil or humus under turf on banks, in hollows under roots of trees, and on the lower sides of logs. According to Hedenäs (1987), *P. jungermannioides* belongs to the Plagiotheciaceae and is not closely related to *P. confervoides*.

M. J. WIGGINTON

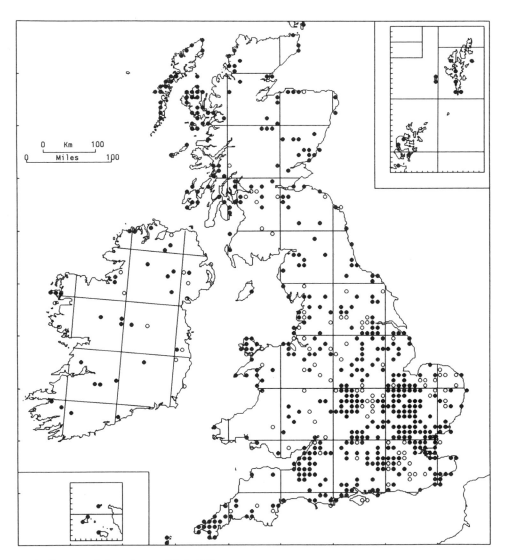

143/1. **Drepanocladus aduncus** (Hedw.) Warnst.

Typically occurs in still, more-or-less base-rich pools, where it may form extensive floating masses with the erect shoot-tips often breaking the surface, or on moist ground in wet fen meadows and eutrophic dune-slacks. It favours rather fertile sites, and is thus frequent in many clayey agricultural areas of the Midlands and S.E. England. Lowland. GB 501+103*, IR 29+13*.

Dioecious; sporophytes very rare.

Widespread in the Northern Hemisphere, occurring from the Arctic south to N. Africa, Sichuan (China) and northern Mexico. Australia, New Zealand.

A very variable species, which has been much subdivided by some authors.

M. C. F. PROCTOR

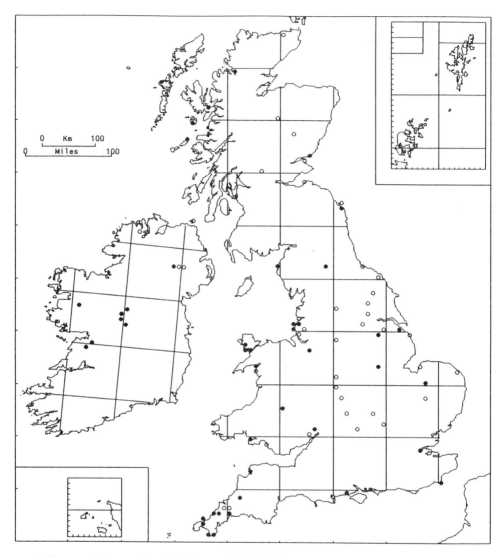

143/2. Drepanocladus sendtneri (Schimp. ex H. Müll.) Warnst.

A rare and local plant of seasonally flooded, calcareous dune-slacks and ecologically similar inland habitats in old marl-pits or natural hollows in limestone flooded in winter. It favours less fertile and ecologically less mature sites than *D. aduncus* and is more strictly limited to highly calcareous conditions. Lowland. GB 36+38*, IR 8+4*.

Dioecious; sporophytes rare.

A mainly northern plant, occurring from the Arctic south to Italy. N. and C. Asia, Sichuan (China), northern N. America, Greenland, Kerguelen Island.

Somewhat plastic in size and habit, but much less variable than *D. aduncus*.

M. C. F. Proctor

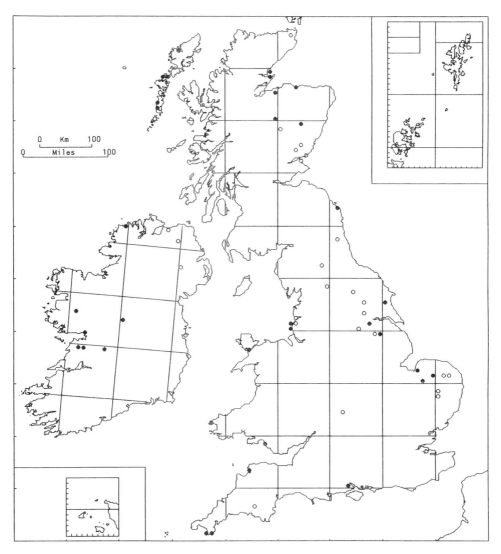

143/3. **Drepanocladus lycopodioides** (Brid.) Warnst. (*Scorpidium lycopodioides* (Brid.) Paul, *Pseudocalliergon lycopodioides* (Brid.) Hedenäs)

Although rare, it may be locally abundant in seasonally flooded, calcareous dune-slacks and in ecologically similar seasonal pools in limestone country, including some turloughs in the Burren. Typical associates are *Carex demissa, C. nigra, C. panicea, Potentilla anserina* and *Salix repens*. It apparently favours infertile sites, but with less annual water-table fluctuation than *D. sendtneri*; the moss carpet is often shallowly submerged in early spring, and dried out at the surface but moist underneath in summer. Occasionally in other wet calcareous sites, including some in the mountains. Mainly lowland, to 520 m (Creag an Dail Bheag). GB 27+23*, IR 7+5*.

Dioecious; sporophytes very rare.

W., C. and N. Europe, east to the Ukraine and north to Iceland and Finnish Lapland. Its occurrence outside Europe requires confirmation (Hedenäs, 1990).

The genus *Drepanocladus* as conventionally delimited is now regarded as an unnatural assemblage. *D. lycopodioides* is placed by Tuomikoski & Koponen (1979) in *Scorpidium* and by Hedenäs (1990) in *Pseudocalliergon*.

M. C. F. PROCTOR

281

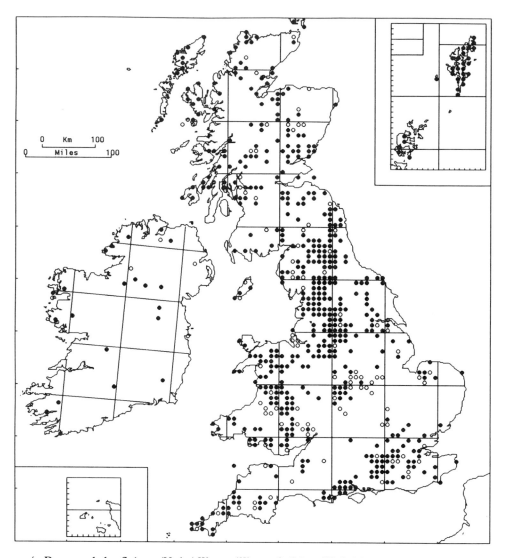

143/4. **Drepanocladus fluitans** (Hedw.) Warnst. (*Warnstorfia fluitans* (Hedw.) Loeske)

On wet heaths and on ombrogenous peats, especially in upland blanket-bogs, usually where these have been degraded by burning, grazing, erosion etc. It grows in shallow acid pools and seepages, probably almost always in base-poor water with less than 5 mg/litre calcium and pH less than 5.0, and will tolerate extremely acid conditions such as at Dersingham Fen, where the pH may fall below 3.0 in summer. 0–870 m (Lake District). GB 521+105*, IR 21+6*.

Autoecious; sporophytes occasional, summer and autumn.

Circumboreal, from the Arctic south to the Azores, Caucasus and Virginia (U.S.A.). Southern Africa, Ecuador, New Guinea, Tasmania, New Zealand, Kerguelen Island.

Variable; there seems to be no clear and consistent morphological or ecological separation between var. *fluitans* and var. *falcatus* (Sanio ex C. Jens.) Roth (*D. h-schulzei* (Limpr.) Loeske, *D. schulzei* Roth). Hedenäs (1993) treats *D. fluitans* and *D. h-schulzei* as synonyms.

M. C. F. PROCTOR

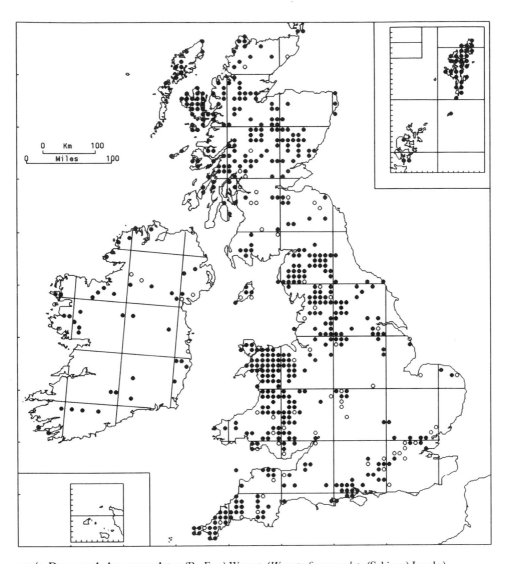

143/5. **Drepanocladus exannulatus** (Br. Eur.) Warnst. (*Warnstorfia exannulata* (Schimp.) Loeske)

It occupies a rather wide range of habitats which have in common moderately acid pH (often 5.0–6.0), low nutrient status, and limited competition. These include seasonally exposed, shelving shores of mildly acid heathland ponds and lakes, open poor-fen and flush communities (with, e.g., *Calliergon stramineum* and base-tolerant sphagna), and base-poor springs (with, e.g., *Calliergon sarmentosum, Dicranella palustris, Philonotis fontana* and *Sphagnum auriculatum*). 0–1335 m (Ben Nevis). GB 512+64*, IR 42+8*.

Dioecious, rarely autoecious; sporophytes rare, summer.

Circumboreal, extending in Europe from Iceland and Svalbard south to Italy, and in other continents from the Arctic south to Kashmir, Yunnan (China) and California (U.S.A.). Colombia, Falkland Islands, New Zealand.

Var. *rotae* (De Not.) Loeske has been shown to have a genetical basis, but not all material can be named satisfactorily so no attempt has been made to map infraspecific taxa. Following Tuomikoski & Koponen (1979), *D. exannulatus* is placed by some recent European and American authors in the genus *Warnstorfia* Loeske.

M. C. F. PROCTOR

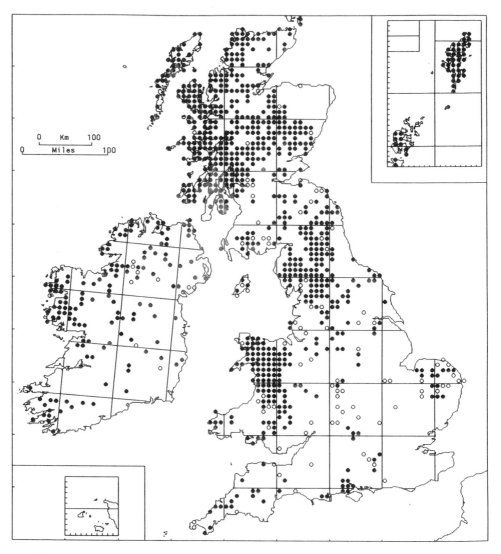

143/6. **Drepanocladus revolvens** (Sw.) Warnst. (*Limprichtia revolvens* (Sw.) Loeske, *Scorpidium revolvens* (Sw.) Hedenäs)

A characteristic component of mildly to strongly calcareous flushes and small-sedge fens, with *Aneura pinguis*, *Campylium stellatum*, *Scorpidium scorpioides* etc. The pH almost always exceeds 6.0. Ascends to 975 m (Beinn a'Chaorainn and Cairn of Claise). GB 749+95*, IR 139+12*.

Dioecious or autoecious; sporophytes uncommon, mature summer.

Circumboreal, extending from the High Arctic south to the mountain ranges of W., C. and E. Europe, Caucasus, C. Asia and northern U.S.A. Northern S. America, New Guinea, New Zealand.

Variable in size and colour. Hedenäs (1989a) places *D. revolvens* in *Scorpidium*, and elevates the slender golden-green var. *intermedius* (Lindb.) R. Wils. to specific rank as *S. cossonii* (Schimp.) Hedenäs. Hedenäs's revision has not yet been applied to British and Irish material, and the two taxa are mapped together.

M. C. F. PROCTOR

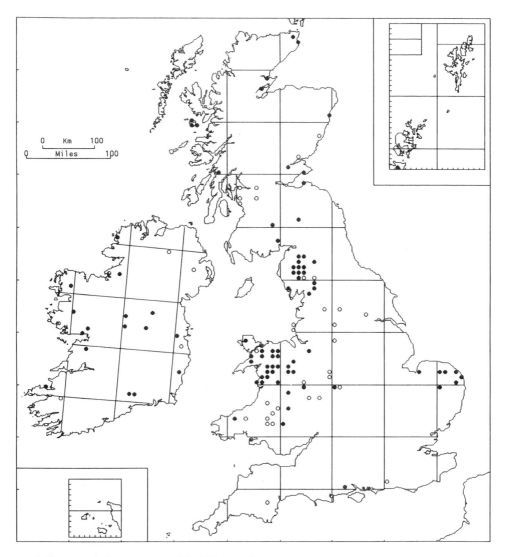

143/7. **Drepanocladus vernicosus** (Mitt.) Warnst. (*Hamatocaulis vernicosus* (Mitt.) Hedenäs)

Mainly a species of somewhat base-rich springs in upland districts, intermediate in its requirements between *Drepanocladus exannulatus* and *Cratoneuron commutatum*. In the lowlands, where it is rare, it generally occurs in spring-influenced sites in mildly basic small-sedge fens. 0–450 m (Snowdon). GB 70+34*, IR 16+5*.

Dioecious; sporophytes very rare, summer.

Circumboreal, ranging from the Arctic south to W., C. and E. Europe, Turkey, Caucasus, C. Asia and northern U.S.A. Disjunct in Dominican Republic.

Perhaps under-recorded, especially in Scotland. Transferred to the new genus *Hamatocaulis* by Hedenäs (1989a), who also gives a map of its distribution in Fennoscandia.

M. C. F. Proctor

285

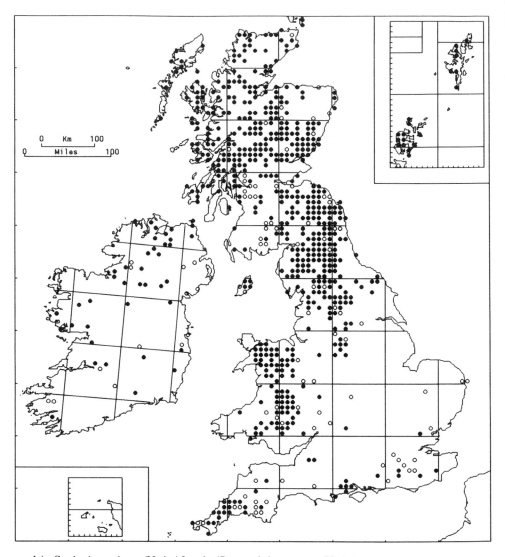

143A/1. **Sanionia uncinata** (Hedw.) Loeske (*Drepanocladus uncinatus* (Hedw.) Warnst.)

It occurs in a range of rather disparate habitats, especially on moist (often shaded) limestone or other calcareous rocks, and on the bases of sallows and other bushes in calcareous or acid carr. It also grows occasionally on the ground, particularly on litter and amongst other mosses in old calcareous fen-meadow. 0–1230 m (Aonach Beag). GB 572+86*, IR 40+13*.

Autoecious; sporophytes frequent, maturing spring and summer.

Widely distributed in Arctic and boreal regions of the Northern Hemisphere, and in the northern part of the broad-leaved forest zone; also in mountain ranges further south, e.g. in C. Europe. Its occurrence in the Southern Hemisphere requires confirmation (Hedenäs, 1989b).

It shows little variation over its British and Irish range. The only likely source of confusion is with the newly-recognized but apparently very rare *S. orthothecioides*.

M. C. F. PROCTOR

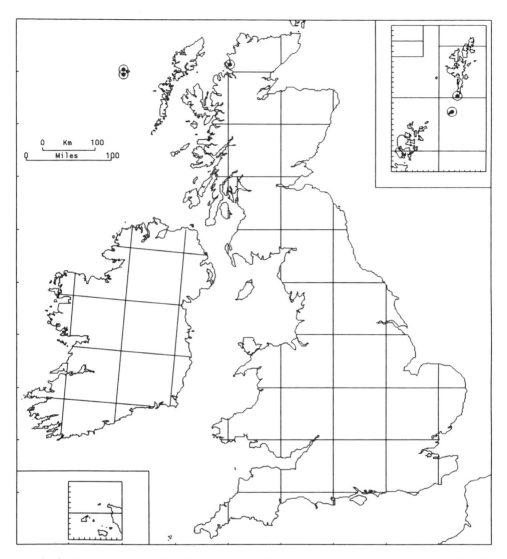

143A/2. **Sanionia orthothecioides** (Lindb.) Loeske (*Drepanocladus uncinatus* ssp. *orthothecioides* (Lindb.) Amann)

On St Kilda it is locally abundant in mossy turf on steep E.-facing slopes close to the sea, associated with *Dicranum scoparium*, *Hylocomium splendens*, *Hypnum jutlandicum*, *Isothecium myosuroides* var. *brachythecioides*, *Pseudoscleropodium purum*, *Rhytidiadelphus loreus* and *R. squarrosus*. It occurs both as relatively pure stands and as isolated stems amongst the other mosses, particularly on well-drained spots overlying partly-buried boulders, and in mossy 'cornices' on small vertical banks. In Shetland, the plant was found in an old quarry, mixed with *Frullania tamarisci*, *Hypnum jutlandicum*, *Polytrichum juniperinum* and *Rhytidiadelphus squarrosus*. In its one site in mainland Scotland it was found in coastal turf. Sea-level (Shetland) to 350 m (St Kilda). GB 5.

Autoecious; sporophytes not yet seen in Scottish material, but a St Kilda specimen bears an old perichaetium. Coasts of Iceland, Faeroes, Svalbard and mainland N. Europe, especially the Arctic coast; in Iceland, it reaches about 50 km inland. Siberia (Yenisey), Alaska, Canada, Greenland, Jan Mayen.

For further information refer to Hedenäs (1989b) and Long (1992, 1993).

D. G. LONG

287

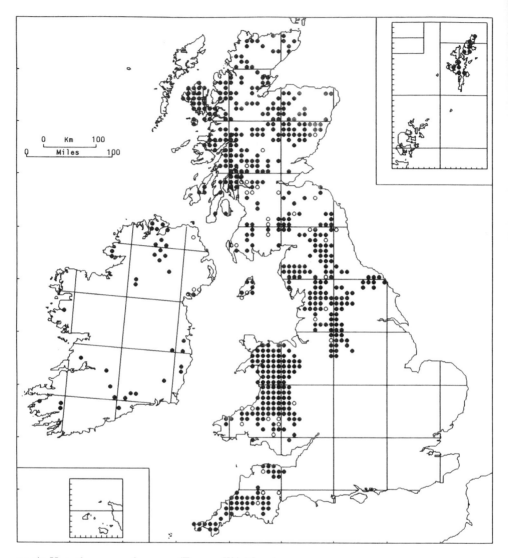

144/1. Hygrohypnum ochraceum (Turn. ex Wils.) Loeske

A calcifuge, forming green or yellow-green, flaccid mats on rocks in flowing water and beside waterfalls. It is often associated with *Scapania undulata*, *Brachythecium plumosum*, *Racomitrium aciculare* and *Rhynchostegium riparioides*. 0–1200 m (Ben Macdui). GB 547+41*, IR 36+5*.

Dioecious; capsules rare.

Circumboreal, from the Arctic south to the mountains of Spain, Greece, Caucasus, Kashmir, Korea, Japan and southern U.S.A. (Colorado, N. Carolina).

It is probably under-recorded in the Scottish Highlands.

G. P. ROTHERO

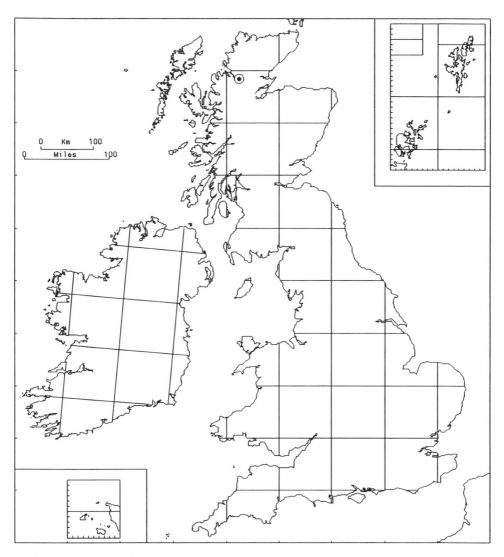

144/2. **Hygrohypnum polare** (Lindb.) Loeske

Known from only one site where it occurs as green cushions on basic, Moine schist boulders at the margin of a small lochan subject to marked fluctuations in water-level. 670 m (Beinn Dearg). GB 1.

Dioecious; capsules unknown.

Circumpolar, found mainly in the Arctic and in the mountains of C. Asia, with few sites south of the Arctic Circle in Europe and N. America. The North American distribution is mapped by Steere (1978).

It was first collected in Scotland in 1952, but the record was not published till twenty years later (Wallace, 1972).

G. P. ROTHERO

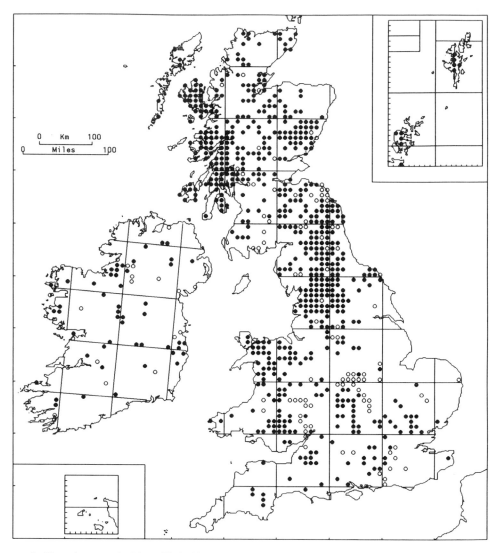

144/3. **Hygrohypnum luridum** (Hedw.) Jenn.

This species forms green or brownish cushions, often denuded below, in flowing water and beside streams, canals and waterfalls where the water is calcareous or at least mildly basic. In southern England it is also found occasionally away from water in situations where it would be inundated only during rain, for example in gutters by churches and on flat tombstones at ground level. 0–910 m (Coire an t-Sneachda). GB 619+98*, IR 49+20*.

Autoecious; capsules common, summer.

Circumboreal, with a wide distribution in the Northern Hemisphere, from the Arctic south to S. Europe, Turkey, Caucasus, Himalaya, China (Yunnan), Japan and southern U.S.A. (Colorado and West Virginia).

Var. *subsphaericarpon* (Schleich. ex Brid.) C. Jens. is a robust form that has very rarely been recorded in recent years. It is not mapped separately.

G. P. ROTHERO

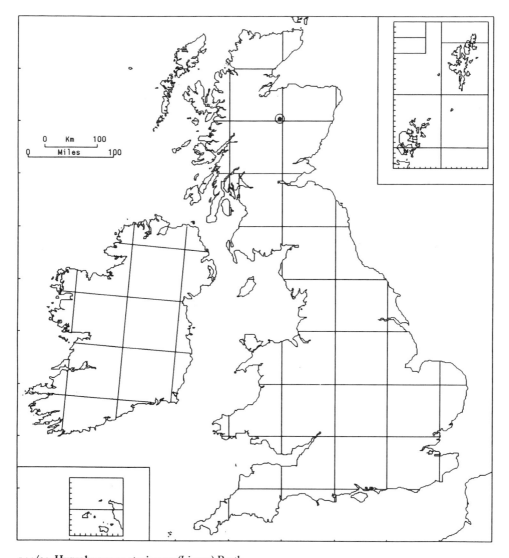

144/3a. **Hygrohypnum styriacum** (Limpr.) Broth.

Known from only one British locality, where it occurs on irrigated crevices of granite crags with *Bryum pseudotriquetrum*, *Hygrohypnum molle* and *Philonotis fontana*. 1075 m (Coire an t-Sneachda). GB 1.

Monoecious; several capsules observed.

Boreal-montane, very rare in Norway, Sweden and Iceland, more frequent in the eastern Alps and the Tatra and Carpathian Mts. Outside Europe known only in N. America from the Rocky Mts, from California to British Columbia.

H. styriacum has newly been added to the British list. For further details, refer to Corley & Rothero (1992).

G. P. Rᴏᴛʜᴇʀᴏ

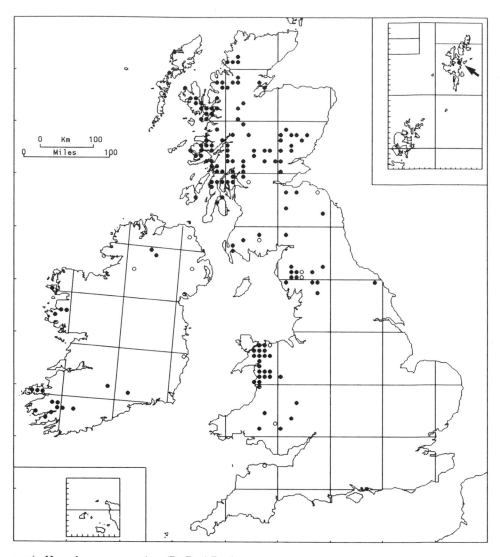

144/4. **Hygrohypnum eugyrium** (Br. Eur.) Broth.

This plant forms neat, flat, often brownish patches on siliceous rocks in rapids and waterfalls, usually where the substrate is at least slightly basic. The habitat and associates are similar to those of *H. luridum* and *H. ochraceum*. 0–800 m (Coire Cheap). GB 137+9*, IR 19+6*.

Autoecious; capsules frequent, spring.

W. and C. Europe north to C. Scandinavia and east to Poland and Romania. E. Asia, Japan, eastern N. America.

G. P. ROTHERO

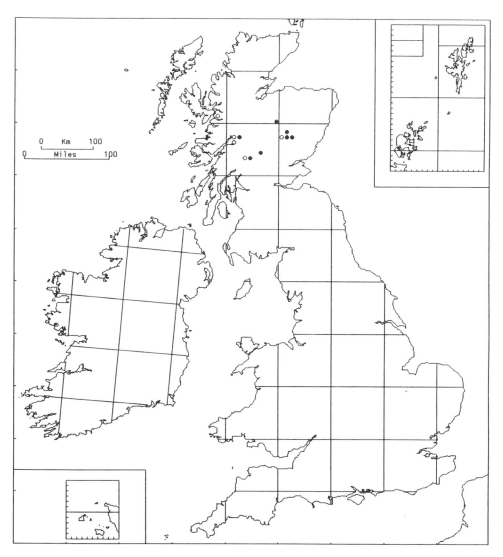

144/5. **Hygrohypnum smithii** (Sw.) Broth.

This species forms mats composed of stiff, green or dark green stems, often denuded below, on rocks in mountain streams. Most of its British sites are noted for their exposures of calcareous rocks and the plant seems to require at least a moderately basic substrate or run-off. 730 m (Aonach Beag) to 1000 m (Coire an t-Sneachda). GB 7+5*.

Autoecious; at least one population produces abundant sporophytes.

Arctic and N. Europe south to the Pyrenees and the mountains of C. and E. Europe. Caucasus, W. Siberia, western and eastern N. America, Greenland.

G. P. Rothero

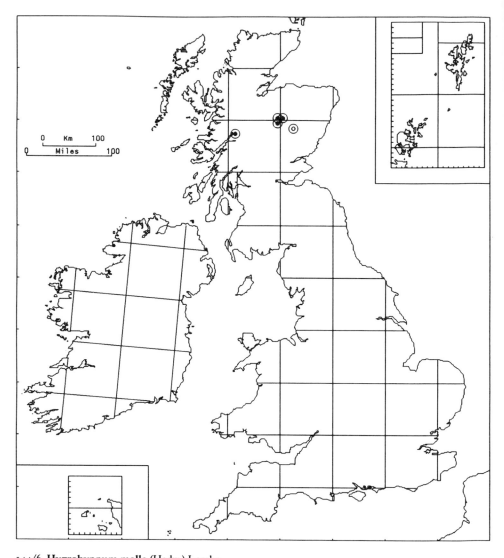

144/6. Hygrohypnum molle (Hedw.) Loeske

A plant of rocky streams and wet rocks at high altitude in the mountains, usually in areas of run-off from late snow-lie. The soft green to yellow-green cushions can be locally extensive in streams, often with *Hygrohypnum ochraceum* and *Rhynchostegium riparioides*. 950 m (Braeriach) to 1250 m (Ben Macdui). GB 4+1*.

Autoecious; capsules not known in Britain.

Arctic Europe and in mountains of C. and W. Europe east to the Carpathians. Siberia, C. Asia, Japan, N. America.

The world distribution is unclear because some authors (notably Crum & Anderson, 1981) treat it as a synonym of *H. dilatatum*.

G. P. ROTHERO

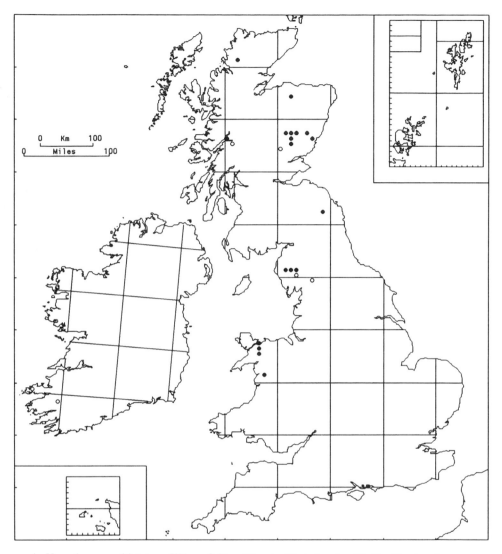

144/7. **Hygrohypnum dilatatum** (Wils. ex Schimp.) Loeske (*H. duriusculum* (De Not.) Jamieson)

This plant forms green or yellow-green tufts on rocks in fast-flowing streams and rivers, usually in montane areas but sometimes at low altitude. Although generally rare, it is locally abundant, especially in the Angus glens. 60 m (Edzell) to 490 m (Twll Du). GB 18+5*, IR 2*.

Dioecious; capsules very rare.

Circumboreal, from the Arctic south to the Pyrenees, Alps, Caucasus, C. Asia, Japan and U.S.A.

There is disagreement about the taxonomic status of *H. dilatatum*. It is treated as a synonym of *H. molle* by Crum & Anderson (1981), and as a synonym of *H. duriusculum* by Corley *et al.* (1981). Anderson *et al.* (1990) list both *H. duriusculum* and *H. molle* as good species, with *H. dilatatum* as a synonym of *H. molle*. A. C. Crundwell (pers. comm.) has written that the nomenclature of Corley *et al.* (1981) is attributable to D. H. Jamieson and he sees no reason to disagree with it.

G. P. Rothero

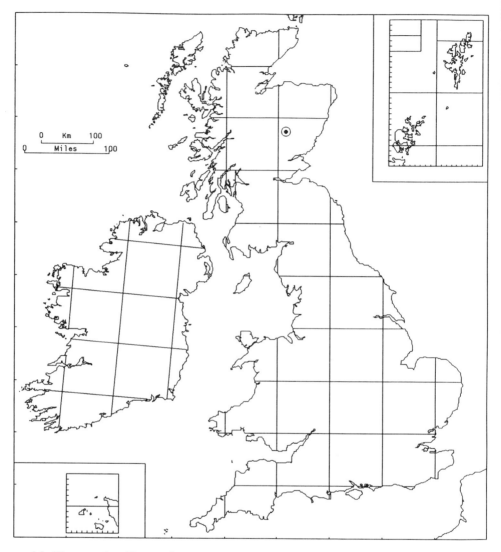

144A/1. Pictus scoticus Townsend

An epiphyte, occurring tightly appressed to the bark of a small tree, probably *Sorbus aucuparia*, on a dry W.-facing limestone outcrop at 500 m altitude. GB 1.

Autoecious; capsules probably common, maturing late summer. No specialized means of vegetative dispersal is known.

The species and genus are as yet unknown elsewhere.

A very rare species, thought to belong to the Amblystegiaceae and still known only from the type collection, made in 1979 (Townsend, 1982). Its survival must be considered precarious as only one small tuft has ever been found. It may perhaps have been overlooked as *Pterigynandrum filiforme*.

F. J. RUMSEY

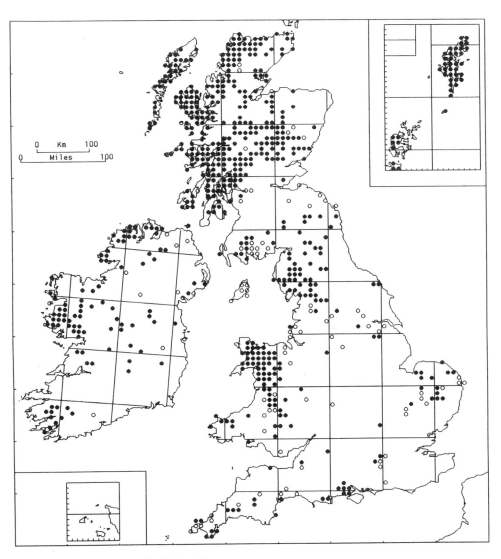

145/1. **Scorpidium scorpioides** (Hedw.) Limpr.

This plant of wet base-rich, often calcareous conditions occurs most frequently in 'rich-fens' dominated by *Carex hostiana*, *C. panicea*, *C. rostrata*, *Eriophorum latifolium* or *Schoenus nigricans*, growing in waterlogged hollows, in soligenous soaks and runnels draining from springs, and at the infilled margins of small low-lying lakes. It is commonly associated with *Aneura pinguis*, *Campylium stellatum*, *Drepanocladus revolvens*, *Fissidens adianthoides* and *Sphagnum contortum*. It also occurs in upland flushes with *Carex demissa*, *C. flacca*, *C. panicea*, *Pinguicula vulgaris*, *Saxifraga aizoides* and *Blindia acuta*, in 'mud-bottom' stands within communities dominated by *Carex diandra*, *C. lasiocarpa* and *C. limosa*, at the margins of small lakes with *Chara* spp., submerged in pools with *Utricularia* spp., in dune-slacks, and in calcareous spring-fens. More rarely it occurs in old marl-pits, in flushes in lowland wet heaths and valley-bogs, and on flushed rocks and cliff-ledges in the uplands. 0–880 m (Ben Alder range). GB 515+83*, IR 94+12*.

Dioecious; capsules rare, spring and summer.

Circumboreal, reaching the High Arctic; also in the Andes. Widespread in Europe, becoming montane towards the south.

H. J. B. Birks

297

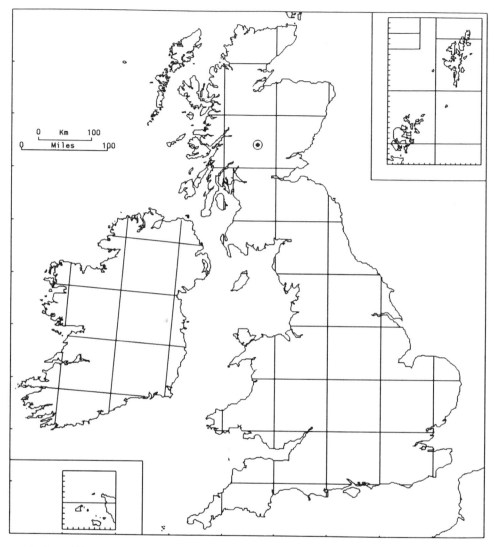

145/2. **Scorpidium turgescens** (T. Jens.) Loeske (*Pseudocalliergon turgescens* (T. Jens.) Loeske)

Locally abundant in *Carex saxatilis*-dominated mires within a small area in the Ben Lawers range. It occurs on gently sloping N.-facing rills and in waterlogged hollows irrigated by snow meltwater, growing with *Carex demissa, C. nigra, Eriophorum angustifolium, Polygonum viviparum, Saxifraga stellaris, Thalictrum alpinum, Aneura pinguis, Scapania undulata, Calliergon sarmentosum, C. trifarium, Drepanocladus revolvens* and *Sphagnum subnitens*. The pH of the mire water is 6.0–6.1. 1000 m. GB 1.

Dioecious; capsules unknown in Britain.

Circumpolar, mainly in sub-Arctic and Arctic regions including the High Arctic, with disjunct occurrences further south. N. Europe and mountains of C. Europe. N. Asia, China (Yunnan), N. America, Ecuador, Bolivia.

A 'glacial relict' today that was more widespread in the British Isles during the last glacial stage and the subsequent late-glacial, as evidenced by fossil remains as far south as Middlesex. Although having a mainly subarctic-alpine distribution in Europe, it is abundant at low elevations by seasonally wet pools on the alvar limestone of the island of Öland, S.E. Sweden. For further details of its habitat in Britain see Birks & Dransfield (1970).

H. J. B. BIRKS

298

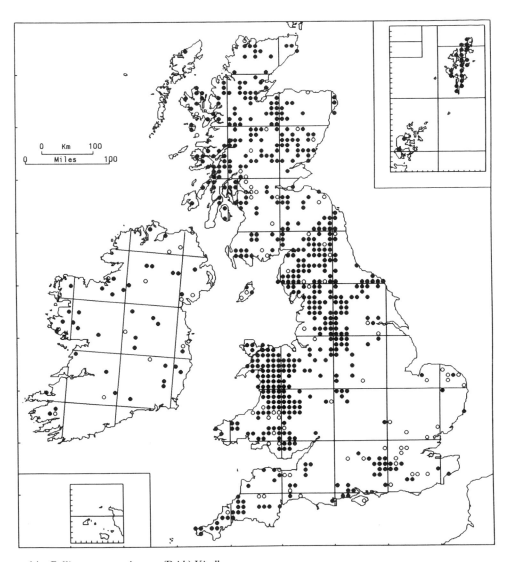

146/1. **Calliergon stramineum** (Brid.) Kindb.

This moss of wet, moderately acid, peaty ground is normally found as scattered stems or small patches among more robust bryophytes, but it also occurs as straggling mats, sometimes semi-submerged, beside lakes and in peaty runnels and ditches. Typical habitats include flushed valley bottoms, acidic fens, lakeside marshes, and ditches and depressions on bogs, especially near their margins or in old peat-diggings. Less often it is found among alder and willows in wet open woodland. It avoids both the most acid and most basic ground, growing with associates such as *Carex echinata, Juncus effusus, Viola palustris, Aulacomnium palustre, Drepanocladus exannulatus, Polytrichum commune* and *Sphagnum recurvum*. It sometimes occurs in stony acid runnels and snow-bed flushes, and has been found, exceptionally, among rocks at the summit of the Cairngorms. 0–1280 m (Braeriach). GB 585+82*, IR 39+13*.

Dioecious; capsules very rare, summer.

Circumboreal, very widespread in cool and cold regions of the Northern Hemisphere north to the High Arctic, reaching south to the mountains of S. Europe, Turkey, the Caucasus and southern U.S.A. (Colorado). Reported from northern S. America.

M. O. HILL

299

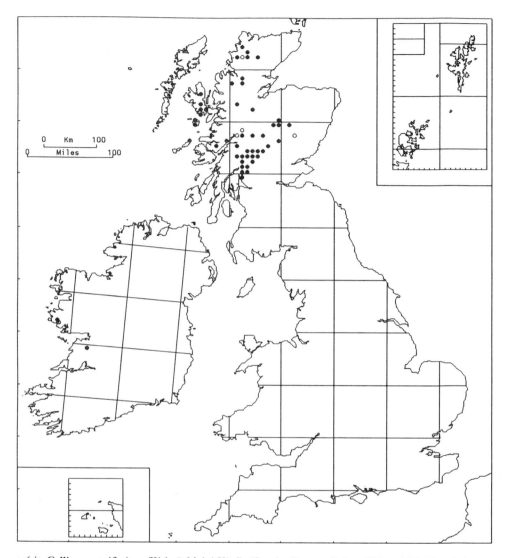

146/2. **Calliergon trifarium** (Web. & Mohr) Kindb. (*Pseudocalliergon trifarium* (Web. & Mohr) Loeske)

A calcicole of wet ground, growing in small-sedge marshes, gravelly flushes and rocky runnels, usually in the mountains. In Scotland, it has been noted especially as an associate of *Carex saxatilis*, *Juncus triglumis* and *Saxifraga aizoides* in montane flushes and beside rills. It also occurs frequently as pure or nearly-pure patches, especially where there is a steady seepage from among rocks. In Ireland it is known from low-altitude calcareous marshes and the shores of seasonally fluctuating lakes, where it grows over bare peat or among other mosses, principally *Scorpidium scorpioides* (Proctor, 1959). 0–960 m (Ben Lawers), descending to sea-level on Skye but rarely below 450 m elsewhere in Scotland. GB 46+4*, IR 2.

Dioecious; capsules not found in Britain or Ireland.

Widespread in boreal and Arctic regions of the Northern Hemisphere, extending southwards in mountains to C. Europe, the Altai and Oregon (U.S.A.). Reported from Haiti and Venezuela.

Like other boreal 'rich-fen' species, *C. trifarium* is declining in the European lowlands. There is a tantalizing record from Whittlesey Mere by Berkeley (1863), but this is unsupported by a specimen.

M. O. Hill

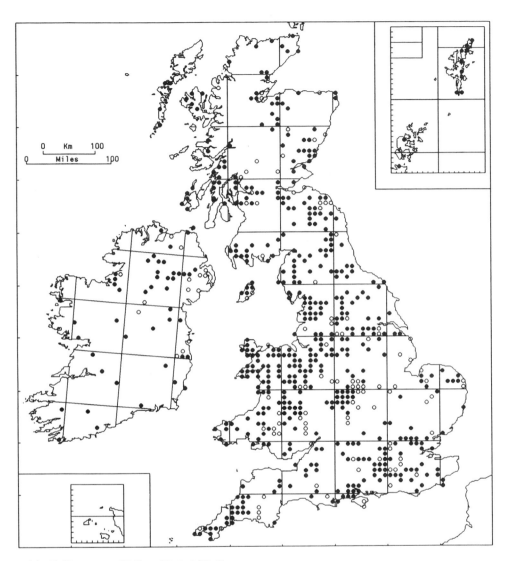

146/3. **Calliergon cordifolium** (Hedw.) Kindb.

This mildly calcifuge moss of wet and sometimes rather squalid places normally grows in the shade of vigorous higher plants, but is also found in rills and swampy flushes in the open. Typical habitats are swamp woodland, especially under alder and willow, lake margins, quaking vegetation rafts, 'moats' at the margins of bogs, ditches on moors, and pools in areas of mineral workings. It often occurs where there are large seasonal fluctuations of the water-table, growing over decaying leaves of trees, grasses or *Typha latifolia*. Its habitat overlaps to some extent with that of *C. stramineum*, but *C. cordifolium* is much more shade-tolerant and tends to occur in places where nutrient levels, but not necessarily the pH, are higher. Mainly lowland, ascending to 910 m (Coire an t-Sneachda). GB 487+101*, IR 46+14*.

Autoecious; capsules occasional, ripe in late spring and early summer.

Circumboreal, rather rare in the Arctic, extending southwards in mountains to S. Europe, Turkey, Caucasus, Nepal and, in N. America, to Colorado and N. Carolina. Reported from New Zealand.

M. O. Hill

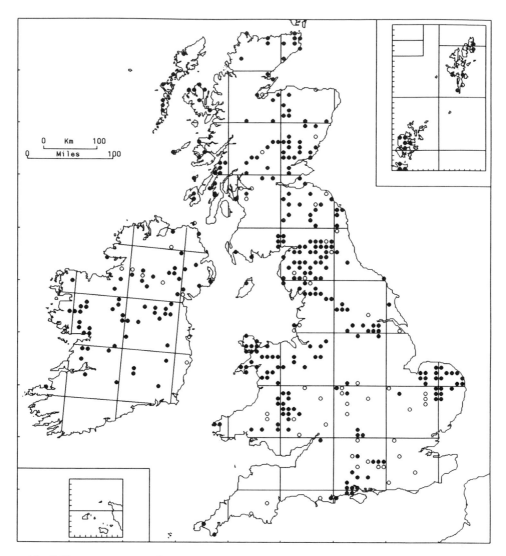

146/4. **Calliergon giganteum** (Schimp.) Kindb.

A plant of wet, open fens, waterlogged hollows, swampy flushes, boggy streamsides and lake margins, confined to sites that are calcareous or at least moderately base-rich. Although sometimes found in bushy places, it generally occurs in the open, often among *Carex rostrata* or among large grasses and herbs in fens. Its other associates are many and varied, including *Carex diandra*, *Potentilla palustris*, *Calliergon cuspidatum*, *Drepanocladus revolvens*, *Scorpidium scorpioides* and *Sphagnum contortum*. 0–910 m (Ben Lawers). GB 283+53*, IR 62+9*.

Dioecious; capsules very rare, summer.

Circumboreal; widespread in cool and cold regions of the Northern Hemisphere north to the High Arctic, extending south in mountains to C. Europe, the Altai and Colorado (U.S.A.).

Like several other 'rich-fen' species, *C. giganteum* is well represented in Pleistocene deposits and was apparently more abundant in the past (Dickson, 1973). It is sometimes confused with *C. cordifolium*; the map shown here must inevitably contain a few errors, but these should be few enough not to distort the picture of either its frequency or its distribution.

M. O. HILL

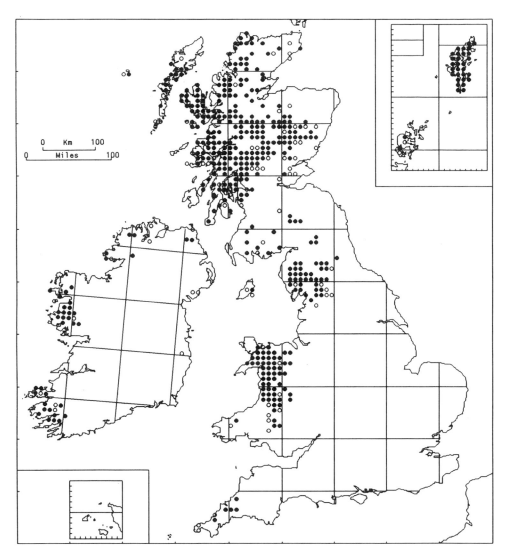

146/5. **Calliergon sarmentosum** (Wahlenb.) Kindb. (*Sarmenthypnum sarmentosum* (Wahlenb.) Tuom. & T. Kop.)

This moss of permanently moist ground in unshaded situations grows in flushes, in stony rills, on dripping cliffs and in bryophyte-dominated springs. It requires some mineral enrichment of the water, and its occurrence on acid heaths and moors is a sure sign of basic seepage. Nevertheless it is not a calcicole and, like *Blindia acuta*, with which it sometimes grows, does not occur on limestone. Typical associates in flushes and on seeping rocks are *Aneura pinguis, Scapania undulata, Campylium stellatum, Drepanocladus revolvens, Scorpidium scorpioides* and *Sphagnum auriculatum*. In springs on mountains it is often found with *Saxifraga stellaris, Dicranella palustris* and *Philonotis fontana*. 0–1070 m (Ben Lawers). GB 382+71*, IR 38+13*.

Dioecious; capsules very rare, summer.

Circumboreal and bipolar, very widespread in cool and cold regions of the world, extending north to the High Arctic and occurring in Antarctica, on subantarctic islands and in New Zealand; scattered on tropical mountains in Africa, Papua New Guinea and South America.

M. O. Hill

303

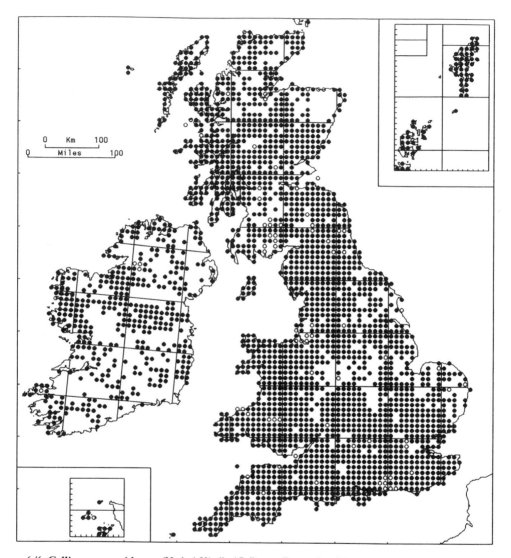

146/6. **Calliergon cuspidatum** (Hedw.) Kindb. (*Calliergonella cuspidata* (Hedw.) Loeske)

It occurs in a wide variety of open or lightly-shaded habitats, especially in turf that is either calcareous or wet, almost always in sites that are too base-rich for growth of sphagnum. In the lowlands it is found by lakes and rivers, in springs, in marshes and fens, in lawns and grassy flushes, in dune-slacks, on undercliffs, by ditches, in old quarries, by tracks and roads, and in woodland rides. It is common and often abundant in dry chalk and limestone grassland where there is a little shelter. It is much less common on mountains, occurring in small-sedge communities, flushed grassland and springs. 0–900 m (Ben Lawers). GB 2126+75*, IR 397+7*.

Dioecious; capsules rare, produced mainly in base-rich marshes and fens, ripe May.

Widespread in cool-temperate regions of the Northern Hemisphere north to the southern boreal zone, absent from most of the Arctic, extending south in mountains of Eurasia to the Himalaya and, in N. America, to California and N. Carolina; also in Australia, New Zealand and the Andes of S. America.

M. O. Hill

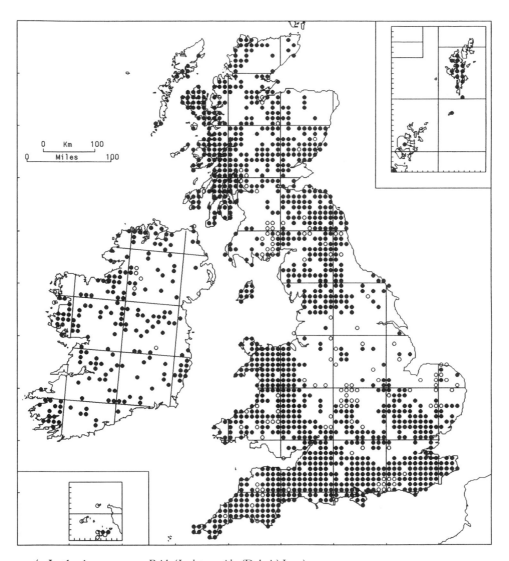

147/1. **Isothecium myurum** Brid. (*I. alopecuroides* (Dubois) Isov.)

This plant is often abundant on tree-roots and boles and on mildly to strongly basic rocks and crags in sheltered situations. It forms deep, robust green wefts, usually in pure stands but also associated with *Brachythecium* spp., *Eurhynchium striatum*, *Hypnum cupressiforme* and *Isothecium myosuroides*. In C. and E. England it is particularly noted from coppice-stools in woodland. 0–650 m (Snowdon). GB 1219+122*, IR 189+9*.

Dioecious; capsules occasional, ripe winter.

Europe north to Iceland and C. Scandinavia. Macaronesia, N. Africa, W. Asia east to Caucasus and Iran; disjunct in Canada (Ontario, a single locality).

G. P. ROTHERO

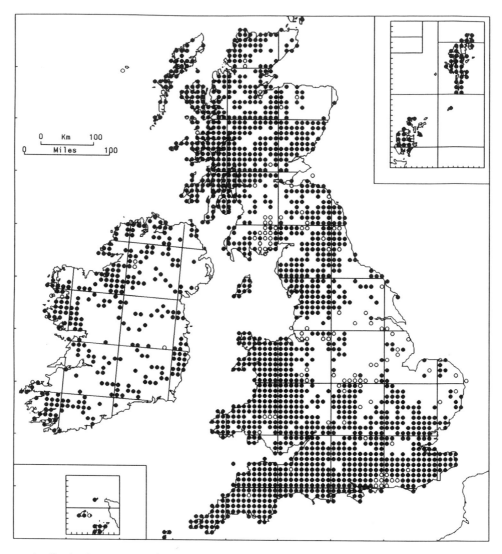

147/2a. **Isothecium myosuroides** Brid. var. **myosuroides**

A shade-tolerant plant of well-drained, hard substrates such as trees, rocks and screes, in woodland and in sheltered to moderately exposed situations on moorland. In the north and west, where it is locally dominant, the dense yellowish to mid-green carpets may cover both rocks and tree-boles to a depth of several centimetres. In rocky western *Quercus petraea* woods, *I. myosuroides* can form a significant part of the ground layer, associated with *Scapania gracilis*, *Dicranum* spp., *Hylocomium brevirostre*, *Hypnum cupressiforme*, *Rhytidiadelphus loreus*, *Thuidium delicatulum* and *T. tamariscinum*, and, in humid sites, *Hymenophyllum wilsonii* and hygrophilous Atlantic bryophytes. It avoids calcareous substrata and in much of C. and S. England is confined to tree-bases. 0–950 m (Coire an t-Sneachda). GB 1620+107*, IR 282+7*.

Dioecious; capsules common in the north and west, rare in parts of S.E. and C. England.

Western Europe north to Iceland and C. Scandinavia (very few records north of the Arctic Circle) and east to the Carpathians. Macaronesia, N. Africa, Turkey, western and eastern N. America.

The map should be broadly correct, but some Shetland records may refer to var. *brachythecioides*.

G. P. ROTHERO

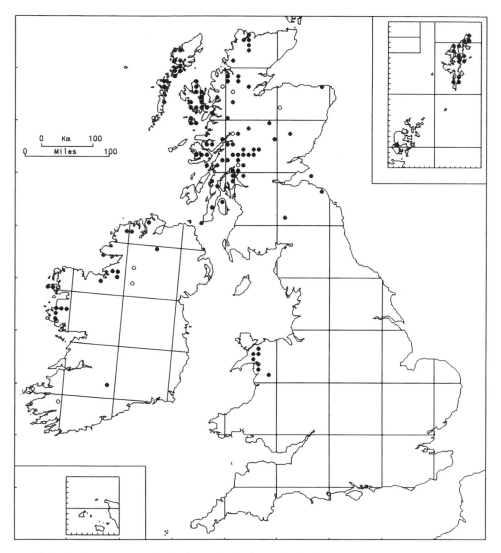

147/2b. **Isothecium myosuroides** Brid. var. **brachythecioides** (Dix.) Braithw.

This plant forms robust, lax, yellow-green patches on vertical or inclined rock-faces, usually in sheltered, humid but not irrigated sites; it also occurs in rock crevices and under overhanging ledges. It is moderately frequent on schistose rocks in the S.W. Highlands of Scotland, where associates may include *Diplophyllum albicans*, *Heterocladium heteropterum*, *Hypnum cupressiforme* var. *cupressiforme* and *Isopterygium pulchellum*. 50 m (Shetland) to 930 m (Cairngorm Mts). GB 102+6*, IR 17+4*.

Dioecious; sporophytes unknown.

North Atlantic seaboard, from N. Spain, British Isles and Faeroes to Iceland and N. Norway. According to Duell (1992) it also occurs on Tenerife (Canary Islands).

G. P. ROTHERO

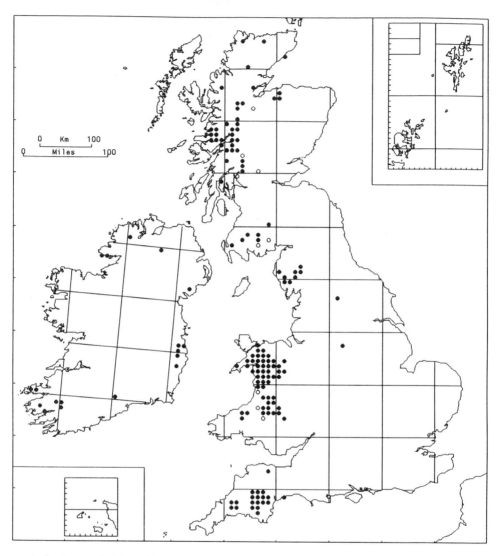

147/3. **Isothecium holtii** Kindb.

A robust moss of rocks, tree-roots and tree-boles in very humid sites in the inundation-zone of fast-flowing rivers and streams, often near waterfalls in deep ravines where spray keeps the humidity level high. Common associates include *Scapania undulata*, *Brachythecium plumosum*, *Hygrohypnum ochraceum*, *Isothecium myosuroides* and *Racomitrium aciculare*. Although a plant of mountainous districts, it is normally found at low elevations, ascending to 550 m in Ireland (Dart Mountain) and 300 m in Scotland (Glen Nevis). GB 126+9*, IR 18.

Dioecious; capsules rare.

Atlantic seaboard of Europe from S. Spain north to S. Norway; rare in France, where it is confined to the Basque country and Brittany; disjunct in the Schwarzwald and Harz. Reported from the Canary Islands and Czechoslovakia.

I. holtii is surprisingly infrequent in the west of Scotland given the extent of apparently suitable habitat there. Similarly, it has a patchy distribution in the Lake District.

G. P. Rothero

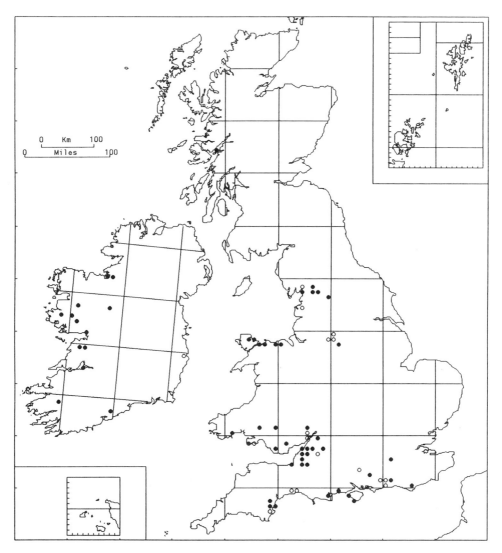

147/4. **Isothecium striatulum** (Spruce) Kindb. (*Eurhynchium striatulum* (Spruce) B., S. & G.)

A strict calcicole, confined to areas of chalk and limestone, where it occurs on dry, shaded stones, rocks, walls and tree-roots. It often forms quite extensive patches, with few associates other than *Porella platyphylla* and common *Brachythecium* and *Eurhynchium* species. Lowland. GB 42+16*, IR 13+2*.

Dioecious; sporophytes occasional.

Europe north to southern Scandinavia and east to the Crimea. Macaronesia, N. Africa, Turkey, Caucasus.

G. P. ROTHERO

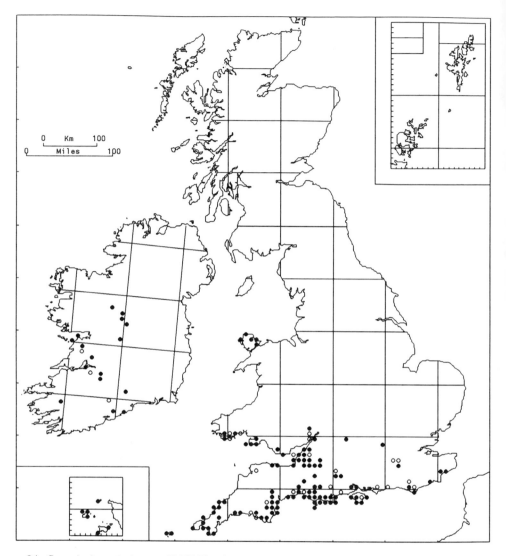

148/1. **Scorpiurium circinatum** (Brid.) Fleisch. & Loeske

In Cornwall and the Channel Islands *Scorpiurium* is found in open turf or around rock outcrops in well-drained, insolated or lightly shaded sites such as banks in sand-dunes, rocky cliffs and earth-covered rocks, walls, Cornish hedges, roadside banks and disused quarries. Although usually growing over non-calcareous rocks its substrates often contain calcium derived from wind-blown sand and are almost always subject to the deposition of salt spray. Elsewhere it is confined to chalk and limestone, being found on cliffs, rocks and limestone walls, often in shaded situations, and on loose stones in woods. It is also recorded from silty tree-bases by calcareous streams. There are isolated occurrences on limestone ornaments in parks, including an old stone seat at Cliveden and a bust of Pan (imported from Italy in the 1920s) at Buscot Park. Lowland. GB 101+23*, IR 16+3*.

Dioecious; sporophytes not known in Britain or Ireland. No specialized means of vegetative propagation is known.

A Mediterranean-Atlantic species, frequent in the Mediterranean countries and reaching its northern limit in the British Isles. Macaronesia, N. Africa, S.W. Asia.

C. D. Preston

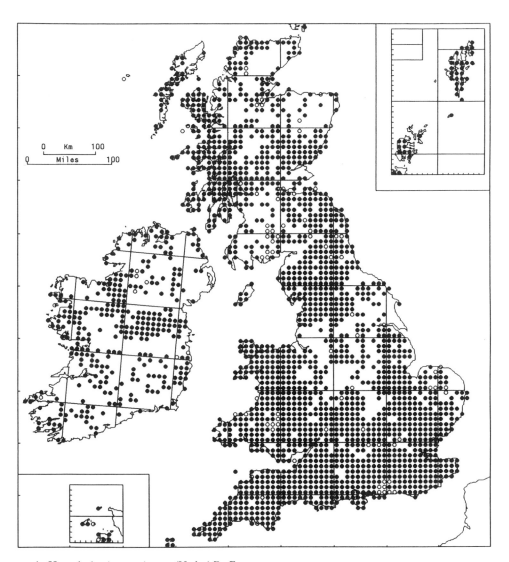

149/1. Homalothecium sericeum (Hedw.) Br. Eur.

This species grows on vertical, horizontal and sloping surfaces, often fully exposed to sun and wind, but best in humid, lightly shaded sites. It occurs on stone or mortared walls, on tiled, thatched or asbestos roofs, and on many rock types, especially limestone but also igneous and siliceous. It is common on living bases, boles and main branches of many tree species in clean-air districts, mostly on less acid bark types (but known on birch and conifers). It is not a pioneer and grows mostly on mature trees. It occurs on the ground in beech-woods on chalk, in quarries, on hard-packed soil, and, rarely, on sand-dunes and sea-cliff turf. Susceptible to aerial pollution, it tolerates mean aerial SO_2 contents of only up to 22 µg S m^{-3} (Gilbert, 1970). 0–610 m (Glen Clova). GB 1843+78*, IR 321+8*.

Dioecious; capsules occasional, especially in sheltered clean-air sites, ripe winter and spring.

All Europe north to Iceland and N. Fennoscandia. Macaronesia, N. and C. Africa, W. Asia, Nepal, Newfoundland.

R. A. Finch

311

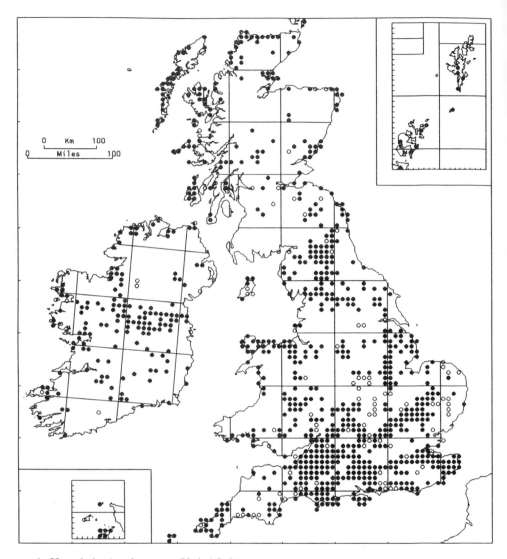

149/2. **Homalothecium lutescens** (Hedw.) Robins.

This is a characteristic and often abundant species of relatively dry calcareous grassland on chalk, limestone, sand, clay, gravel or basic igneous rock. It grows on sunny south-facing cliffs and slopes, but is more luxuriant on ones facing north. It does not tolerate much shade, and in woods is confined to edges, by paths and in open glades. Common on bare soil and chalk in quarries or in species-rich short turf, it rarely grows directly on rock, but in limestone districts is locally frequent on walls and stable screes. On coasts, it grows in cliff-top turf, on fixed dunes, in machair and, rarely, in salt-marsh. 0–490 m (Tailbridge Hill). GB 709+92*, IR 146+9*.

Dioecious; capsules rare but locally numerous, ripe January to April.

Nearly all Europe from the Mediterranean north to Iceland and southern Scandinavia. Macaronesia, N. Africa, W. Asia.

A sterile prostrate form, blackening when old and resembling *H. sericeum* in habit, grows on limestone screes and walls in Derbyshire and Yorkshire. It may be var. *fallax* (Philib.) Bertsch., which extends from Norway through C. Europe to Turkey and has distinct sporophyte characters.

R. A. FINCH

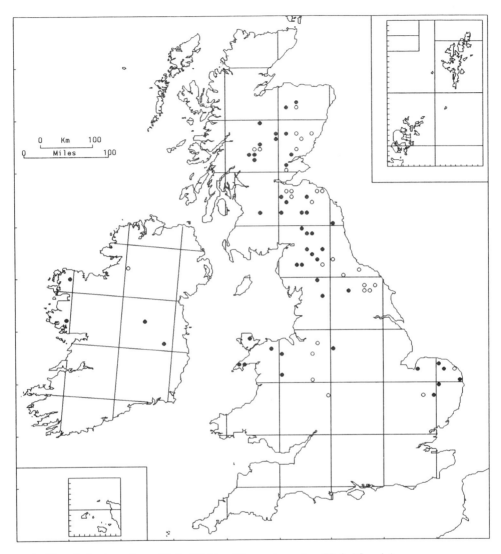

149/3. **Homalothecium nitens** (Hedw.) Robins. (*Tomentypnum nitens* (Hedw.) Loeske)
This species is found in open calcareous mires and flushes with gently flowing water of pH 5.8 or more. Its lowland sites are often species-rich fens or bog-fen transition mires, with sedges, rushes and a hypnoid moss carpet developed over a peaty or mineral substrate bare in places and with few tall plants. At upland sites it occurs in wet fields and flushes, often on gentle slopes flushed by water from steeper ones. In the Scottish Highlands it is mainly in mesotrophic peaty alpine mires dominated by *Carex* spp. and 'brown mosses'. It reaches 840 m on Meall nan Tarmachan. GB 45+28*, IR 4+1*.

Dioecious; capsules very rare, ripe summer.

Circumpolar, occurring widely in the Arctic and northern boreal zones, extending south in mountains to Spain, C. Europe, Caucasus, C. Asia, N.E. China and southern U.S.A. (New Mexico).

Declining in W. and C. Europe and rare in fruit throughout its range, so recent finds of fruiting plants in Ireland (Lockhart, 1987) are remarkable. *H. nitens* was locally common in British mires from 10,000 to 6,000 B.P., often with *Helodium blandowii* and *Paludella squarrosa*. Natural processes of soil acidification caused its decline in highland areas.

R. A. Finch

313

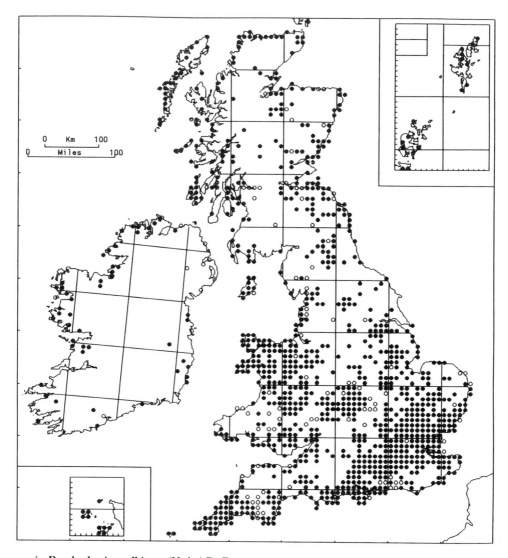

150/1. **Brachythecium albicans** (Hedw.) Br. Eur.

A plant of neutral or acidic, base-deficient, usually sandy or gravelly soils in unshaded situations. It is particularly prominent in open acidic *Agrostis-Festuca* grassland where associates include *Ceratodon purpureus*, *Hypnum cupressiforme* and *Polytrichum piliferum*. It is also characteristic of certain types of thin, stony, slightly leached, calcareous soils in Breckland, associated with *Barbula* spp. and *Dicranum scoparium*, in conditions where the relatively continental climate leads to summer drought. On the coast, it reaches its maximum abundance on sand-dunes, being tolerant of low rates of deposition of wind-blown sand, and is often found with *Tortula ruralis* ssp. *ruraliformis*. It also occurs on tracks, by roadsides, on wall-tops, on sandy heaths, in quarries and on mine-waste. Mainly lowland, but ascending to at least 350 m by moorland roads in N. Wales. GB 974+95*, IR 48+9*.

Dioecious; sporophytes rare, maturing autumn to winter.

Circumboreal, extending north to the Arctic and south to Macaronesia, Turkey, California and eastern Canada; very widespread in Europe. Introduced in Australia and New Zealand.

R. D. PORLEY

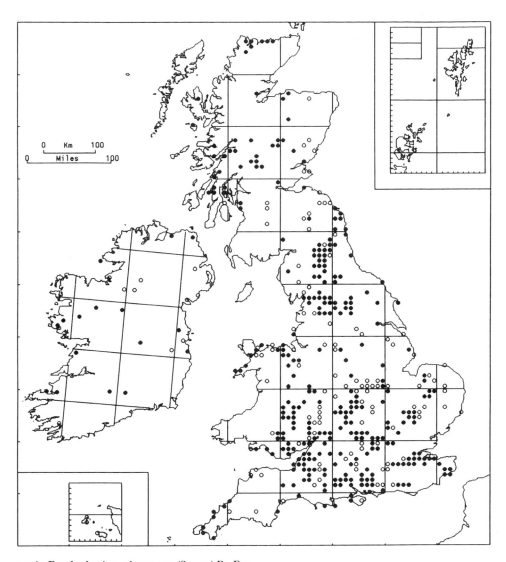

150/2. **Brachythecium glareosum** (Spruce) Br. Eur.

This species occurs in a variety of habitats, most characteristically on dry calcareous substrates. It is most frequent in chalk or limestone turf, on banks and in disused quarries. Commonly encountered associates include *Ctenidium molluscum*, *Homalothecium lutescens* and *Pseudoscleropodium purum*. It tolerates some shading, and is found on dry banks at the edge of woodlands, and on tracks in open woodland. It is occasionally found on sand-dunes, on rocks and, more rarely, on tree-bases. 0–1065 m (Aonach Beag). GB 300+111*, IR 16+10*.

Dioecious; sporophytes rare, maturing in winter.

Throughout Europe north to Arctic Fennoscandia. Widespread in northern Asia, from Turkey and the Caucasus through C. Asia to China and Japan.

R. D. PORLEY

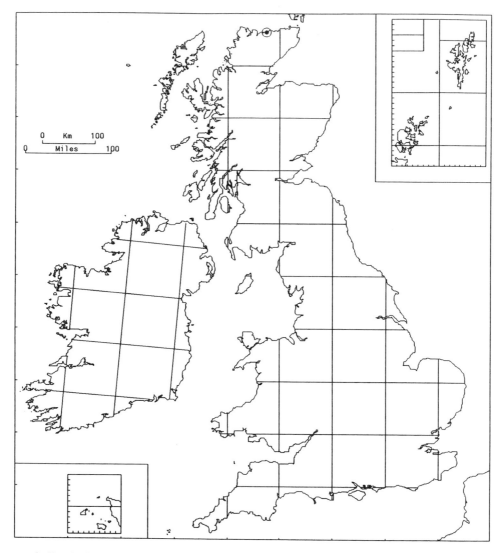

150/3. **Brachythecium erythrorrhizon** Br. Eur.

On slopes of low sand-hummocks in calcareous coastal dunes, with *Dryas octopetala*, *Ditrichum flexicaule*, *Homalothecium lutescens* and *Tortula ruralis* ssp. *ruraliformis*. Lowland. GB 1.

Dioecious; capsules unknown in Britain.

Circumboreal-montane. Arctic and N. Europe, mountains of C. and E. Europe. Turkey, Siberia, C. and E. Asia, western and eastern N. America.

First found in Britain at Bettyhill, a locality well known for northern calcicoles, in 1948 (Barkman, 1955); it was refound there in 1956 and 1988.

G. P. ROTHERO

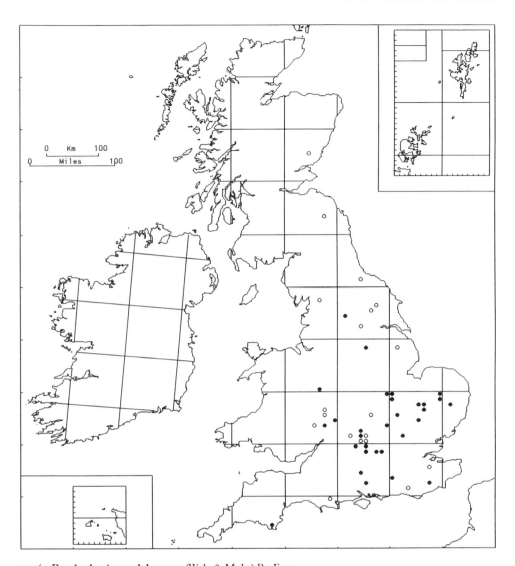

150/4. **Brachythecium salebrosum** (Web. & Mohr) Br. Eur.

Primarily a species of rotting wood in wet shady woodlands. It is most characteristic of alder carr, where it is perhaps overlooked as *B. rutabulum*, which is a common associate in this habitat. It occurs more rarely in wet *Molinia* grassland, on wet ditch-banks, on leaf-litter, tree-bases and tracks in damp woodland, and on rock. It has also been recorded on stone in disused railway cuttings and on clinker in open woodland. Lowland. GB 30+19*.

Autoecious, rarely synoecious; sporophytes frequent to common, maturing in winter.

Circumboreal, from the Arctic south to the northern deciduous forest zone, and in mountains south to Macaronesia, N. Africa, C. Asia and central N. America (Colorado and Tennessee). Southern Australia, Kerguelen Island.

Brachythecium salebrosum and *B. mildeanum* have often been confused in Britain. Only records checked by M. O. Hill for a vice-comital revision (Corley & Hill, 1981) and those made subsequently have been mapped. Both species are probably under-recorded.

R. D. PORLEY

317

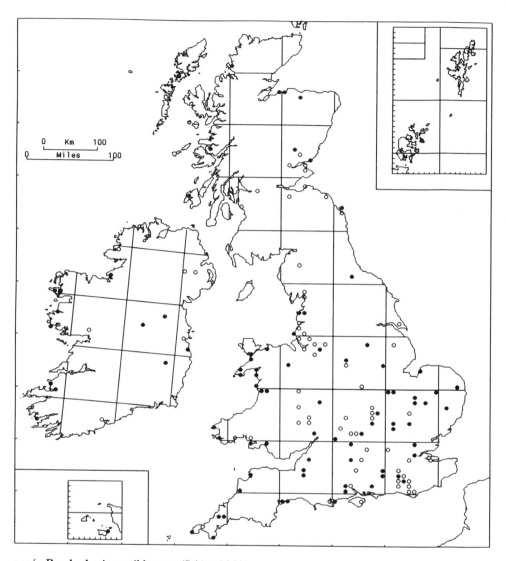

150/5. **Brachythecium mildeanum** (Schimp.) Milde

A species primarily of damp, open, calcareous ground. It occurs amongst grasses around ponds and pools, on river-banks, in tall-herb fen, and on damp, clayey woodland tracks, often associated with *Calliergon cuspidatum*. In old water-meadows it is sometimes found with *Eurhynchium speciosum*. It is often locally abundant in seasonally damp dune-slacks, typically with *Riccardia chamedryfolia*, *Campylium polygamum* and *Drepanocladus aduncus*. It also behaves as a weedy colonist of man-made habitats such as tracks, roadsides, car-parks, railway ballast and disturbed sandy soil by the coast. Lowland. GB 75+60*, IR 12+6*.

Autoecious; sporophytes occasional, maturing in winter.

Widespread in Europe north to about the Arctic Circle. Macaronesia, W. and C. Asia, western N. America.

B. mildeanum has often been confused with *B. rutabulum* and *B. salebrosum*. The map is based mainly on herbarium specimens checked by M. O. Hill, and the high proportion of old records reflects the fact that many specimens collected before 1950 have been examined. There is no reason to think that *B. mildeanum* has declined.

R. D. PORLEY

318

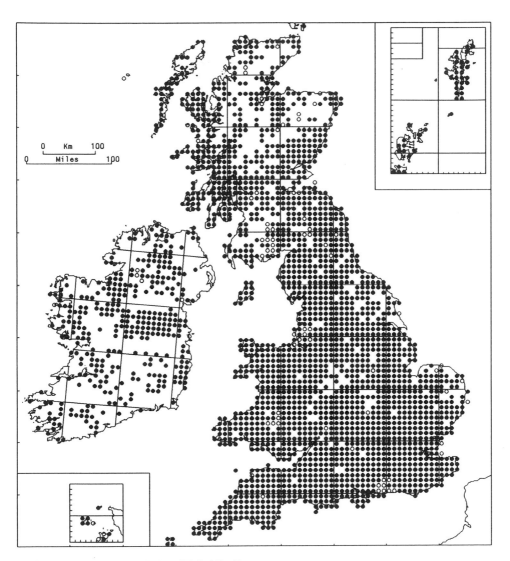

150/6. **Brachythecium rutabulum** (Hedw.) Br. Eur.

This species occurs in a very wide range of habitats, often growing luxuriantly in eutrophic situations, but scarce or absent on more acid and oligotrophic substrates. It is shade-tolerant and reaches its finest development in wet woodland on base-rich soils. It is often present in tall-herb communities such as *Urtica dioica* stands, where few other bryophytes can survive. In autumn and winter it rapidly colonizes the last season's accumulation of stem litter, having a higher relative growth rate than many vascular plants (Furness & Grime, 1982). It is ubiquitous in most types of woodland, including conifer plantations of the uplands, growing on tree-bases, stumps, rotting wood, soil and rocks. Elsewhere it occurs on roadsides, in hedges, by streams and rivers, in grassland, on heaths, in quarries and on cliffs. It is a rapid colonist. 0–1344 m (Ben Nevis). GB 2159+73*, IR 343+7*.

Autoecious; sporophytes common, maturing late autumn to spring.

Very widely distributed in the North temperate zone, south to Macaronesia, N. Africa, Himalaya, Japan and central N. America (Iowa, N. Carolina); rare in and largely absent from the Arctic. Hawaii, Guatemala, Australia, New Zealand.

R. D. PORLEY

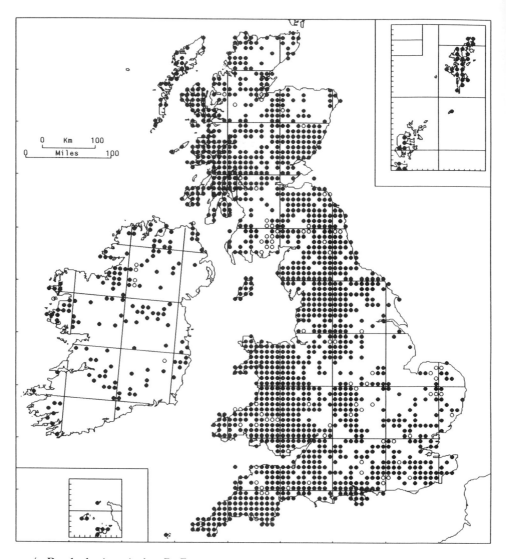

150/7. **Brachythecium rivulare** Br. Eur.

A species of moist or wet, eutrophic or mesotrophic habitats. It is characteristic of wet woodland, particularly alder carr, where a frequent associate on the ground is *Brachythecium rutabulum*, and it occurs in flushes in many other types of woodland, sometimes with *Trichocolea tomentella*. It is also common in and beside streams and rivers, particularly those with fast-flowing water, growing on rocks, banks, tree-bases, concrete and rotting wood, with associates such as *Conocephalum conicum*, *Cratoneuron filicinum* and *Thamnobryum alopecurum*. It is prominent in certain types of upland springs, either base-poor with *Philonotis fontana* or calcareous with such associates as *Aneura pinguis*, *Pellia endiviifolia*, *Cratoneuron commutatum* and *Philonotis calcarea*. It is sometimes present in wet grassland. 0–960 m (Ben Lawers). GB 1488+83*, IR 144+8*.

Dioecious; sporophytes occasional, maturing autumn to spring.

Very widespread in the Northern Hemisphere, from the Arctic south to Macaronesia, N. Africa, S.W. Asia, Himalaya, Taiwan and southern U.S.A. (New Mexico, N. Carolina). Rare in the Southern Hemisphere, reported from Chile, Australia (New South Wales) and Kerguelen Island.

R. D. PORLEY

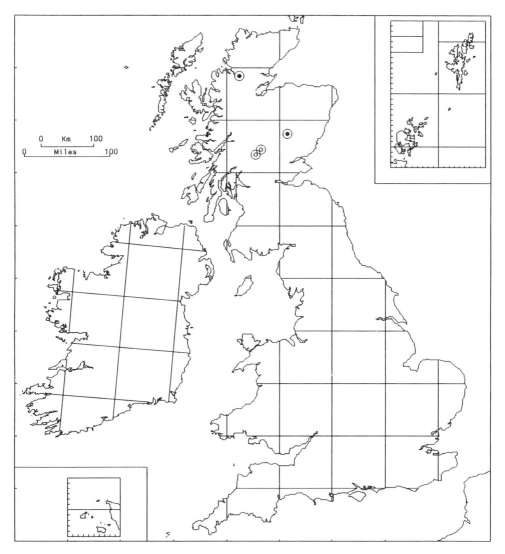

150/8. **Brachythecium starkei** (Brid.) Br. Eur.

It has been found growing in the interstices of calcareous mountain rocks, on a boulder beside a mountain loch, and in flushed *Agrostis vinealis* grassland below a snow-bed. 500 m (Seana Bhraigh) and 960 m (above Caenlochan Glen). GB 2+2*.

Autoecious; capsules abundant at the Caenlochan site, not found elsewhere in Scotland.

Circumpolar arctic-alpine. Widespread in N. Europe and in mountains further south. Turkey, Caucasus, C. and E. Asia, eastern N. America.

This very rare plant was poorly understood in Britain until Scottish specimens were revised by Crundwell (1959a). It has not been seen in the Ben Lawers area since 1941.

G. P. ROTHERO

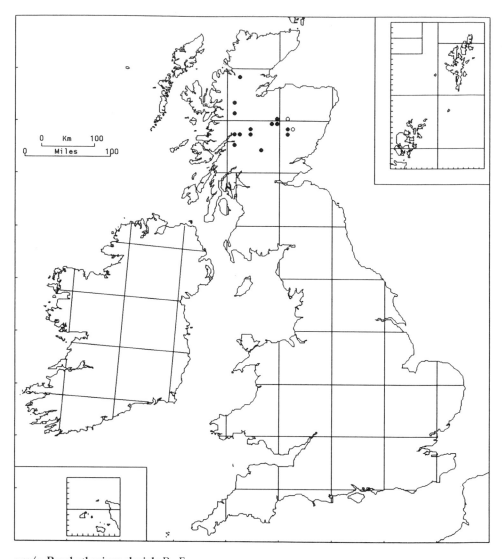

150/9. **Brachythecium glaciale** Br. Eur.

This moss occurs as straggling, yellow-green stems in the crevices of sheltered block-scree in the mountains. It is particularly characteristic of fern-dominated communities with *Athyrium distentifolium* in N.- and E.-facing corries where snow patches linger well into the summer, and is often to be found growing on the fern litter. Frequent associates are *Barbilophozia floerkei, Lophozia sudetica, Hylocomium umbratum, Hypnum callichroum* and *Plagiothecium* spp. Where snow-cover is especially prolonged, it may occur in scree and on more open rock-faces, with *Diplophyllum taxifolium, Lescuraea patens* and *Kiaeria starkei*. 710 m (Aonach Mor) to 1200 m (Ben Nevis). GB 14+2*.

Autoecious; capsules generally rare but more frequent where snow-cover is very persistent.

Bipolar arctic-alpine. Widespread in Europe. N. and C. Asia, N. America (rare), Greenland, southern S. America, S. Georgia.

G. P. Rothero

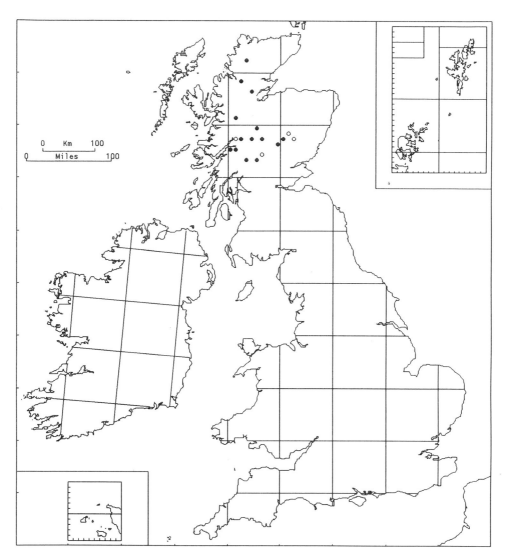

150/10. **Brachythecium reflexum** (Starke) Br. Eur.

This inconspicuous species occurs in the interstices of block-scree in sheltered montane habitats. Like *B. glaciale*, it is commonly found on litter of *Athyrium distentifolium* in fern-dominated snow-bed communities. It usually occurs sparingly in a mixed weft of larger bryophytes such as *Diplophyllum albicans*, *Hylocomium umbratum*, *Hypnum callichroum* and *Plagiothecium* spp. 700 m (Glen Duror) to 1080 m (Aonach Beag). GB 14+4*.

Autoecious; capsules are uncommon but may be locally abundant.

Circumboreal, widespread in the Northern Hemisphere south in the mountains to Turkey, Caucasus, Kashmir, Japan and central N. America (Oregon, N. Carolina).

In northern Europe, *B. reflexum* occurs commonly in broad-leaved woodland, and it has recently appeared in numerous woods in the Netherlands (Touw & Rubers, 1989; Bremer & Ott, 1990). Its absence from woodland in Britain is rather surprising.

G. P. ROTHERO

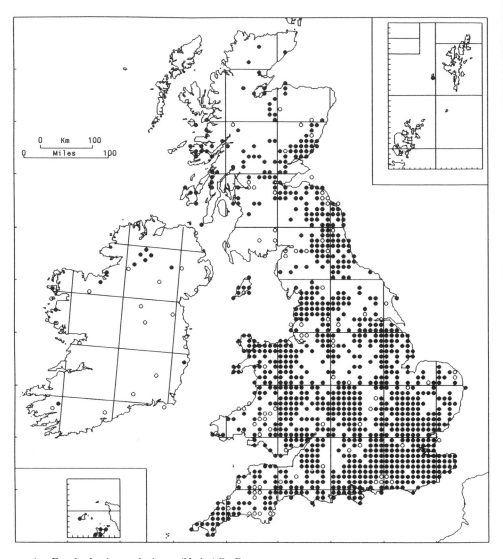

150/11. **Brachythecium velutinum** (Hedw.) Br. Eur.

This species of well-drained, shaded habitats is common in calcareous districts as well as on sands and gravels. It occurs most typically on tree-bases, stumps and logs in woodland, with *Amblystegium serpens* and the superficially similar *Rhynchostegium confertum*. It also occurs on the bark of elders, and on banks, rocks and soil beside lanes, hedges, streams and tracks, especially where the substratum is leached or slightly acid. It is frequent in towns and cities, often growing with *Bryum capillare*, *Grimmia pulvinata* and *Tortula muralis* in dry, quite exposed places on walls, stones and brickwork. Lowland. GB 1038+90*, IR 12+19*.

Autoecious; sporophytes common, maturing winter to spring.

Widespread in the Northern Hemisphere, from the Arctic (where it is rare) south to Macaronesia, N. Africa, Caucasus, C. Asia, Japan and northern U.S.A. (Oregon, New England). Reported from Campbell Island and New Zealand.

R. D. PORLEY

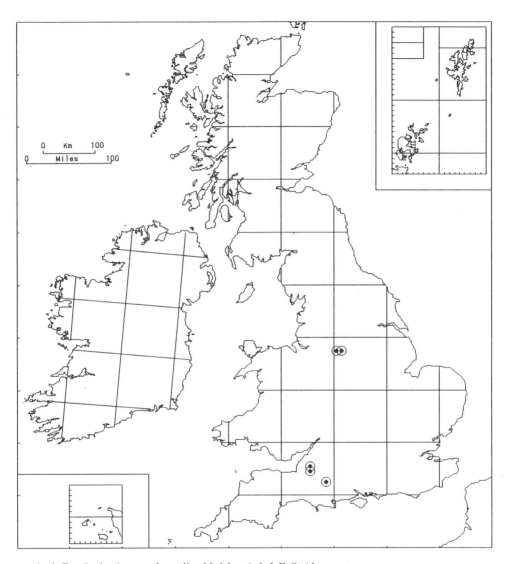

150/11A. **Brachythecium appleyardiae** McAdam & A. J. E. Smith

On shaded calcareous rocks, typically on a thin layer of soil overlying rock-ledges in wooded valleys, where associates include *Eurhynchium praelongum*, *Scorpiurium circinatum* and *Thamnobryum alopecurum*. All but one of the colonies are on Carboniferous Limestone, but the Wiltshire colony is anomalous in being on an exposed sandstone wall with much *Homalothecium sericeum*. Wall-mortar may in this case provide the necessary calcareous substrate. Lowland. GB 5.

Dioecious; gametangia and sporophytes are unknown in the wild; Wiltshire plants have produced sporophytes in cultivation.

Not yet discovered outside Britain.

B. appleyardiae was first collected by Mrs J. Appleyard in Somerset in 1967 and tentatively identified as *Eurhynchium pulchellum* (Appleyard, 1970). It was later described as a species of *Brachythecium*, but its generic placement remains problematical (McAdam & Smith, 1981).

R. D. PORLEY

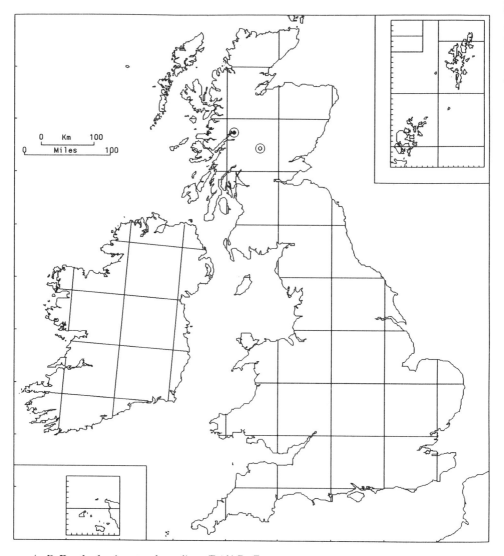

150/11B. **Brachythecium trachypodium** (Brid.) Br. Eur.

On calcareous mountain rocks. In its western locality it was low on the near-vertical north side of a large craggy block in an E.-facing gully, mixed with *Cephalozia bicuspidata*, *Diplophyllum albicans*, *Tritomaria quinquedentata*, *Amphidium lapponicum*, *Blindia acuta*, *Ctenidium molluscum*, *Kiaeria starkei*, *Pohlia ludwigii* and *Racomitrium fasciculare*. At its eastern site it was found growing among *Encalypta rhaptocarpa*. 1080 m (Aonach Beag). GB 1+1*.

Autoecious; capsules not found in Britain.

A circumpolar arctic-alpine species known from Europe, Algeria, Turkey, Caucasus, Afghanistan, Siberia, Alaska, Canada and Greenland.

Found on Ben Lawers in the 19th century, but the determination was regarded as doubtful until the specimen was re-examined following the discovery of the species on Aonach Beag in 1989 (Corley, 1990).

G. P. ROTHERO

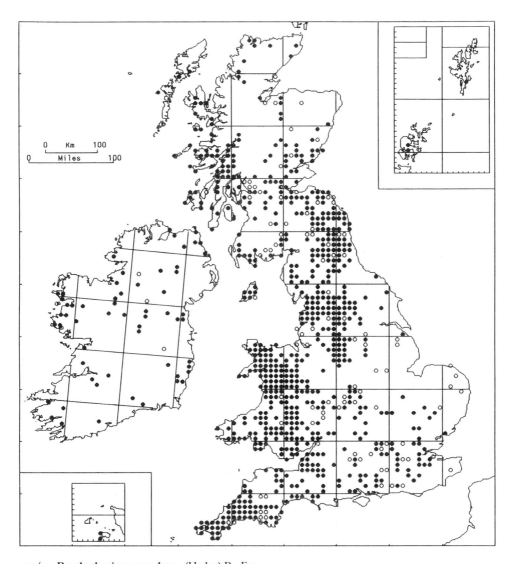

150/12. **Brachythecium populeum** (Hedw.) Br. Eur.

A species of well-drained substrates in shaded places, often in woodland and hedgerows. It occurs on tree-roots, tree-bases, stumps and soil-banks, frequently associated with *Amblystegium serpens*, *Eurhynchium praelongum* and *Hypnum cupressiforme*. It is equally common on rocks, walls and concrete. Occasionally, it is found in more exposed habitats such as stones in fields, flints overlying chalk, and rocks in quarries and on cliffs. 0–580 m (Seana Bhraigh). GB 671+101*, IR 58+5*.

Autoecious; sporophytes frequent, maturing autumn to spring.

Widespread in the cool-temperate Northern Hemisphere, from the boreal zone south to Macaronesia, N. Africa, Himalaya, Japan, British Columbia (Canada) and N. Carolina (U.S.A.); it is almost completely absent from the Arctic.

R. D. PORLEY

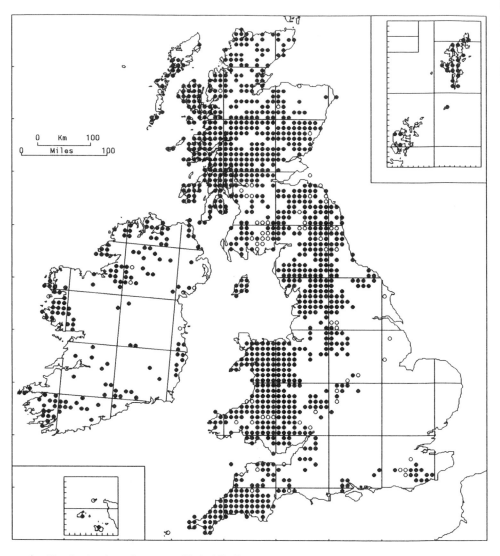

150/13. **Brachythecium plumosum** (Hedw.) Br. Eur.

This is a characteristic species of swiftly flowing rivers and streams, where it is as least periodically inundated. It is weakly calcifuge and can tolerate shade. It is most common on siliceous boulders subject to submergence by mountain torrents, where it is frequently associated with *Scapania* spp., *Hygrohypnum* spp. and *Racomitrium aciculare*. It also occurs on tree-roots and tree-bases. Oligotrophic mountain springs are another habitat, which it shares with *Scapania undulata*, *Dicranella palustris* and *Philonotis fontana*. It is also found by lakes and pools, and occasionally on rocks and rock-ledges away from water. 0–1205 m (Ben Lawers). GB 1053+68*, IR 149+7*.

Autoecious; sporophytes frequent, maturing in winter.

Very widespread in the Northern Hemisphere, from Arctic Eurasia south to N. Africa, Himalaya and S.E. Asia and from boreal N. America south to Oregon and Arkansas. Scattered in the tropics and subtropics; New Zealand.

R. D. PORLEY

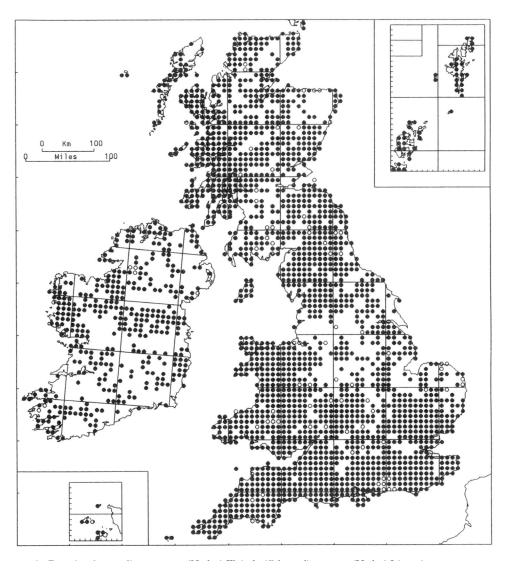

151/1. **Pseudoscleropodium purum** (Hedw.) Fleisch. (*Scleropodium purum* (Hedw.) Limpr.)

Forming coarse wefts in grassland and heaths on a variety of substrates from highly calcareous to mildly acidic, but avoiding the most acid soils. It also avoids deep shade and very wet ground. Grassland habitats include meadows, dunes, marshes, quarries, churchyards, woodland rides, roadsides, banks and cliff-tops. In these, it is often associated with *Brachythecium rutabulum*, *Hypnum cupressiforme*, *Rhytidiadelphus squarrosus* and other large bryophytes. In dry or seasonally damp *Calluna* heathland it commonly grows with *Hypnum jutlandicum* and *Pleurozium schreberi*. At higher altitudes it occurs on ungrazed ledges, usually where there is some basic influence. Predominantly lowland, ascending to 1000 m (Ben Lawers). GB 1909+79*, IR 365+8*.

Dioecious; sporophytes occasional in Cornwall (Paton, 1969) but generally rare, ripe in autumn and winter.

Widespread in Europe north to Iceland and southern Scandinavia. Macaronesia, Algeria, Turkey, Caucasus. Widely introduced in other parts of the world, including western N. America, eastern N. America (perhaps casual only), Hawaii, Australia and New Zealand.

N. G. HODGETTS

329

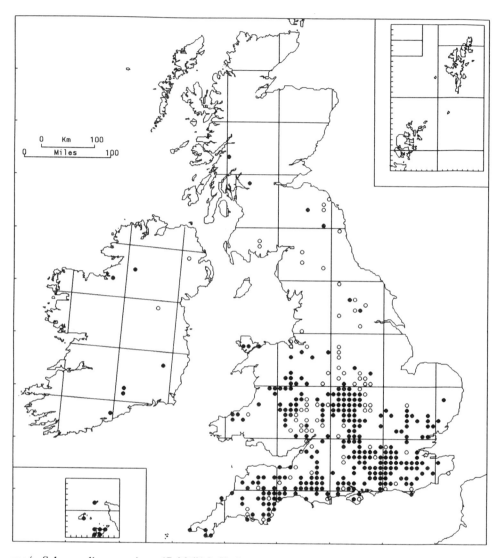

152/1. **Scleropodium cespitans** (C. Müll.) L. Koch

Predominantly a riparian species occurring, often in abundance, on tree-bases, soil, stones and masonry which are occasionally inundated by calcareous or silt-laden stream water. It is intolerant of continuous immersion and possibly benefits from periods of desiccation. It also commonly occurs away from watercourses, particularly on flat concrete and tarmac surfaces with slight soil accumulation. Lowland. GB 274+86*, IR 6+2*.

Dioecious; sporophytes extremely rare, ripe winter.

Atlantic zone of Europe, extending north to Scotland, east to Holland, France and Italy, and south to Corsica and Sardinia. Turkey, western N. America (British Columbia to California).

Perhaps sometimes confused with other species of similar appearance and probably under-recorded in parts of central southern England. It has apparently increased in built-up areas in recent years.

J. W. Bates

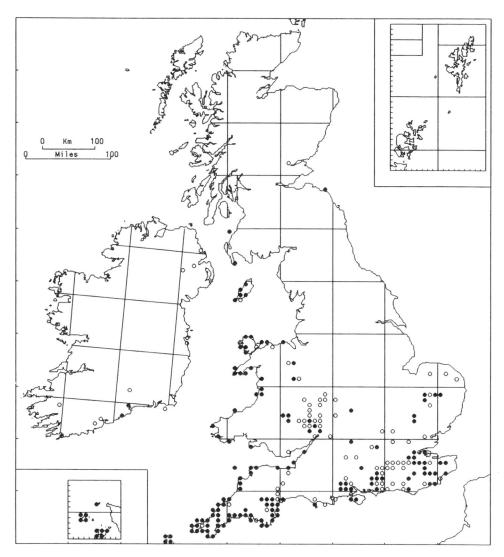

152/2. **Scleropodium tourettii** (Brid.) L. Koch

A xerophyte of well-drained soils and rocks in warm situations on both calcareous and lime-free substrata. The wormlike shoots are most characteristic of cliffs, stable dunes, banks, paths and walls where drought, trampling or rabbit grazing produce a sparse, low turf. It has a predilection for S.-facing aspects but is not limited to them. A significant number of records are from banks in woodland. Lowland. GB 144+78*, IR 2+11*.

Dioecious; sporophytes rare.

Mediterranean, C. and W. Europe north to Scotland and Denmark. Macaronesia, N. Africa, S.W. Asia, western N. America.

Its disappearance from a significant number of inland sites probably results from reduced rabbit-grazing following myxomatosis and from changes in land use.

J. W. BATES

331

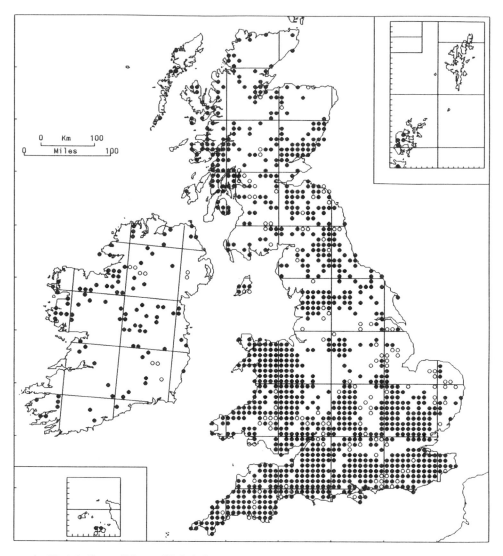

153/1. **Cirriphyllum piliferum** (Hedw.) Grout

A moss of mesic, shaded situations on basic or neutral soils. It frequently grows amongst grasses and scrub on river-banks, and is often abundant in the moister parts of hazel, ash and maple woods on clay and limestones. It is also recorded from shady churchyards, mesotrophic fens, and, rarely, basic mountain ravines and cliffs. Typical associates are *Eurhynchium striatum*, *E. swartzii* and *Rhytidiadelphus triquetrus*. Mostly lowland but ascending to 800 m (Seana Bhraigh). GB 1066+132*, IR 97+9*.

Dioecious; sporophytes are rare.

Widespread in northern Eurasia, from the boreal zone south in the mountains to Spain, Italy, Caucasus, Iran, Mongolia and Japan. N. Africa, Alaska, eastern N. America, Greenland.

J. W. BATES

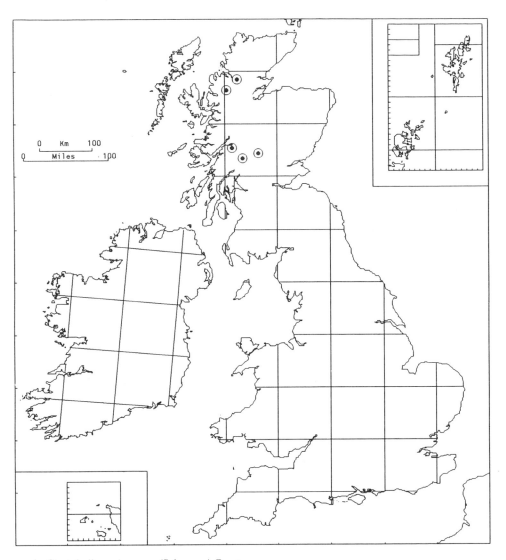

153/2. **Cirriphyllum cirrosum** (Schwaegr.) Grout

On strongly basic mountain rocks, usually on N.- or E.-facing cliffs. 730 m (Inverlael Forest) to 1170 m (Ben Lawers). GB 5.

Dioecious; sporophytes have not been reported in Britain.

Common in the High Arctic and found in many of the mountain ranges of the Northern Hemisphere south to the Atlas, Himalaya and Colorado (U.S.A.).

J. W. Bates

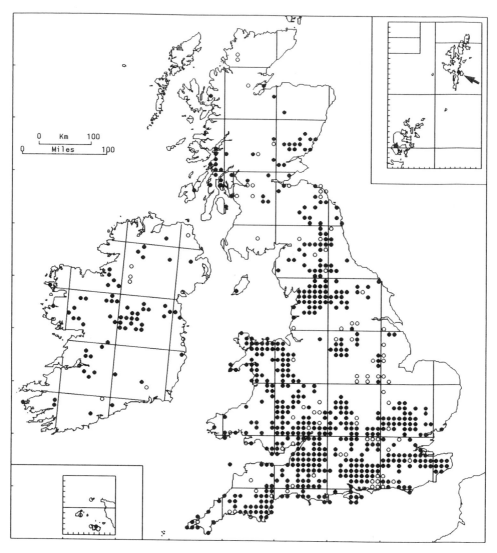

153/3. **Cirriphyllum crassinervium** (Tayl.) Loeske & Fleisch. (*Eurhynchium crassinervium* (Wils.) Schimp.)

A highly characteristic member of the woodland flora on calcareous soils in the lowlands. It forms dense patches on tree-bases, especially of beech, and on stones, frequently in association with *Barbula sinuosa, Isothecium myurum* and *Thamnobryum alopecurum*. Commonly found at the bases of old walls and other masonry where calcareous stone or lime mortar has been used. It also occurs frequently as a member of the riparian flora of tree-roots and stones by calcareous streams, often accompanied by *Amblystegium* spp., *Anomodon viticulosus, Leskea polycarpa, Scleropodium cespitans* and *Tortula latifolia*. Lowland. GB 571+84*, IR 65+9*.

Dioecious; sporophytes occasional, ripe winter.

Europe north to Scandinavia. Macaronesia, Algeria, Turkey, Caucasus, Iran.

J. W. Bates

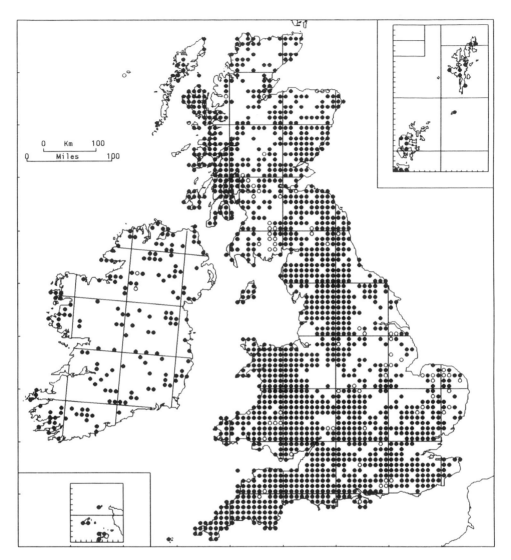

154/1. **Rhynchostegium riparioides** (Hedw.) C. Jens. (*Platyhypnidium riparioides* (Hedw.) Dix.)

On rocks, tree-roots, concrete or brickwork, either submerged or subject to submergence in streams, rivers, ditches and culverts, favouring nutrient-rich water. It is often abundant in swiftly flowing calcareous water on weirs and in sluices and locks. Less frequently it occurs in calcareous springs, in sewage-beds, and in wet places on cliffs and rock-faces. Associates include *Brachythecium rivulare* and *Fontinalis antipyretica*, and, usually in more montane areas, *Brachythecium plumosum* and *Hygrohypnum luridum*. Predominantly lowland, to 950 m (Braeriach). GB 1565+84*, IR 165+4*.

Autoecious; sporophytes frequent in exposed plants, late summer to winter.

Widespread in Europe north to Iceland and southern Scandinavia. Macaronesia, N. Africa, Asia, N. and C. America, northern S. America.

A variable species, smaller in drier habitats, larger when submerged or floating.

N. G. HODGETTS

335

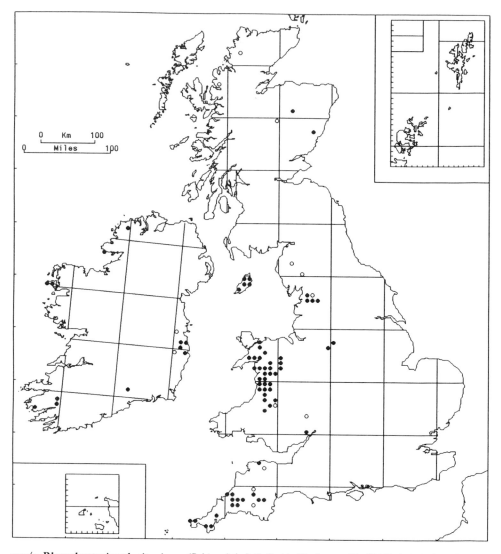

154/2. **Rhynchostegium lusitanicum** (Schimp.) A. J. E. Smith (*R. alopecuroides* (Brid.) A. J. E. Smith)

A calcifuge, occuring on submerged rocks and stones in fast-flowing streams and rivers, often associated with *Brachythecium plumosum*, *B. rivulare*, *Hygrohypnum ochraceum* and *Rhynchostegium riparioides*. Generally at rather low altitudes, ascending to 450 m (Berwyn Mountains). GB 60+11*, IR 12+2*.

Autoecious; sporophytes rare in some districts but frequent in North Wales (Hill, 1988), winter.

A European endemic known from Portugal, Spain, France and Germany (Eifel, Schwarzwald).

N. G. HODGETTS

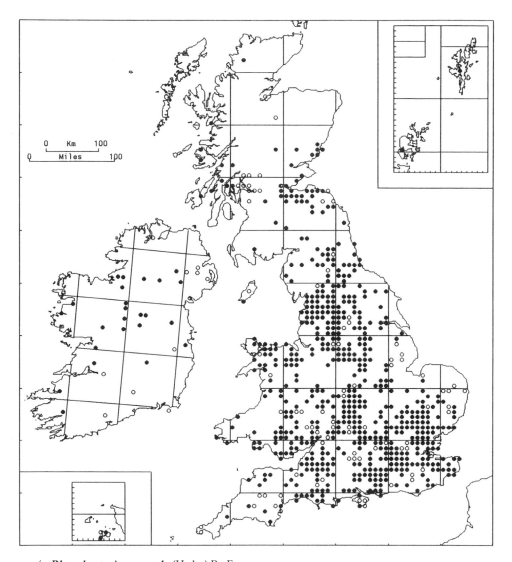

154/3. **Rhynchostegium murale** (Hedw.) Br. Eur.

A calcicole, characteristic of shaded rocks and walls on limestone. It also occurs on tree-boles and bare soil in calcareous areas, and, more rarely, on damp brickwork and basic rocks elsewhere. Occasionally it grows in more open, better-lit, sheltered situations, especially if there is some temporary water-flow. Associates include *Brachythecium rutabulum*, *Bryum capillare*, *Homalothecium sericeum* and other common saxicolous and wall-top bryophytes. Mainly lowland, ascending to 730 m (Knock Fell and Mickle Fell). GB 584+87*, IR 24+11*.

  Autoecious; sporophytes common, winter to spring.

  Most of Europe north to Iceland and Scandinavia. Madeira, Canaries, Algeria, S.W. Asia, Afghanistan, Japan.

N. G. HODGETTS

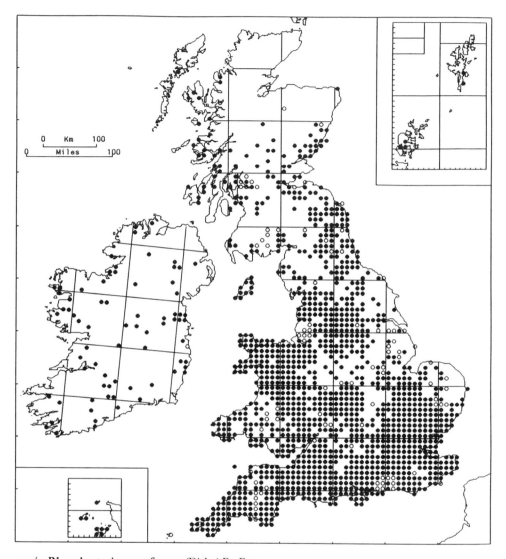

154/4. **Rhynchostegium confertum** (Dicks.) Br. Eur.

A mildly calcicolous moss of well-drained substrata in mainly shaded, often damp places. It grows in patches on rocks, stones, walls, banks, soils, concrete, roofs and base-rich bark, where it is particularly characteristic of elders. Frequent associates include *Amblystegium serpens*, *Brachythecium rutabulum*, *Bryum capillare*, *Orthotrichum diaphanum* and *Tortula* spp. Mainly lowland, ascending to 350 m (Colt Park Wood) but reported from 950 m in Westmorland (Wilson, 1938). GB 1279+92*, IR 72+2*.

Autoecious; sporophytes common, winter.

Europe north to Iceland and southern Scandinavia. Macaronesia, Algeria, S.W. Asia.

N. G. HODGETTS

338

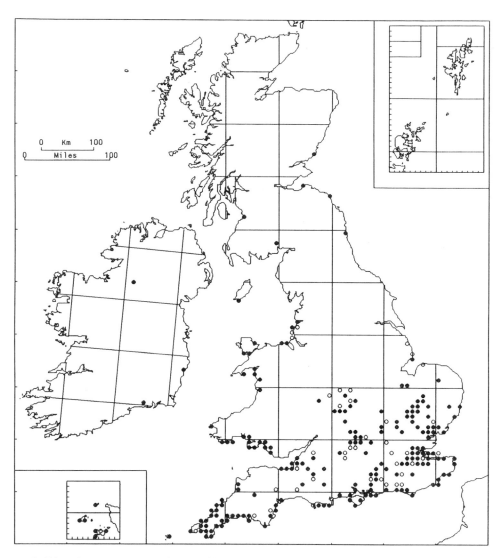

154/5. **Rhynchostegium megapolitanum** (Web. & Mohr) Br. Eur.

A calcicole forming straggling patches on well-drained soils. It occurs on stable sand-dunes, sandy and chalky banks, roadsides, cliff turf and grass heaths, and in chalkpits. Lowland. GB 189+40*, IR 3+1*.

Autoecious; sporophytes frequent, autumn to winter.

Western and southern Europe. Macaronesia, N. Africa, S.W. Asia.

This species is probably under-recorded, as it is easily overlooked for *Brachythecium rutabulum*, particularly when sterile.

N. G. HODGETTS

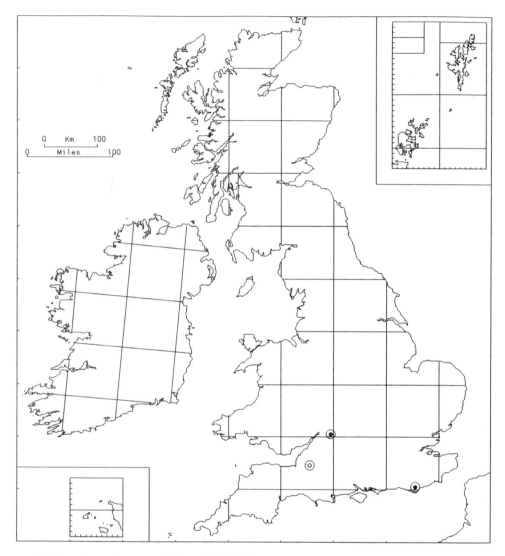

154/6. **Rhynchostegium rotundifolium** (Brid.) Br. Eur.

On shaded boles and exposed roots of ash and field maple trees in hedgerows, with *Amblystegium serpens*, *Anomodon viticulosus* and *Neckera complanata*. It also occurs on sheltered and shaded limestone rocks and on damp stone at the base of a limestone wall, with *Amblystegium serpens*, *Barbula sinuosa*, *Brachythecium rutabulum*, *Cirriphyllum crassinervium*, *Eurhynchium praelongum*, *Homalothecium sericeum*, *Rhynchostegium confertum* and *Thamnobryum alopecurum*. A lowland species, to 230 m (Gloucestershire). GB 2+1*.

Autoecious; sporophytes common, winter.

Scattered in C. and S. Europe. Caucasus, Japan.

It is possible that this species still occurs at its third British station, 'near Wells', where it was found in 1887 by C. H. Binstead, but the exact locality is not known. According to Duell (1992), most localities in C. Europe are in secondary habitats, especially in the neighbourhood of old castles, giving rise to a suspicion that it may be introduced there.

N. G. Hodgetts

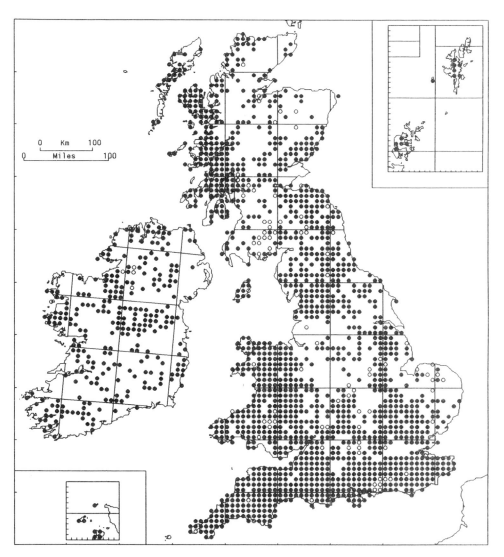

155/1. **Eurhynchium striatum** (Hedw.) Schimp.

Frequent on neutral or basic soil and rocks. It grows on shaded sandstone and limestone rock, on calcareous cliff-ledges, on soil and boulders in woodland and scrub, in hedges, on roadside and streamside banks, in sheltered grassland over basalt, chalk and limestone and in turf on sea-cliffs. It is particularly conspicuous in lowland woods, where it often forms extensive carpets on soils derived from chalk, limestone and calcareous boulder clay in both primary and secondary woodland and in scrub. It sometimes spreads to tree-bases. 0–780 m (Seana Bhraigh). GB 1450+92*, IR 273+6*.

Dioecious; sporophytes uncommon in S.E. England but more frequent in the west, maturing in winter.

Widespread in Europe, extending north almost as far as the Arctic Circle in coastal Norway; rare in S. Europe, where it is restricted to mountainous areas. Macaronesia, N. Africa, Asia from Turkey to Japan.

The three commonest *Eurhynchium* species, *E. praelongum*, *E. striatum* and *E. swartzii*, overlap in their ecological requirements. *E. striatum* and *E. swartzii* are less catholic in their habitat preferences than *E. praelongum*, but tend to replace it on highly calcareous ground. *E. striatum* is more restricted to stable, permanent habitats than *E. swartzii*.

C. D. PRESTON

341

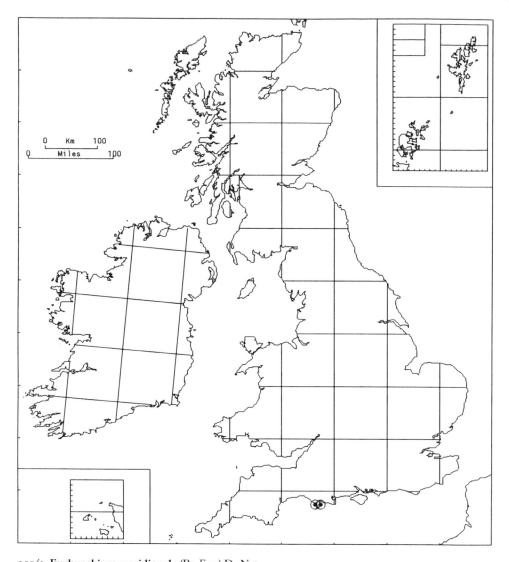

155/2. **Eurhynchium meridionale** (Br. Eur.) De Not.

Confined to coastal limestone, where it grows on sheltered rocks and boulders, tracks and stony banks in disused quarries and on cliff-slopes. It has also been recorded from a roadside wall. Associated species include *Fissidens cristatus* and *Trichostomum brachydontium*. Lowland. GB 2.

Dioecious; sporophytes not known in Britain but occasional in the Mediterranean region, where they mature in winter and spring.

S. Europe from Portugal and Spain to Greece; Switzerland. Macaronesia, N. Africa, Israel, Turkey.

The Isle of Portland represents an isolated, northerly outpost of this predominantly Mediterranean species, which in France occurs no further north than the department of Landes.

C. D. PRESTON

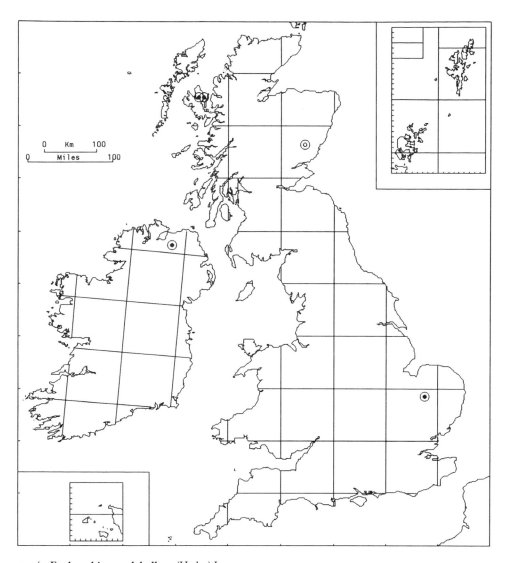

155/3. **Eurhynchium pulchellum** (Hedw.) Jenn.

A highly disjunct species in Britain and Ireland, found in dry basic habitats in areas notable for their rich flora. In Skye and N.E. Ireland it is confined to basalt, growing on rock-faces, decomposed rock and mossy cliff-ledges. In E. Scotland it was found on sandstone rocks. In the Suffolk Breckland, it was discovered in 1950 in open vegetation on calcareous soil in an area from which turf had been removed during the 1939–45 War. By 1980 natural recolonization and a reduction in grazing had resulted in a longer turf in which *E. pulchellum* was reduced to a single small patch; in 1990 it could not be refound. 0–500 m (The Storr). GB 3+1*, IR 1.

Dioecious, with small budlike males. Sporophytes have not been observed in Britain or Ireland.

Circumboreal, south to N. Africa, Afghanistan and northern S. America. Widespread in Europe north to Svalbard.

It is rare in the Atlantic zone of Europe, and, like *Entodon concinnus*, *Rhytidium rugosum* and *Thuidium abietinum*, has probably been much reduced since the Late Glacial. Elsewhere it commonly grows on rotten tree-stumps and humic soil in woodland. Hill (1993) concluded that material from all British and Irish localities was referable to var. *diversifolium* (Br. Eur.) C. Jens.

C. D. PRESTON

343

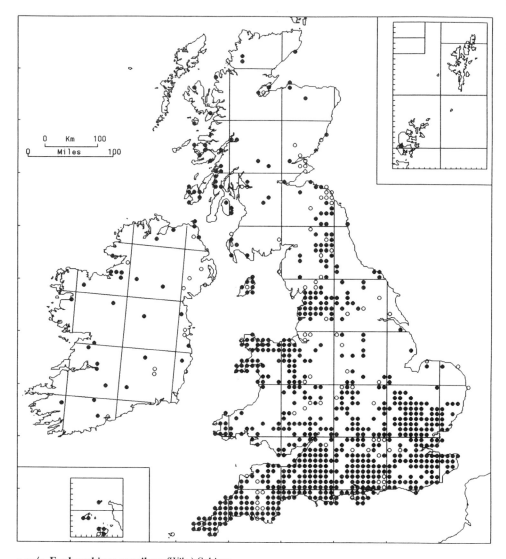

155/4. **Eurhynchium pumilum** (Wils.) Schimp.

The prostrate stems or low mats of *Eurhynchium pumilum* characteristically grow on heavily shaded basic substrates where competition from other bryophytes is low and litter does not accumulate. Its habitats include woodland banks, hedges, stream-banks, ditchsides, soil on the walls of chalkpits and silt-covered brickwork by water. It is often found on the banks which mark the boundaries of ancient woods, which are shaded by trees but blown free of leaves. *E. pumilum* favours friable or compacted soil derived from alluvium, chalk, limestone or calcareous boulder clay, from which it spreads onto tree-bases, exposed tree-roots and stones. Towards its northern limit it grows on damp limestone and basic sandstone in low-lying ravines and sea-caves. Lowland. GB 717+81*, IR 29+12*.

Dioecious; sporophytes infrequent, maturing from November to March.

S., W. and C. Europe north to Scotland and S.W. Norway. Macaronesia, N. Africa, Turkey. Introduced to Australia.

C. D. PRESTON

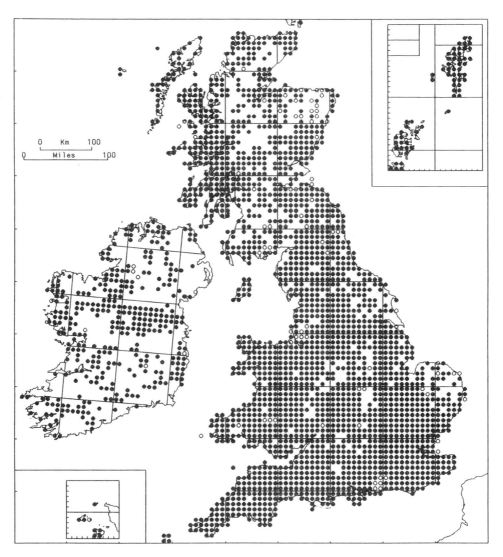

155/5. **Eurhynchium praelongum** (Hedw.) Br. Eur.

On fertile, acidic or neutral substrates in a wide range of open to deeply shaded, damp or dry habitats. It grows on rotting wood (both logs and branches), tree-bases, soil and boulders in woodland and scrub; in hedges and on roadside banks; on the sides of rivers, streams and ditches; in turf in pastures, salt-marshes, sand-dunes and sea-cliffs; in *Juncus*-dominated marshes; on boulders in streams and soil-covered rocks; on earthy cliff-ledges and on sandstone rocks. It also occurs in artificial habitats such as damp, shaded walls, arable and fallow fields and on waste ground and rubble. At high altitudes it grows in stabilized block-scree, in montane tall-herb stands and in fern-dominated snow-bed communities. It is usually absent from highly calcareous substrates and from peaty soils in moorland and heathland, but it can occasionally be found as straggling stems in sphagnum bogs. 0–1050 m (Beinn a' Bhuird), most frequent below 500 m. GB 2172+79*, IR 338+9*.

Dioecious; sporophytes frequent on well grown plants in moist habitats, maturing in winter.

Widespread in cool-temperate parts of the Northern Hemisphere south to N. Africa, Japan and south-west U.S.A. (California); very rare in eastern N. America. S. America, Australia (perhaps introduced), New Zealand.

Var. *stokesii* has not been recorded consistently and is not mapped separately.

C. D. Preston

345

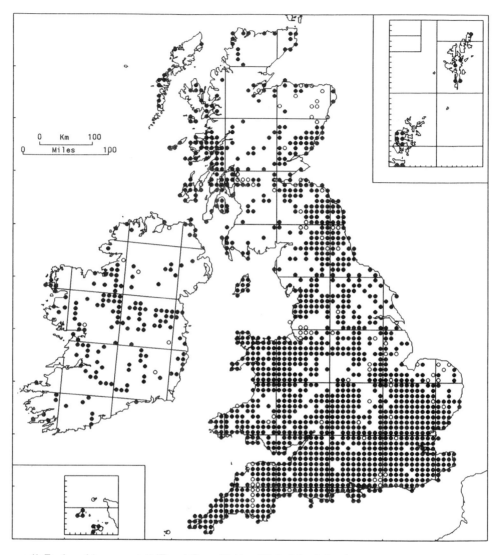

155/6. **Eurhynchium swartzii** (Turn.) Curn. (*E. hians* (Hedw.) Sande Lac.)

On neutral or basic, fertile soils in a variety of habitats. It is frequent in woodland, scrub and hedgebanks over chalk, limestone and calcareous boulder clay; in chalk and limestone grassland and disturbed ground in chalkpits; in damp pastures, moist streamsides and river-banks; and in stubble and fallow fields. It also grows on waste ground, railway banks, damp walls, damp concrete and occasionally on wet upland rocks. 0–450 m (Moel Hebog). GB 1431+91*, IR 154+7*.

Dioecious; sporophytes rather rare, maturing in winter.

Widespread in temperate parts of the Northern Hemisphere. In Europe from the Mediterranean to Iceland and N. Norway. Reported from St Helena.

It is strange that *E. swartzii*, which fruits rather rarely and has no specialized means of vegetative reproduction, should be frequent in disturbed habitats such as arable fields. The map includes all records, including those of the rather ill-defined var. *rigidum* (Boul.) Thér.

C. D. Preston

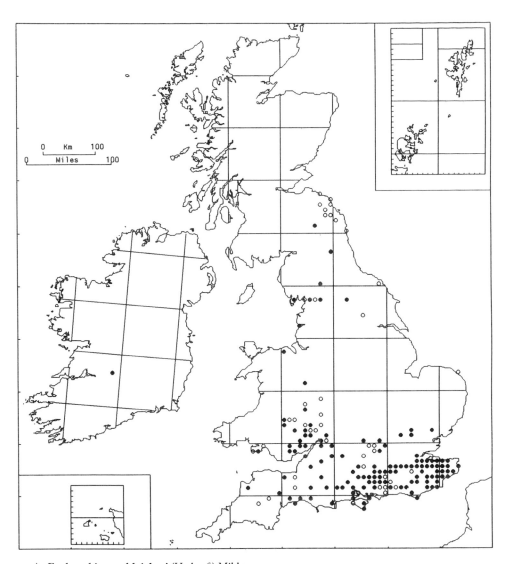

155/7. **Eurhynchium schleicheri** (Hedw. f.) Milde

On friable, well-drained, basic or slightly acidic soil on shaded banks in woodland and by roads and sunken lanes; occasionally on streamsides and river-banks. It is found on soil derived from limestone, chalk and clay, and on sandy soils over substrata such as Greensand and Eocene sands. In some areas it apparently grows directly over chalk, but more often it avoids such soils, occurring where a chalkpit or road has been dug through an acid soil layer, so that the chalk is amended with acid soil. As a result it typically grows with a mixture of calcicoles and calcifuges such as *Ctenidium molluscum*, *Dicranella heteromalla* and *Tortula subulata*. Lowland. GB 125+44*, IR 1.

Dioecious; sporophytes infrequent, maturing from December to March.

Widespread in Europe, north to southern Scandinavia. Macaronesia, Turkey, Iran.

C. D. PRESTON

347

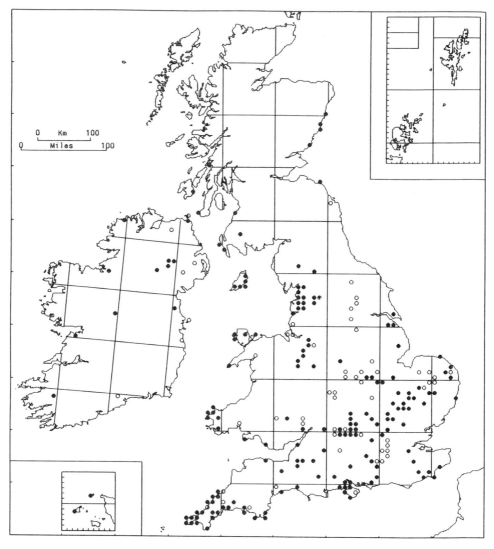

155/8. **Eurhynchium speciosum** (Brid.) Jur.

In wet, mesotrophic or eutrophic, often shaded habitats. It grows on soil, stones, exposed tree-roots and tree-bases in wet woodland and fen carr, in fens, flushes and marshy ground, in turf by ponds and turloughs, on the sides of canals, streams and ditches, on wet rocks, banks and paths, damp ground at the base of cliffs and damp walls by water. At the northern edge of its range it is confined to sea-cliffs, where it occurs in damp hollows, in wet rock crevices, on wet rocks and on the dripping walls of caves. Lowland. GB 161+53*, IR 11+7*.

Autoecious or synoecious; sporophytes frequent, maturing in winter.

S., W. and C. Europe, north to Scotland and S. Scandinavia. Macaronesia. Reported from N. Africa, Saudi Arabia and Iran.

E. speciosum is under-recorded; it closely resembles *E. swartzii* (from which is is thought to have evolved by autopolyploidy) and can also be passed over as *Brachythecium rutabulum*. It is almost certainly commoner in S.E. England and Ireland than shown on the map, but in S.W. England and Wales it is genuinely restricted to coastal sites.

C. D. PRESTON

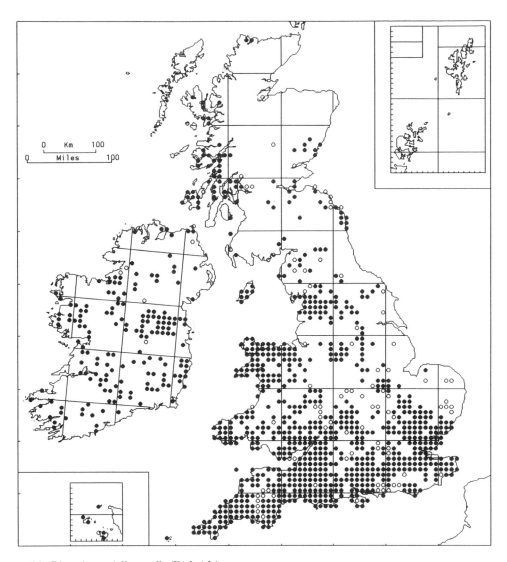

156/1. **Rhynchostegiella tenella** (Dicks.) Limpr.

A species of dry but usually sheltered base-rich rocks. It is common in natural habitats on chalk, limestone and other basic strata in southern Britain, especially in woods and often in strong shade, growing on chalk and flint pebbles, on stones, boulders and cliffs, and rarely on tree-roots and stumps. It is also found widely on sheltered brick and stone walls and in churchyards, e.g. at the base of gravestones; in some districts these are the commonest habitats. In coastal areas it occurs on sea-cliffs and in sea-caves. Towards the north it is increasingly confined to sheltered lowland habitats. 0–350 m (Peak District). GB 802+92*, IR 166+7*.

Autoecious; capsules frequent, ripe winter to spring.

Most of Europe north to S. Scandinavia, commoner in the south. Macaronesia, N. Africa, W. Asia.

Most records are of var. *tenella*. Var. *litorea* (De Not.) Rich. & Wall. is known from scattered localities in southern England, occurring on stumps, trunks and stones.

T. L. BLOCKEEL

349

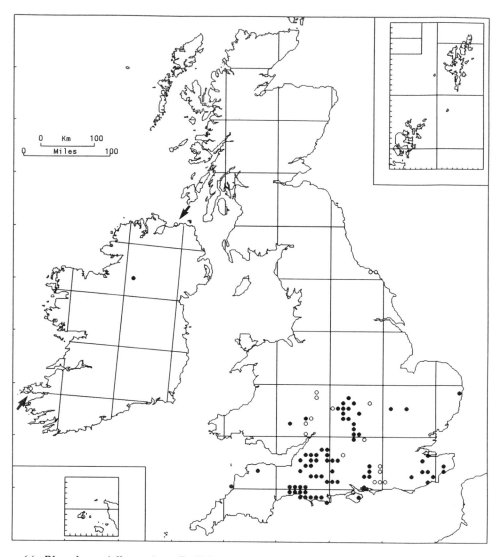

156/2. **Rhynchostegiella curviseta** (Brid.) Limpr.

A species of damp stones and tree-roots, especially on wooded stream-banks. It also occurs in man-made habitats such as bridge supports and retaining walls by rivers, canals and lakes, and on the banks of shaded lanes and cuttings. Natural substrata include sandstone and limestone, but it is absent from very acid situations. Lowland. GB 73+17*, IR 1+2*.

Autoecious; capsules frequent, maturing in winter and spring.

S. and W. Europe, north to Belgium, Germany, Czechoslovakia and Romania. Macaronesia, N. Africa, Cyprus, Turkey, Palestine, Iraq.

T. L. BLOCKEEL

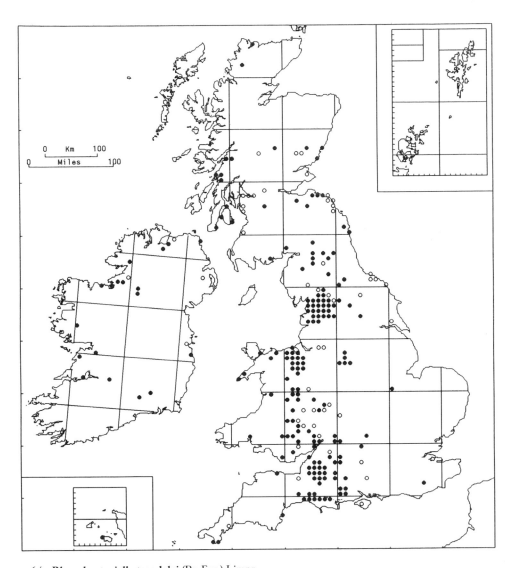

156/3. **Rhynchostegiella teesdalei** (Br. Eur.) Limpr.

A species of wet rocks, masonry and wood by streams and waterfalls, often close to flowing water. It is tolerant of shade and may occur in dark hollows, on ravine walls and in caves. It is frequent on Carboniferous Limestone in northern England, but is also widespread in other parts of the country on a variety of substrata, including sandstone and shales, but always requiring some base enrichment. Lowland. GB 165+47*, IR 18+5*.

Autoecious; capsules frequent, maturing in winter and spring.

S. and W. Europe north to southern Norway and Sweden. Macaronesia, Caucasus.

Similar in appearance and habitat to *R. curviseta* but typically occurring in wetter places. It has a more northerly distribution and is much less common than *R. curviseta* in the Mediterranean region.

T. L. BLOCKEEL

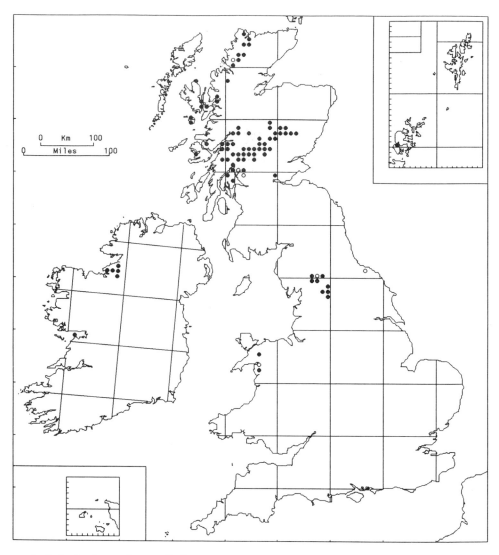

157/1.**Orthothecium rufescens** Br. Eur.

Occurs most frequently as dense vinous-red silky patches on damp, periodically flushed, N.- to E.-facing steep, often vertical sheltered montane cliff-faces of limestone, calcareous mica-schist, basalt, or other basic volcanic rocks, growing with *Saxifraga aizoides, S. oppositifolia, Selaginella selaginoides, Aneura pinguis, Anoectangium aestivum* and *Blindia acuta*, and on damp, shaded limestone walls of low-lying ravines, growing with *Asplenium viride, Cystopteris fragilis* and *Fissidens cristatus*. More rarely it occurs on the sides of turfy tufaceous hummocks in calcareous 'rich-fens' and springs, associated with *Thalictrum alpinum, Tofieldia pusilla, Leiocolea alpestris* and *L. bantriensis*. 10–1250 m (Breadalbane). GB 76+6*, IR 6.

Dioecious; capsules rare, summer.

A boreal-montane plant occurring in W., C. and N. Europe, including Iceland. Siberia, Japan, Alaska, Greenland.

H. J. B. BIRKS

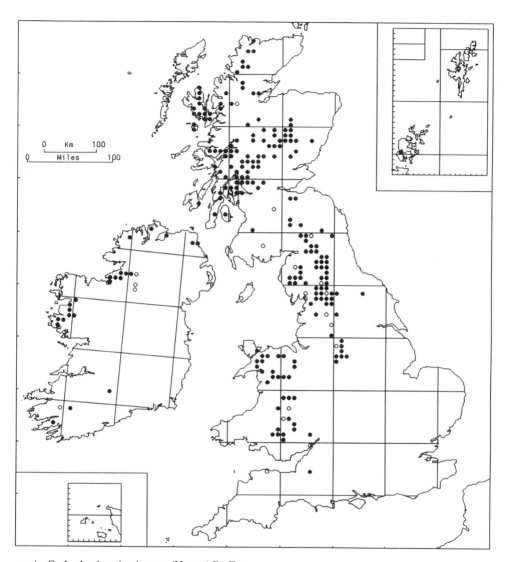

157/2. **Orthothecium intricatum** (Hartm.) Br. Eur.

Usually present in small quantity in dry or damp, shaded crevices and under sheltered overhangs in N.- to E.-facing basic cliffs, often growing with *Ctenidium molluscum, Mnium marginatum, M. stellare* and *Pohlia cruda*, and in crevices in basic rock-walls of low-lying wooded ravines. Confined to calcareous rocks and commonest on limestone and calcareous mica-schists. 0–1180 m (Ben Lawers). GB 192+18*, IR 23+4*.

Dioecious; capsules very rare, summer.

A boreal-montane species widespread in Europe north to Iceland and Svalbard. N. Africa, Turkey, Caucasus, Himalaya, N. Asia, N. America, Greenland.

A form confined to the Yoredale limestones of N. England has been distinguished as var. *abbreviatum* Dix. Biometrically it is not distinct and appears to be a very stunted form of var. *intricatum*.

H. J. B. BIRKS

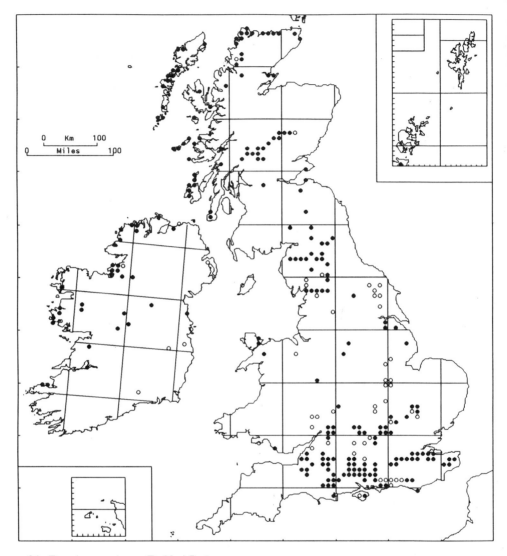

158/1. **Entodon concinnus** (De Not.) Paris

A calcicole of well-drained open situations, typically in species-rich short turf on well-illuminated, especially S.-facing, slopes. It grows, usually in small quantity, in long-established grassland on chalk downs and ancient trackways, in quarries, on limestone rock-ledges and scree, and on sandy banks, coastal dunes and machair. Rare in the mountains, on well-drained brown-earth soil or in dry or periodically damp *Dryas* heath. 0–1175 m (Ben Lawers). GB 196+46*, IR 31+7*.

Dioecious; capsules unknown in the British Isles.

Discontinuously circumpolar, in cold steppes and the Arctic to about 70°N, and in mountains further south. Widespread in Eurasia and western N. America, south to Caucasus, C. Asia, Himalaya, Japan and southern U.S.A. (New Mexico); very rare in eastern N. America, known from Newfoundland and N. Carolina. Disjunct in alpine habitats in Ecuador and New Guinea.

R. A. FINCH

354

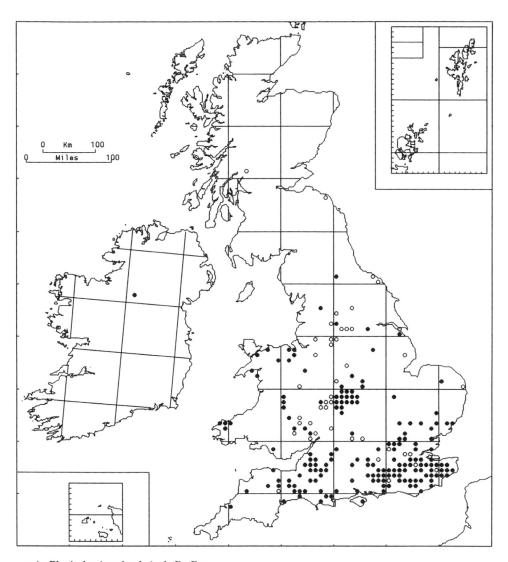

159/1. **Plagiothecium latebricola** Br. Eur.

A species of decomposing vegetable matter in damp shady places. It is particularly characteristic of fern and tussock-sedge stools in swampy woodland, usually in deep shade, but also occurs on decaying logs. Less often it grows on soil and living tree-trunks. Frequent associates are *Eurhynchium praelongum* and *Plagiomnium* spp. Mainly lowland, ascending to 350 m (edge of Mynydd Hiraethog). GB 171+43*, IR 1.

Dioecious; sporophytes rare, spring. Fusiform gemmae, often produced in leaf-axils and on leaf-tips, are presumably the normal means of spread; similar gemmae are produced on the protonema in culture (Whitehouse, 1987).

Widespread in Europe north to central Scandinavia, east to Estonia and the Carpathians. Turkey, Japan, eastern N. America.

N. G. HODGETTS

355

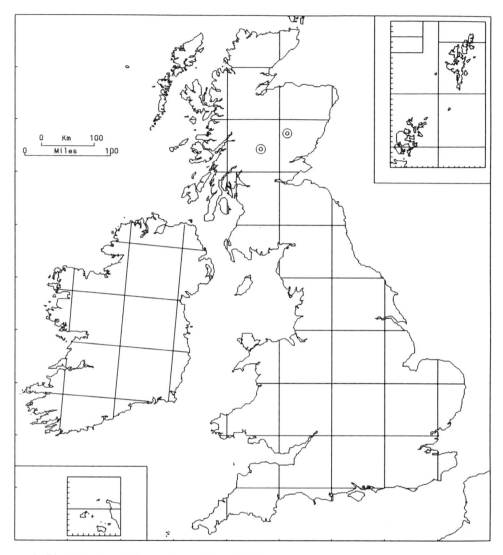

159/2. **Plagiothecium piliferum** (Sw. ex Hartm.) Br. Eur.

On shaded rock-ledges at about 800 m (Ben Lawers and Caenlochan Glen). GB 2*.

Autoecious; sporophytes common, summer. Gemmae unknown.

C. and N. Europe, Italy (Sardinia, Tuscany). Siberia, Far East, western N. America.

Last seen in Britain at Caenlochan Glen in 1939. Although it was generally reported as preferring acidic rock-ledges, the sites at which it occurred are especially noted for their calcareous nature. Presumably it grew only where the substrate was leached. Outside the British Isles it grows at lower elevations in subalpine woodlands and on acid-barked trees and rotten wood in swampy woodland.

N. G. HODGETTS

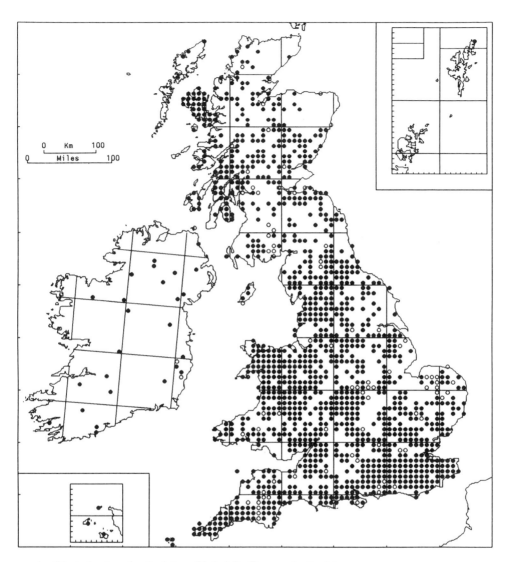

159/3a. **Plagiothecium denticulatum** (Hedw.) Br. Eur. var. **denticulatum**

A generally calcifuge plant of moist shady places, especially in woodland, where it grows on tree-stumps, tree-bases, rotting logs, banks, soil and rocks. It also occurs in marshes and reed-swamps, on shaded stone walls and on stream-banks. In montane regions it grows in wooded ravines and, at higher altitudes, in shady recesses between boulders in stable, block-scree and in earthy crevices in cliffs, often where there is some calcareous influence. 0–1085 m (Ben Alder). GB 1195+81*, IR 25+4*.

Autoecious; capsules common, late spring to summer. Fusiform axillary gemmae occasional.

Very widely distributed in the Northern Hemisphere, including almost the whole of Europe. Reports from the Southern Hemisphere are probably erroneous (Ireland, 1992).

In the lowlands it is a strict calcifuge, more so than *P. nemorale. Plagiothecium ruthei* Limpr. is doubtfully distinct from *P. denticulatum* and has not been distinguished by most recorders; it is not mapped separately. It tends to grow in wetter places.

N. G. HODGETTS

357

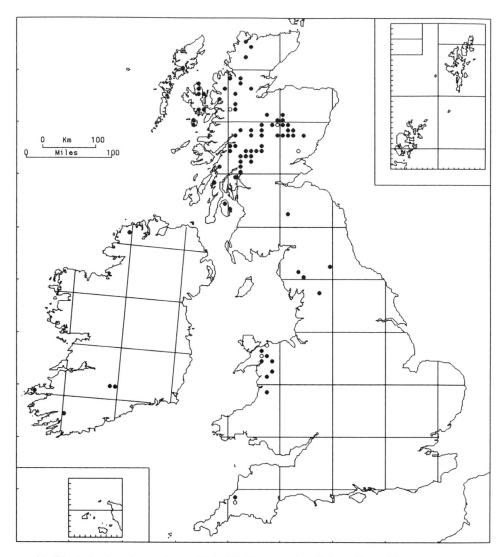

159/3b. **Plagiothecium denticulatum** (Hedw.) Br. Eur. var. **obtusifolium** (Turn.) Moore

It grows among boulders and in rock crevices on mountains, where it is often a component of the vegetation of high-level boulder fields and late-lying snow patches, along with species such as *Lophozia opacifolia*, *Moerckia blyttii*, *Pohlia ludwigii* and large oceanic-montane liverworts. It also occurs in crevices on calcareous rock-ledges, usually on N.- or E.-facing cliffs, with *Ctenidium molluscum*, *Distichium capillaceum*, *Isopterygium pulchellum* and other less widespread montane species, and in recesses between boulders in stable basalt screes. 330 m (Skye) to 1205 m (Ben Lawers), mostly above 600 m. GB 75+6*, IR 5.

Autoecious; sporophytes common, autumn. Gemmae not recorded.

N. and C. Europe. Widespread elsewhere in cool parts of the Northern Hemisphere. Reports from the Southern Hemisphere are probably erroneous (Ireland, 1992).

A taxonomically difficult segregate requiring further study (Smith, 1978).

N. G. HODGETTS

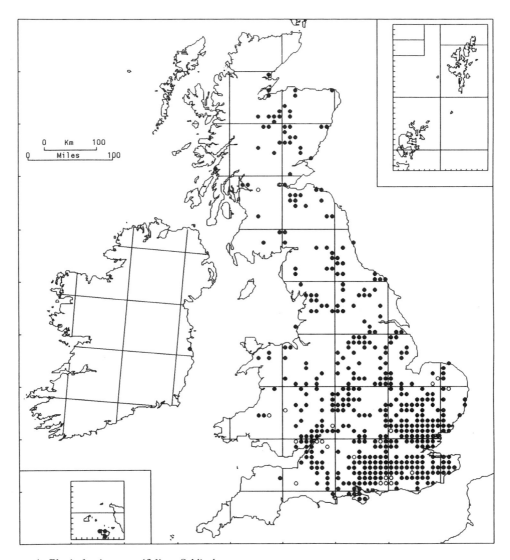

159/5. **Plagiothecium curvifolium** Schlieph.

A calcifuge species of tree-stumps, tree-bases, exposed roots, fallen logs, litter and humus-rich soil in woodland, growing particularly luxuriantly on soil covered with decaying conifer needles, often in deeply shaded conifer plantations. Associated species may include *Eurhynchium praelongum* and *Isopterygium elegans*. Predominantly lowland, ascending to 400 m (Clocaenog Forest). GB 459+22*, IR 1.

Autoecious; sporophytes common, late spring or early summer. Fusiform axillary gemmae occasional.

Most of Europe, becoming scarcer in the north and south. Japan, N. America.

Probably increasing as a result of the widespread planting of conifers. It is very closely related to and often hard to distinguish from *P. laetum*; in many countries the two are regarded as conspecific. The map shown here must contain some errors but these are unlikely to obscure the overall distribution.

N. G. HODGETTS

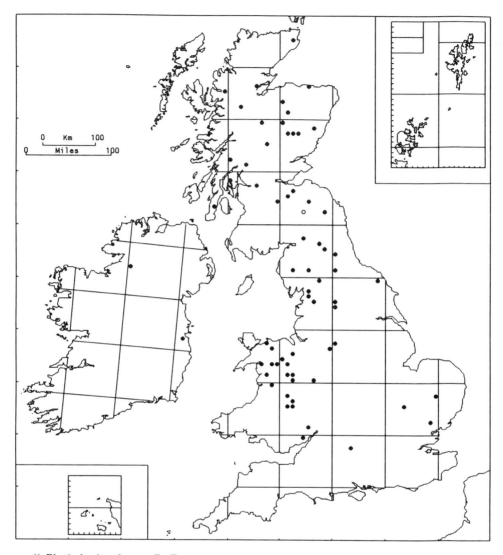

159/6. **Plagiothecium laetum** Br. Eur.

A calcifuge found on tree-bases, stumps, rotten logs and soil in broadleaved woodland; seldom under conifers. It also occurs on banks and among boulders in the mountains. Mainly at low altitudes, ascending to 420 m (Tarnbrook Fell). GB 64+1*, IR 2.

Autoecious; sporophytes common. Fusiform axillary gemmae occasionally produced on leaf-tips and in leaf-axils.

Most of Europe, montane in the south. Caucasus, Siberia, N. America, Greenland.

Probably somewhat overlooked. Although collected in 1928 it was not recognized as a distinct species in Britain for another 30 years (Crundwell, 1959b). It is very closely related to *P. curvifolium* and occasional plants are hard to identify and appear to be intermediate.

N. G. Hodgetts

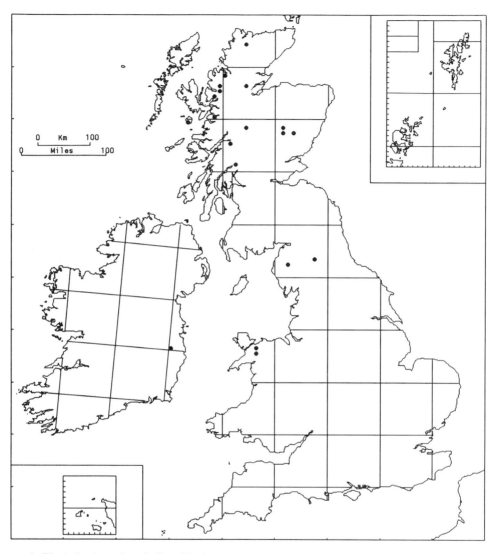

159/7. **Plagiothecium platyphyllum** Mönk.

A plant of wet, shaded habitats in montane areas, growing in mountain flushes, in rock crevices on cliffs, and by waterfalls and streams. Ascends to 850 m (Glen Coe). GB 18, IR 1.

Autoecious; sporophytes frequent, summer. Gemmae unknown.

Widespread but scattered in European mountains north to central Scandinavia. Turkey, Caucasus, Iran, N. America.

Probably under-recorded. It is easily confused in the field with *P. succulentum*, which sometimes grows in similar habitats and from which it can be distinguished with certainty only by microscopic characters. In the most recent American checklist (Anderson *et al.*, 1990) both *P. platyphyllum* and *P. succulentum* are treated as synonyms of *P. cavifolium* sensu lato.

N. G. HODGETTS

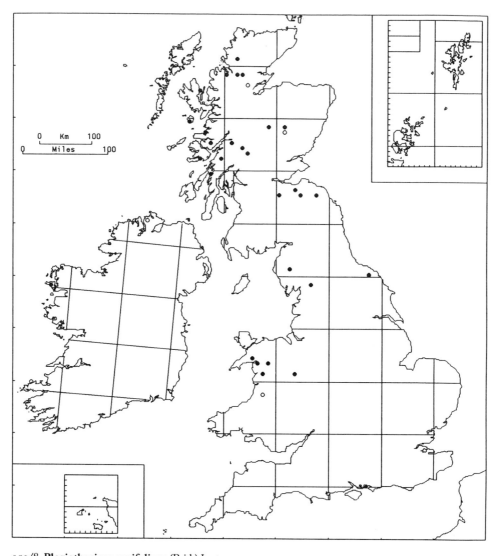

159/8. **Plagiothecium cavifolium** (Brid.) Iwats.

On moist, basic, mountain rock-ledges, where it may be associated with a wide variety of other calcicoles, including *Ctenidium molluscum*, *Encalypta ciliata*, *Fissidens adianthoides*, *Hypnum callichroum*, *Isopterygium pulchellum* and *Plagiobryum zieri*. Less often it is found on siliceous rock-ledges at lower altitudes, sometimes where the basic influence is only slight. 0–1205 m (Ben Lawers). GB 28+3*, IR 1+2*.

Dioecious; sporophytes rare. Fusiform axillary gemmae sometimes present.

Widespread in N. Europe and in mountains further south. N. Asia, Japan, N. America, Greenland. Reported from the Falkland Islands.

A critical species, closely related to *P. succulentum* and sometimes appearing to intergrade with it. North American authors generally regard the two taxa as conspecific.

N. G. Hodgetts

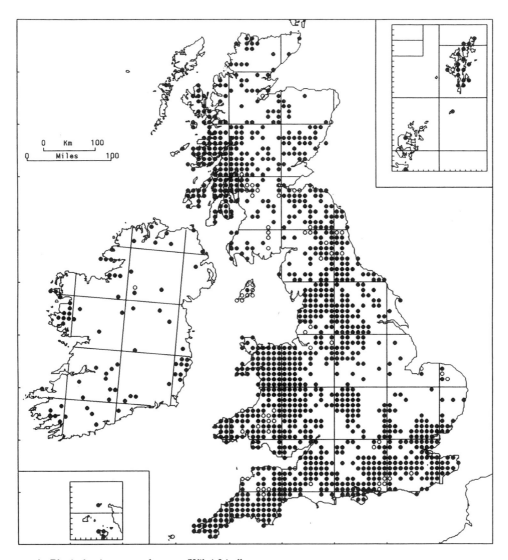

159/9. **Plagiothecium succulentum** (Wils.) Lindb.

On soil-banks, tree-bases or, more rarely, stone walls and decaying wood, in woods and hedgerows, often near streams. Here it is a calcifuge, frequently growing in deep shade on shallow humus overlying neutral sandy or clayey soils, but avoiding peat. In chalky areas it is confined to well-leached substrata. In the mountains it is found among boulders and on rock-ledges, most frequently where there is some basic influence. It sometimes also forms tufts in mountain springs. 0–1160 m (Ben Lawers). GB 1172+69*, IR 73+1*.

Normally dioecious; sporophytes occasional, summer. Occasionally producing fusiform axillary gemmae.

Widespread throughout Europe. Macaronesia, N. Africa, Siberia, N. America.

Plagiothecium succulentum and *P. nemorale* are hard to separate in the field and are sometimes regarded as conspecific (Hemerik, 1989). In general, *P. succulentum* tends to be the commoner species in upland areas, though this is apparently not the case in Skye (Birks & Birks, 1974).

N. G. HODGETTS

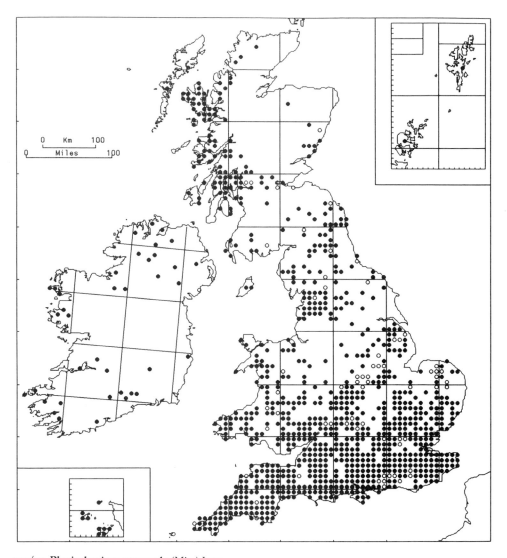

159/10. **Plagiothecium nemorale** (Mitt.) Jaeg.

On earthy banks, tree-bases and streamsides in woodlands and hedgerows. On tree-bases it tends to grow in a zone above *Mnium hornum*, where it is associated with many common lowland woodland bryophytes such as *Lophocolea heterophylla*, *Eurhynchium praelongum*, *Hypnum cupressiforme* and other *Plagiothecium* spp. On soil it may be associated with *Fissidens* spp. and other terrestrial woodland mosses. Mainly lowland, to 460 m (Skye). GB 877+68*, IR 38+1*.

Dioecious; sporophytes occasional, summer to autumn. Occasionally producing fusiform axillary gemmae.

Widespread in Europe north to C. Scandinavia. Macaronesia, N. Africa, Turkey, Caucasus, Iran, Japan, N. America.

P. nemorale is commoner than *P. succulentum* in most lowland areas, particularly on slightly leached, calcareous soils. In upland areas, it must frequently be overlooked, because of its similarity to that species.

N. G. HODGETTS

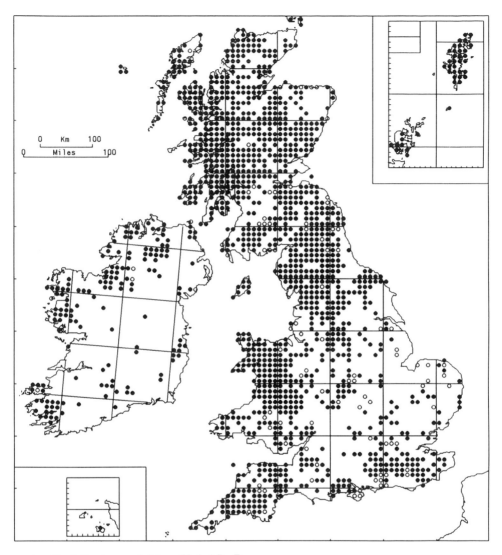

159/11. **Plagiothecium undulatum** (Hedw.) Br. Eur.

Strongly calcifuge, forming extensive patches in acid woodland, including conifer plantatons, and on wet and dry heath. It is often a major component of the bryophyte layer in heather moorland and in dwarf-shrub heath on mountain slopes, particularly favouring north and east aspects and growing among boulders in block-scree. In Scottish birch- and pine-woods it grows among *Calluna* and *Vaccinium* along with *Dicranum scoparium*, *Hylocomium splendens*, *Pleurozium schreberi*, *Ptilium crista-castrensis*, *Rhytidiadelphus loreus* and *Sphagnum capillifolium*. It is also found on rotten logs, in flushed grassland, in bogs, and on acid cliff-ledges. 0–1100 m (Ben Alder). GB 1295+89*, IR 151+6*.

Dioecious; sporophytes frequent in the north and west, rare in the drier south-east, spring to summer. Gemmae unknown.

W., C. and S. Europe, north to 70°N in western Norway, montane in the southern part of its range. Turkey, Iran, western N. America.

N. G. HODGETTS

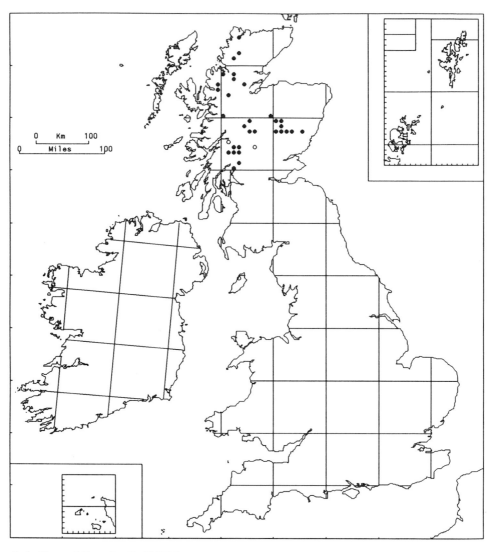

160/1. **Herzogiella striatella** (Brid.) Iwats.

A generally rare, but locally frequent, plant of moist peaty soils in shaded habitats in lowland to alpine regions. It grows on soil in crevices of rock screes or amongst stones and boulders, often over schist. It sometimes occurs in species-rich alpine *Racomitrium lanuginosum* heath in the N.W. Highlands of Scotland, where it is associated with such species as *Carex bigelowii*, *Festuca ovina*, *Ptilidium ciliare*, *Aulacomnium turgidum*, *Dicranum fuscescens*, *Hylocomium splendens* and *Cetraria islandica*. It also occurs in *Vaccinium myrtillus*-dominated snow-bed communities with *Hylocomium splendens*, *Pleurozium schreberi*, *Polytrichum alpinum* and *Rhytidiadelphus loreus*. At lower altitudes it occurs in woods, sometimes as an epiphyte on tree-roots, though it is scarce in that habitat. Lowland to 950 m (Aonach Beag). GB 31+3*.

Autoecious; sporophytes common, summer.

Widespread in northern and montane Europe, with a subarctic-subalpine distribution. N. Asia, N. America, Greenland.

M. J. Wigginton

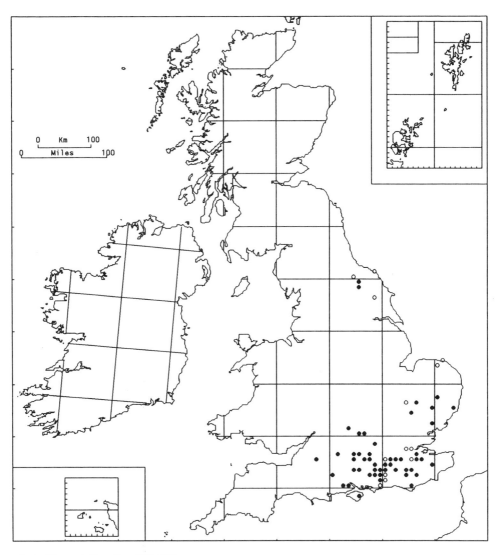

160/2. **Herzogiella seligeri** (Brid.) Iwats.

It occurs mainly on rotting logs and tree-stumps in shaded places in broad-leaved woodland. It has also been found on bases and roots of living trees. It very rarely grows on soil, though where it does the substrate is likely to be highly organic, perhaps representing the last remains of a tree-stump. It is recorded mainly on non-native tree species such as sweet chestnut, particularly in managed plantations in the Weald. Lowland. GB 49+13*.

Autoecious; usually abundantly fertile, sporophytes mature in spring and summer.

Widespread in Europe from Portugal to northern Scandinavia. N. Africa, Turkey, Caucasus, C. Asia, China, north-western N. America.

Although it was first discovered in England in 1843, its tendency to occur on exotic trees has led to the suggestion that it may not be truly native (Whitehouse, 1964).

M. J. Wigginton

367

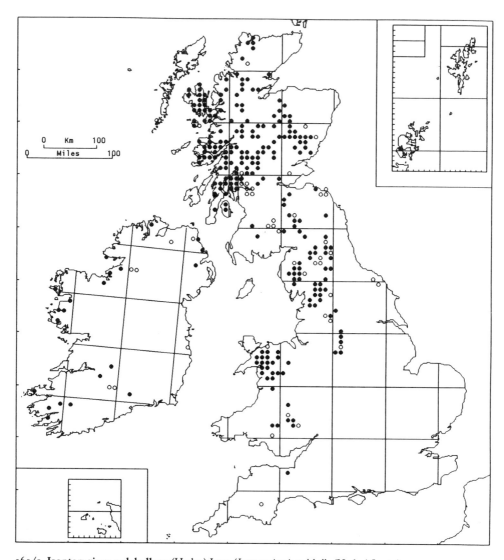

161/1. **Isopterygium pulchellum** (Hedw.) Jaeg. (*Isopterygiopsis pulchella* (Hedw.) Iwats.)

A species of moist ledges and crevices on a variety of rock types, mostly in upland and mountainous regions. It rarely occurs directly on the rock surface, usually creeping on accumulated humus or through other bryophytes. It is not a particularly strong calcicole, often occurring for example on leached pockets of soil on limestone ledges, but it always requires at least slight base enrichment. Habitats include wooded valleys and ravines, rocks on stream-banks, and montane cliffs and crags. 0–1170 m (Ben Lawers). GB 228+39*, IR 21+8*.

Autoecious; capsules common, maturing in spring and summer. Vegetative propagation is by means of axillary gemmae, but these are apparently rare.

Circumboreal arctic-alpine, extending north to the High Arctic and occurring widely in mountain ranges further south. Reported from New Zealand.

T. L. BLOCKEEL

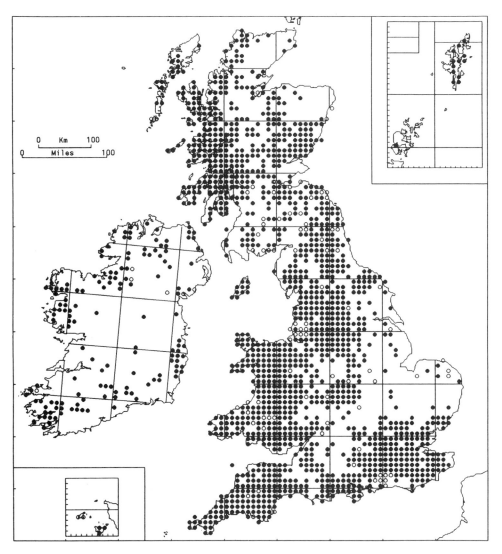

161/2. **Isopterygium elegans** (Brid.) Lindb. (*Pseudotaxiphyllum elegans* (Brid.) Iwats.)

A species of acid habitats in woods and on moorland. It occurs on loamy, sandy and peaty soil in woods and on sheltered banks, and commonly also on and among rocks, both on raw humus and colonizing bare surfaces. It may also be found on tree-stumps and old logs. In southern and eastern districts it is nearly confined to shaded habitats, but it may occur in more open places in the west and north. It is tolerant of heavy shade and sometimes grows in extensive patches under dense shrub cover; it is able to grow, in attenuated form, in deep dark holes in block-scree and on rocky banks. *Diplophyllum albicans*, *Dicranella heteromalla* and *Mnium hornum* are regular associates. 0–980 m (Ladhar Bheinn). GB 1417+84*, IR 143+6*.

Dioecious; capsules very rare, maturing in spring and summer. Propagules, in the form of reduced axillary branchlets, are frequent and sometimes abundant.

Widespread in Europe north to the Arctic Circle, most common in the north-west and becoming progressively rarer to the east and south. Macaronesia, N. Africa, W. Asia, western and eastern N. America, Hawaii.

T. L. Blockeel

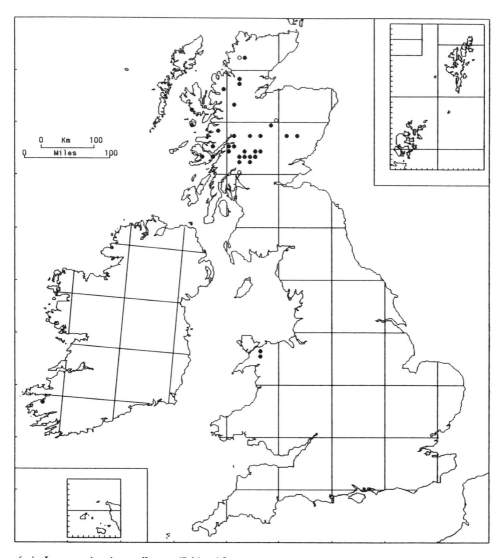

162/1. **Isopterygiopsis muellerana** (Schimp.) Iwats.

It occurs as pale, white or yellow-green patches or scattered stems in very sheltered crevices in block-scree or at the base of crags where there is either periodic irrigation or persistently high humidity. The substrate is usually at least mildly basic. Populations always seem to be small and fragmented. Most sites in the Scottish Highlands are at the base of large crags with a north to north-east aspect where snow probably lies late but at least one colony occurs in a N.-facing wooded ravine. The most frequent of a long list of associates are *Diplophyllum albicans*, *Jungermannia* spp. and *Philonotis fontana*. 200 m (Loch Beoraid) to 1070 m (Aonach Beag). GB 31+3*, IR 1.

Dioecious; sporophytes unknown in Britain or Ireland.

A montane species, widespread in the Northern Hemisphere but rare in the Arctic (Alaska and Greenland). Mountains of W., C. and E. Europe north to S. Norway. Turkey, Caucasus, Himalaya, China, Japan, Russian Far East, N. America, Greenland, New Zealand.

G. P. ROTHERO

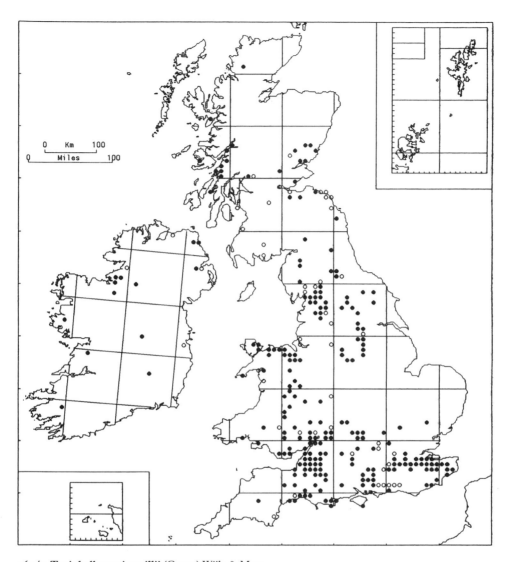

163/1. **Taxiphyllum wissgrillii** (Garov.) Wijk. & Marg.

This calcicolous species is found mainly in chalk and limestone areas, growing on well-drained substrata in rather deep shade, typically in woodland or wooded gullies. It grows on soil, on tree-bases and roots, on basic rocks and stones (including calcareous sandstone, shales and igneous rocks), and on banks. On tree-bases, associated species may include *Brachythecium velutinum*, *Heterocladium heteropterum* var. *flaccidum* and *Isothecium myurum*; on rocks they may be *Cirriphyllum crassinervium* and *Eurhynchium striatum*. Mainly lowland but reaching 400 m (near Malham Tarn). GB 216+46*, IR 13+4*.

Dioecious; sporophytes very rare.

W., C. and E. Europe north to Iceland and C. Scandinavia; absent from most of the Mediterranean region. N. Africa, Turkey, Caucasus.

M. J. WIGGINTON

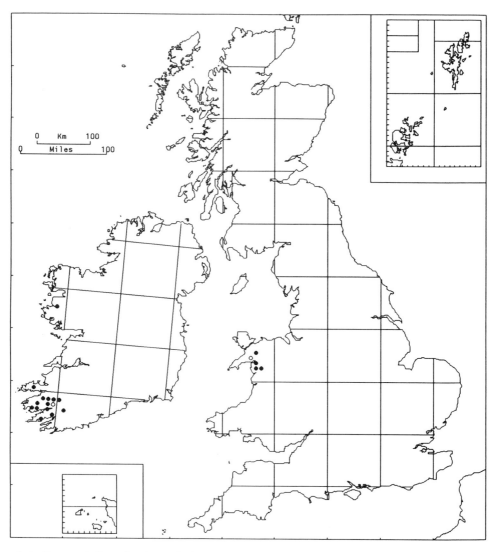

164/1. **Sematophyllum demissum** (Wils.) Mitt.

A species of lightly shaded, humid habitats in low-lying natural or semi-natural mixed deciduous woodland (mainly oak, elm, ash or birch). Most often it occurs on flat or gently sloping terrain, growing on acid or mildly basic slabs and bedded blocks facing south-east to west and with a sparse bryophyte cover where water seeps intermittently. Common associates include *Harpanthus scutatus*, *Marsupella emarginata*, *Scapania umbrosa*, *Tritomaria exsecta*, *Brachythecium plumosum* and *Heterocladium heteropterum*. More rarely it occurs on sloping flushed blocks in small rivulets and streams on wooded slopes. 0–330 m (Kerry). GB 4+1*, IR 14+1*.

Autoecious; capsules common, autumn.

W. and C. Europe north to Norway. N. Africa, Turkey, Japan, eastern N. America. For a map of its European distribution see Schumacker & de Zuttere (1982); this excludes the Norwegian record, which is reported, without exact locality, by Frisvoll & Blom (1992).

It has declined in some localities as a result of woodland clearance or of excessive shade from the spread of rhododendron or the dense growth of bramble.

H. J. B. BIRKS

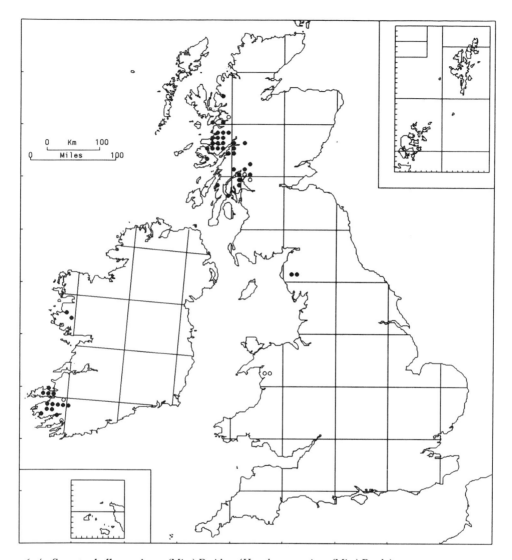

164/2. **Sematophyllum micans** (Mitt.) Braithw. (*Hygrohypnum micans* (Mitt.) Broth.)

Occurs as pure patches or mixed with other bryophytes on periodically irrigated, lightly to moderately shaded, acid or mildly basic, gently or steeply sloping slabs and boulders with an incomplete bryophyte cover. The usual habitat is deciduous woodland (mainly oak or birch) on steep, rocky slopes facing north to east. Common associates include *Harpanthus scutatus*, *Lejeunea patens*, *Grimmia hartmanii*, *Heterocladium heteropterum* and *Hypnum callichroum*. More rarely it occurs on intermittently flushed rocks in N.- or E.-facing wooded ravines and by waterfalls, growing with *Heterocladium heteropterum*, *Hyocomium armoricum* and *Oxystegus tenuirostris*. 0–420 m (Kerry), but ascending only to 330 m in Britain (Cumberland). GB 33+5*, IR 15+2*.

Dioecious; capsules unknown in the British Isles.

France, Germany. British Columbia, eastern N. America, Mexico, S. America.

The ecological differentiation between *S. micans* and *S. demissum* is unclear. In S.W. Ireland, where they occur together, *S. micans* shows a wider ecological tolerance and is less restricted to wooded habitats. *S. micans* has declined through woodland destruction and the spread of rhododendron.

H. J. B. Birks

373

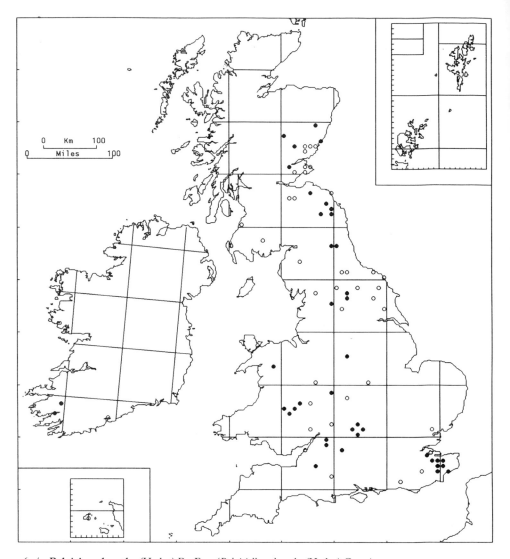

165/1. **Pylaisia polyantha** (Hedw.) Br. Eur. (*Pylaisiella polyantha* (Hedw.) Grout)

An epiphyte on the trunks and branches of a range of broad-leaved trees and shrubs in hedgerows and open woodland, very rarely on stumps or logs. It is most frequently found on hawthorn, ash, elm and lime. Lowland. GB 38+38*, IR 2+1*.

Autoecious; highly fertile and usually fruiting abundantly, capsules mature autumn to winter.

Circumboreal, extending south to the Canaries, Turkey, Caucasus, Kashmir, Japan, southern U.S.A. (Arizona, N. Carolina) and Mexico. Widespread in Europe, not uncommon through most of Scandinavia, becoming rare in the north.

Always uncommon, this species has undoubtedly declined through the wholesale destruction of hedgerows, loss of elms to Dutch elm disease, and bark acidification due to air pollution. It may sometimes be overlooked because of its similarity to the abundant and protean *Hypnum cupressiforme*.

F. J. RUMSEY

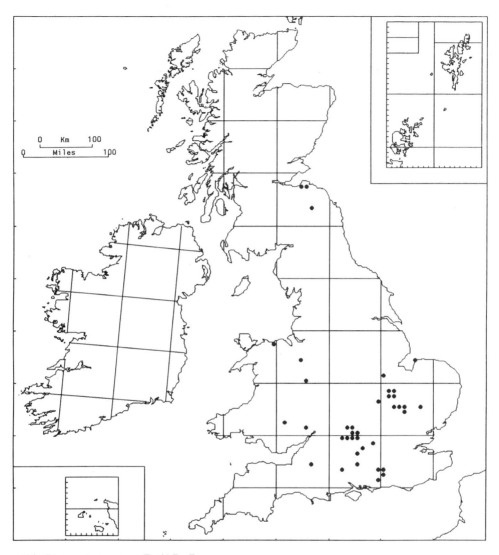

166/1. **Platygyrium repens** (Brid.) Br. Eur.

Occurs as an epiphyte on the lower trunk and branches of a wide range of deciduous trees and shrubs, with a mild preference for less acidic bark types and an inclined surface. It is also found on logs. Lowland. GB 38.

Dioecious, mostly sterile or occasionally male; sporophytes found once in the Wye valley near Monmouth. Tiny shootlets, produced abundantly, are presumably dispersed on the bellies of birds and squirrels.

Widely distributed in the North temperate zone south to North Africa, Caucasus, Mongolia, Japan, western Canada (British Columbia) and south-eastern U.S.A. (Florida). In Europe it is commoner towards the east, and in N. America it is much commoner east of the Rocky Mountains.

Possibly a recent colonist; first collected near Oxford in 1945 but not recognized as a British species until 1962 (Warburg & Perry, 1963). It is easily overlooked as *Hypnum cupressiforme* var. *resupinatum*. At some woodland sites it is abundant, but in most localities it persists for some time as a single colony before disappearing. The scattered distribution and frequently casual status support the view that it is an invading species, spreading from a small number of widely dispersed introductions. The species appears also to be spreading in N. France and Belgium (Jones, 1991).

J. W. BATES

375

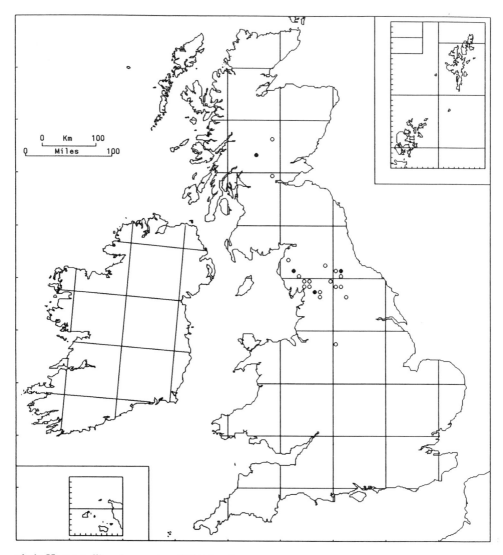

167/1. **Homomallium incurvatum** (Brid.) Loeske

A calcicole, found on rocks, walls or sometimes soil-banks in sheltered and shaded situations in valley woodlands. It occurs particularly on limestone. Lowland. GB 4+18*.

Autoecious; capsules common, ripe June to August.

Widespread in the upper forest zone of all major European mountain ranges and in the southern boreal zone of N. Europe, extending north of the Arctic Circle in Norway. Widespread also in Asia, from Turkey, Caucasus and Iran east through Siberia and Mongolia to Japan and the Russian Far East; very rare in N. America (Kentucky).

The reasons for the apparent decline are not known, but *H. incurvatum* may simply have been overlooked in recent years because of its similarity to *Hypnum cupressiforme* var. *resupinatum*. In other parts of its range it frequently grows on tree-trunks and tree-roots.

N. F. Stewart

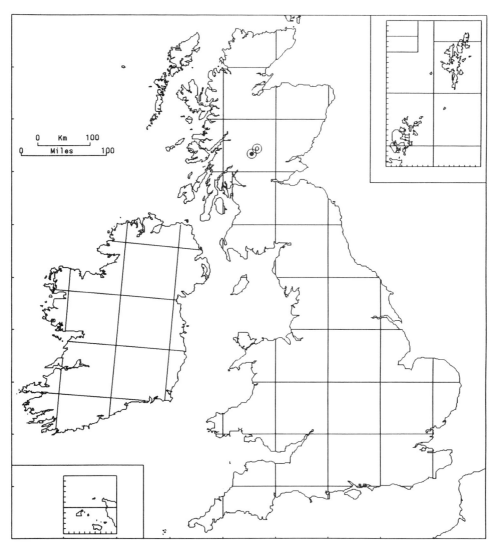

168/1. Hypnum revolutum (Mitt.) Lindb.

In thin to dense mats or tufts on basic mountain rocks at approximately 1000 m altitude (Ben Lawers). GB 1+1*.
Dioecious; capsules not known in Scotland.

A bipolar species distributed throughout the Northern Hemisphere in Arctic and montane habitats, in Europe from the Pyrenees, Alps and Caucasus north to Spitsbergen. Argentina (Patagonia), South Island of New Zealand, Antarctica.

A variable species in which Ando (1973) recognizes var. *revolutum* and var. *dolomiticum* (Milde) Mönk., but admits that they are linked by puzzling intermediates. One Scottish gathering resembles var. *revolutum*, one var. *dolomiticum*, but the third is intermediate (Smith, 1978). Nevertheless, Corley & Crundwell (1991), following the opinion of some Alpine bryologists, treat *H. dolomiticum* Milde as a distinct species.

A. J. E. SMITH

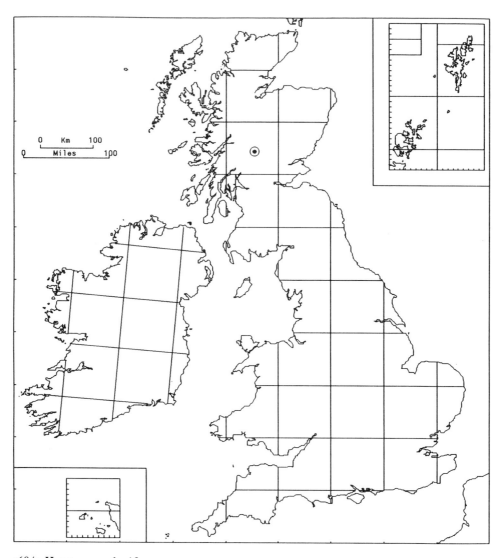

168/2. Hypnum vaucheri Lesq.

A calcicolous species found only once in the British Isles, growing in small patches on calcareous mica-schist rocks on an E.-facing slope at about 730 m (Creag an Lochain). GB 1.

Dioecious; sterile in Scotland, and capsules extremely rare throughout the range of the species.

Boreal and montane parts of the Northern Hemisphere north to the High Arctic; in Europe from the Picos de Europa, Alps and Caucasus to Spitsbergen.

Found in Scotland in 1962 (Perry & Fitzgerald, 1963).

A. J. E. SMITH

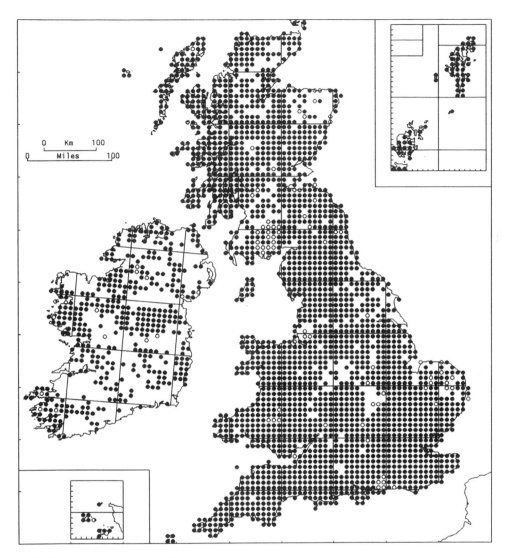

168/3. Hypnum cupressiforme Hedw.

The commonest pleurocarp in the British Isles, being found on trunks and lower branches of trees, on logs, stumps, rock outcrops, walls, roofs, tombstones, rarely on soil, usually in dry acidic exposed or sheltered habitats, sometimes forming extensive patches. It extends into urban areas, tolerating average annual levels of SO_2 pollution of up to 0.016 p.p.m. (about 21 µg S m⁻³) (Gilbert, 1969). The var. *lacunosum* Brid. occurs on basic rocks, soil, sand-dunes and in calcareous turf. 0–760 m (Skye). GB 2283+78*, IR 396+16*.

Dioecious; capsules frequent to common in the north and west, rare to occasional elsewhere, ripening in late autumn or winter. Vegetative propagules unknown but leaf and stem fragments readily develop into new plants.

More-or-less cosmopolitan, occurring in temperate, montane, subarctic and subantarctic regions.

Because of past confusion, and because until recently allied species have been treated as varieties of *H. cupressiforme*, some mapped records may belong to *H. jutlandicum* or *H. mammillatum*. However, *H. cupressiforme* is so common that it will not be mapped in any square in which it does not occur. The infraspecific taxa recognized by Smith (1978), namely var. *cupressiforme*, var. *lacunosum* Brid. and var. *resupinatum* (Tayl.) Schimp., are not mapped separately because the records are unreliable.

A. J. E. SMITH

379

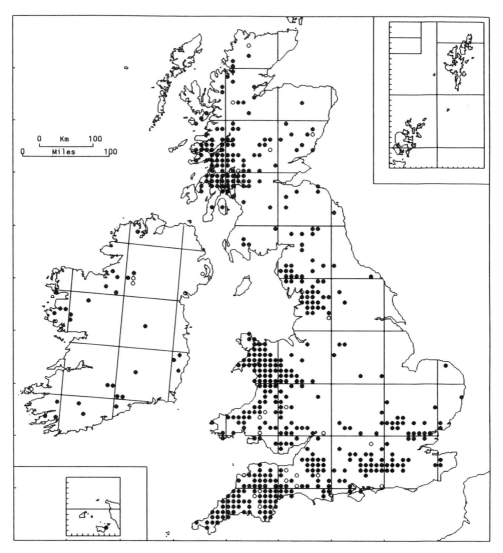

168/4. **Hypnum mammillatum** (Brid.) Loeske (*H. andoi* A. J. E. Smith)

Found, sometimes forming extensive mats, on trunks and lower branches of trees, including conifers, in woodland, less frequently on logs, acidic rocks and wall-tops. Often abundant in sheltered humid habits in the west, rare elsewhere. 0–700 m (Creag an Lochain). GB 479+23*, IR 29+5*.

Dioecious; capsules often abundant, autumn.

W., C. and N. Europe to about 68°N. Macaronesia, north-east N. America (Newfoundland, Nova Scotia, Maine). Reported from Turkey.

In the early years of the Mapping Scheme this species was treated as a variety of *Hypnum cupressiforme* and was recorded as var. *filiforme* Brid., var. *mammillatum* Brid. or simply as *H. cupressiforme* sensu lato. Only records identified as var. *mammillatum* or *Hypnum mammillatum* have been mapped here, because identification of var. *filiforme* was not reliable. The map is therefore incomplete.

A. J. E. SMITH

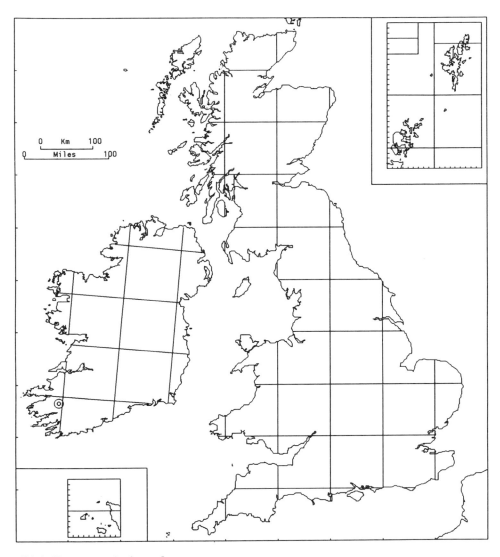

168/4A. **Hypnum uncinulatum** Jur.

Reported from rocks, but details of the Irish habitat of this extremely oceanic species are inadequate. In Macaronesia it occurs most commonly on tree-trunks and logs, occasionally on rocks and earth; rarely it is epiphyllous (Ando, 1986). Lowland. IR 1*.

Dioecious; capsules occasional in Ireland, maturing in late autumn or winter (Ando, 1986).

Macaronesia. Reports from Portugal, Spain, Turkey and Iran require confirmation.

There are only five gatherings from Ireland, where it has not been seen for more than 90 years, very possibly because it has not been looked for. Omitted by Smith (1978) because the author was informed that the record of this species was incorrect, it was restored to the Irish list by Ando & Townsend (1980).

A. J. E. SMITH

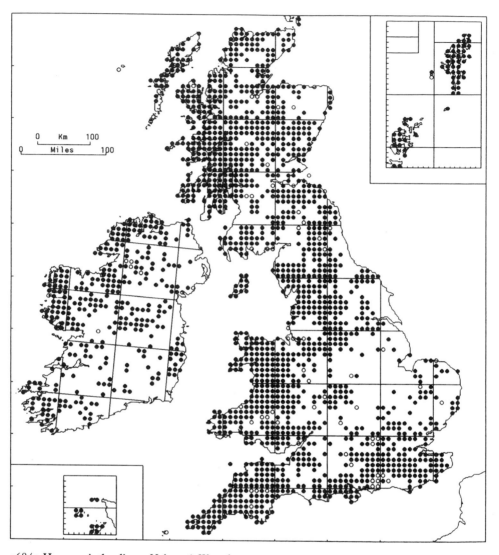

168/5. **Hypnum jutlandicum** Holmen & Warncke

A calcifuge species common and sometimes abundant on dry to damp lowland and upland heath, high-altitude *Racomitrium* heath and drier parts of bogs, particularly associated with *Calluna vulgaris*. It also occurs on humus, rocks and tree-bases in birch woodland and on the ground and stumps in conifer plantations. It has been found as a pendulous epiphyte on *Pinus contorta* in W. Ireland (Doyle, 1986). 0–1040 m (Cairngorms). GB 1512+67*, IR 323+9*.

Dioecious; capsules late winter, occasional in the north and west, very rare elsewhere.

C. and N.W. Europe to N. Fennoscandia. Azores, Canada (Newfoundland) (Ando, 1987). Reported from Cyprus, Turkey and Iran.

In the early years of the Mapping Scheme it bore the name *Hypnum cupressiforme* var. *ericetorum* B., S. & G. but it was often recorded as *Hypnum cupressiforme* sensu lato. The map is therefore somewhat incomplete.

A. J. E. SMITH

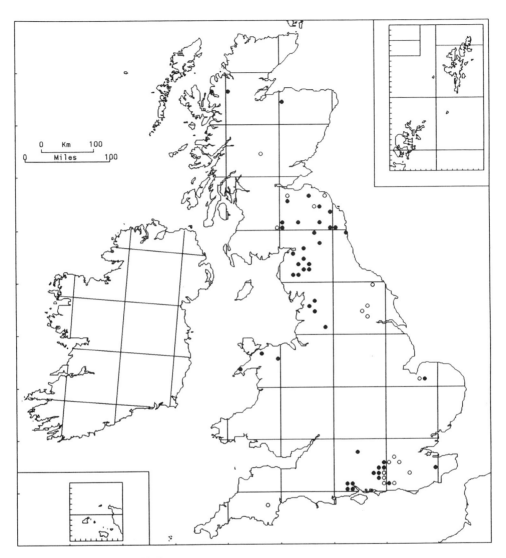

168/6. **Hypnum imponens** Hedw.

An infrequent but sometimes locally common species of wet heath and of raised, valley- and blanket-bogs. 0–600 m (Llyn Anafon). GB 46 + 19*.

Dioecious; capsules very rare.

Temperate and montane Europe from the Pyrenees, Switzerland and Austria to about 67°N. Azores, north-east N. America. Records from other areas have not been confirmed by Ando (1986) and are therefore discounted.

In eastern N. America, where *Hypnum imponens* is very common, it is a woodland plant, occurring mainly on logs and to a lesser extent on banks and rocks. *H. cupressiforme*, by contrast, is relatively uncommon there, occurring in dry places, mainly on calcareous rocks and soil. In Britain, *H. imponens* must often be overlooked, because it is very similar to *H. jutlandicum* and grows in similar places, sometimes mixed with it.

A. J. E. SMITH

383

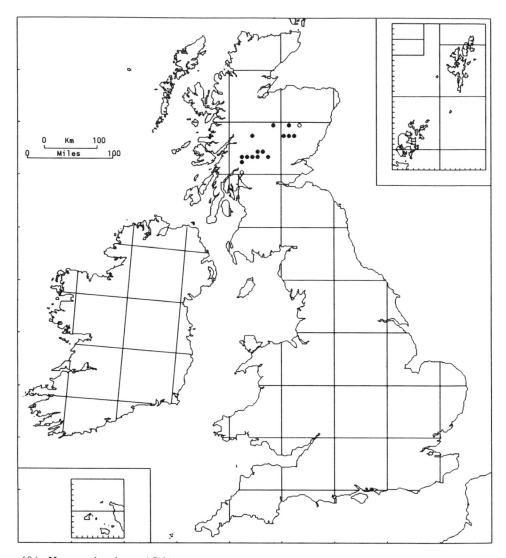

168/7. **Hypnum bambergeri** Schimp.

A calcicole, growing in tufts or patches on sheltered N.- or E.-facing slopes. It has been found on the tops and sides of limestone boulders sunk into turf, in crevices of mica-schist rock, on moist mica-schist rock-faces, and on stony soil at the foot of such rock-faces. The bryophyte communities in which it occurs are often very species-rich, including common calcicoles such as *Plagiochila porelloides*, *Anoectangium aestivum* and *Mnium marginatum* as well as rarities such as *Barbilophozia quadriloba*, *Bryum stirtonii* and *Timmia norvegica*. 900 m (Coire Cheap) to 1175 m (Ben Lawers). GB 14+2*.

Dioecious; capsules unknown in Scotland.

Circumpolar, widespread in the Arctic including the High Arctic; in Europe extending from Italy and Yugoslavia north to Iceland and Spitsbergen. Turkey, Siberia, N. America, Greenland.

A. J. E. Smith

384

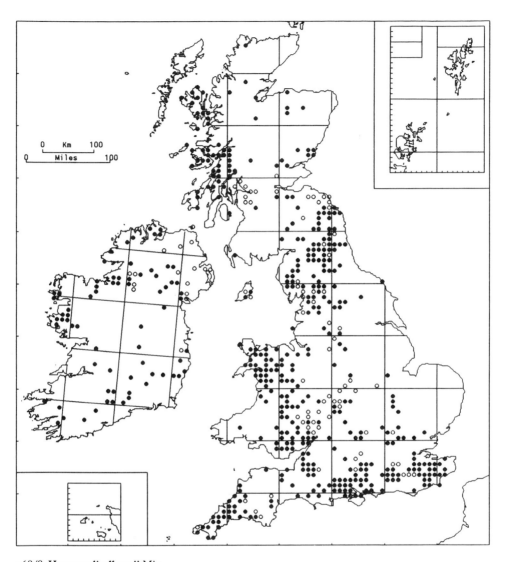

168/8. **Hypnum lindbergii** Mitt.

A mainly lowland species of man-made habitats, but occasionally ascending into the alpine zone. It occurs on clayey to gravelly, slightly acidic soils on roadsides, in ditches, beside woodland rides, in old quarries, gravel-pits and sand-dunes, rarely on soil on mountain rock-ledges. 0–900 m (Caenlochan Glen). GB 392+72*, IR 71+16*.

Dioecious; capsules unknown in the British Isles and exceedingly rare elsewhere.

Circumboreal, very widespread in temperate, Arctic and alpine regions of Eurasia and N. America, in Europe extending to northern Fennoscandia.

A. J. E. SMITH

385

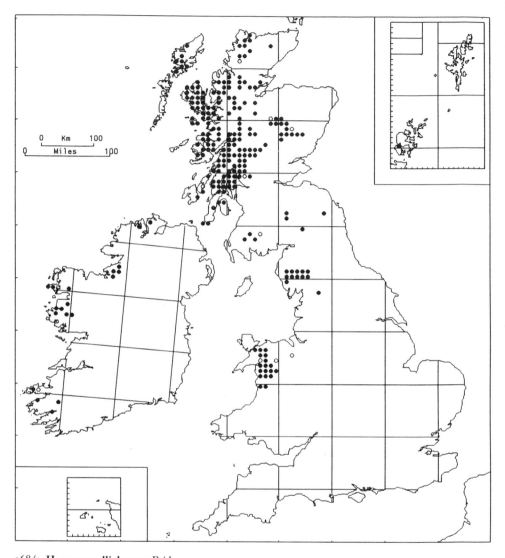

168/9. **Hypnum callichroum** Brid.

It grows on rocks and turf in sheltered situations, both on mountains and in woods and gullies, favouring sites that are intermittently irrigated. Typical habitats are sheltered stream-banks in woods, rock-ledges in ravines, and declivities among blocks below cliffs and in stable screes. In some districts it is confined to substrata that are at least slightly basic, but in W. Scotland it occurs also on granite, quartzose schist and quartzite scree. Exceptionally, it has been found on a rotten log in a limestone ravine. Often descending to near sea-level in the west, perhaps more exclusively a mountain plant in the east. 0–1225 m (Aonach Beag). GB 209+11*, IR 19+1*.

Dioecious; capsules occasional, summer.

Circumpolar arctic-alpine, widespread in the Northern Hemisphere north to the High Arctic; in Europe occurring in all the main mountain ranges north to Iceland, N. Fennoscandia, N. Russia (Kola Peninsula) and Svalbard.

M. O. Hill

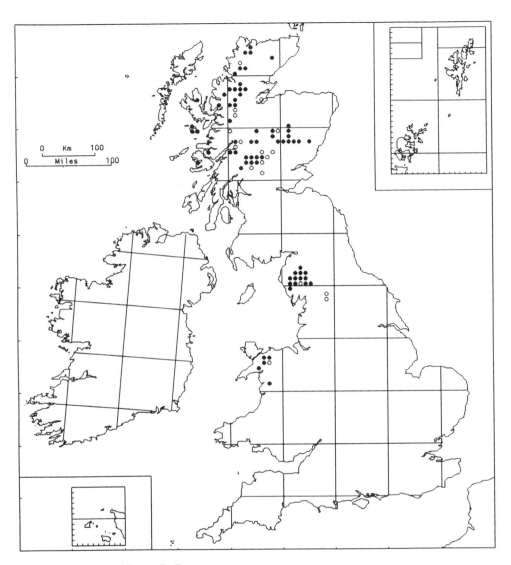

168/10. **Hypnum hamulosum** Br. Eur.

A calcicole of sheltered rocks, typically in cool situations with a north or east aspect. It often grows directly on vertical rock-faces and is also found on rock-ledges, in rock crevices and in turf among rocks. The communities in which it occurs are rich in bryophyte species, but none appears to be a particularly characteristic associate. 200 m (Creag a'Ghaill) to 1000 m (Ben Lawers). GB 68+18*.

Autoecious; capsules occasional, summer.

Circumpolar arctic-alpine. Mountainous and Arctic regions of Europe from the Pyrenees, Alps and Caucasus north to Spitsbergen. Turkey, Siberia, China, Japan, N. America, Greenland.

A. J. E. Smith

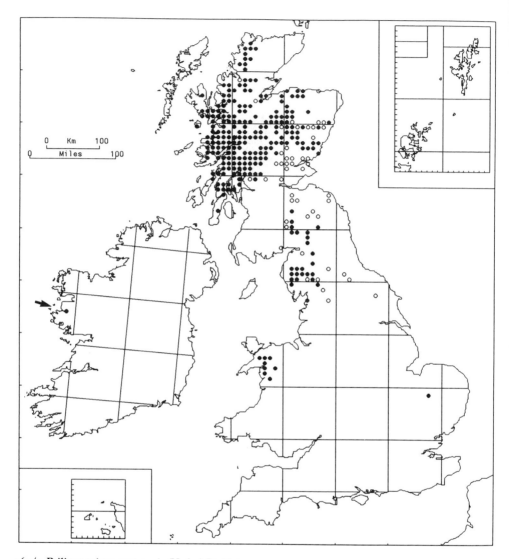

169/1. **Ptilium crista-castrensis** (Hedw.) De Not.

Most frequent on acid, humus-rich soil in shaded situations on the ground and on boulders in natural or semi-natural pine, birch, oak, or birch-hazel woods, growing with *Barbilophozia barbata*, *Hylocomium splendens*, *H. umbratum*, *Plagiothecium undulatum*, *Pleurozium schreberi* and *Rhytiadelphus loreus*. It also occurs under heather on steep N.- to E.-facing block-strewn slopes, associated with *Breutelia chrysocoma*, *Campylopus setifolius* and a wide range of large leafy liverworts. It is less frequent in other damp montane sites such as shaded recesses in stable block-screes, and ledges in low-lying wooded ravines. Its occurrences in East Anglia are in conifer plantations, where it has possibly been introduced by man. 0–800 m (Snowdon). GB 242+49*, IR 1.

Dioecious; capsules very rare, autumn.

Circumboreal, extending north to the Arctic and south in Eurasia to Yunnan (China) and in N. America to Oregon and N. Carolina. Widespread and with a characteristic boreal-montane distribution in Europe.

It resembles some other boreal species such as *Barbilophozia hatcheri* and *Ptilidium ciliare* in being very rare or absent in W. Ireland and the Outer Hebrides.

H. J. B. BIRKS

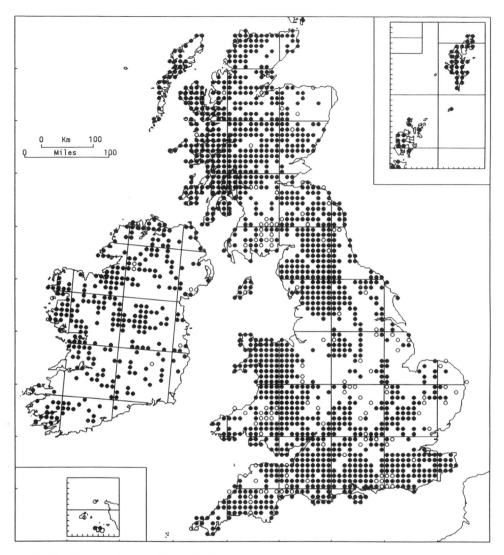

170/1. **Ctenidium molluscum** (Hedw.) Mitt.

This species occurs in diverse mildly to strongly basic habitats which may be dry or periodically irrigated. It grows on rocks and soil, in grassland and woodland, on cliffs and stream-banks, and in basic peaty flushes. Among its many associates the calcicole element is usually strong and may include *Campylium* spp., *Fissidens cristatus*, *Tortella tortuosa* and *Weissia* spp. 0–1205 m (Ben Lawers). GB 1459+122*, IR 297+7*.

Dioecious; capsules rather rare but locally abundant, ripe in spring.

Throughout Europe from the Arctic south to the Mediterranean. Macaronesia, N. Africa, S.W. Asia, N. Siberia, north-west N. America (Alaska, Yukon Territory). (*Ctenidium* from eastern N. America is now assigned to a separate species, *C. malacodes* Mitt.)

The varieties have not been recorded systematically and are not mapped separately. Var. *fastigiatum* (Bosw. ex Hobk.) Braithw. is a poorly marked form of drier basic rocks; var. *condensatum* (Schimp.) Britt. is a more robust, often orange taxon of flushed basic rocks in the mountains, where it may be frequent; the striking var. *robustum* Boul., distinct in habit and size, is a rare plant of irrigated mountain rocks. A form occurring on mildly acid ground in woodland may belong to a distinct taxon.

G. P. ROTHERO

389

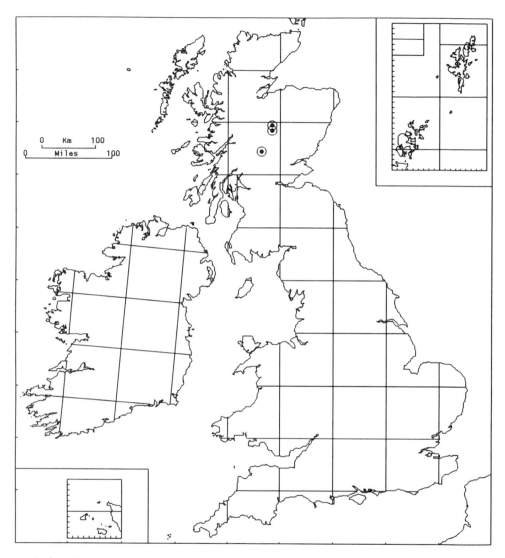

170/2. **Ctenidium procerrimum** (Mol.) Lindb. (*Hypnum procerrimum* Mol.)

It forms dense yellow-brown patches on outcrops of calcareous schist on mountains. Associated species include *Ditrichum flexicaule* sensu lato, *Entodon concinnus*, *Mnium thomsonii* and *Rhytidiadelphus triquetrus*. 450 m (Glen Feshie) and 900 m (Ben Lawers). GB 3.

Dioecious; capsules unknown.

Circumpolar arctic-alpine, widespread in the Arctic, extending south in Eurasia to the mountains of Spain, Alps, Carpathians, Caucasus, C. Asia and China (Yunnan) and in N. America to Alberta and Newfoundland (Canada).

G. P. ROTHERO

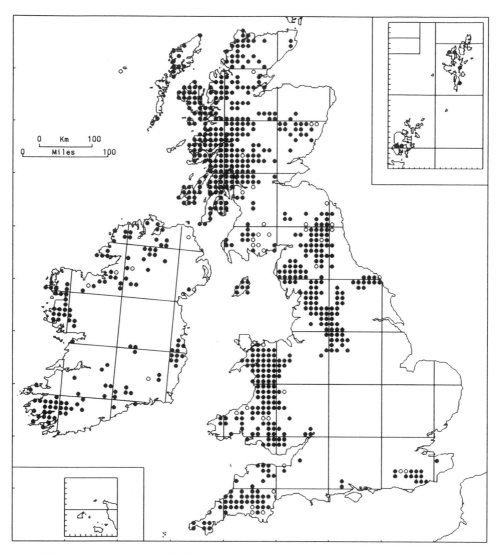

171/1. **Hyocomium armoricum** (Brid.) Wijk & Marg.

Most common close to water, growing on acid or mildly basic boulders and banks of wooded streams and ravines with *Heterocladium heteropterum*, *Hygrohypnum eugyrium*, *Isothecium holtii*, *Rhynchostegium riparioides* and *Thamnobryum alopecurum*. Its sites are often splashed by spray or periodically flooded. It also occurs on wet, shaded flushed rocks, rock outcrops and cliff-faces in natural or semi-natural deciduous woodlands associated with *Marsupella emarginata* and *Heterocladium heteropterum*, and on wet, shaded dripping rocks by waterfalls. More rarely, it can be found on boulders and on banks of streams in open moorland and in damp shaded gullies of low-lying N.- to E.-facing acid or mildly basic cliffs. Generally at low altitudes (0–360 m on Skye), but ascends to 610 m in Kerry and 770 m in E. Scotland (Coire an t-Sneachda). GB 630+40*, IR 127+7*.

Dioecious; capsules very rare, autumn to spring.

Atlantic zone of Europe north to the Faeroes and W. Norway, east to the Massif Central (France) and the Schwarzwald and Eifel (Germany), with isolated stations in the Harz Mts (Germany) and on Corsica. Azores, Turkey, Transcaucasia, Japan.

For a map of its European distribution, see Schumacker *et al.* (1981).

H. J. B. Birks

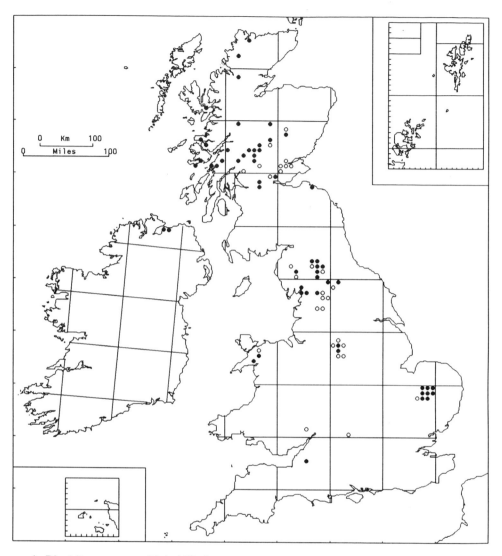

172/1. **Rhytidium rugosum** (Hedw.) Kindb.

A calcicolous species of dry, short and often rather open grassland and turf on well-drained shallow rendzina soils developed over a range of substrates including limestone, calcareous sands, basic andesites and other volcanic rocks, and mica-schists. It also occurs on dry S.- to S.W.-facing rock-ledges of inland cliffs and, more rarely, in dune grasslands in Ireland and in limestone quarries in N. England. Common associates include *Scapania aspera*, *Ditrichum flexicaule*, *Entodon concinnus*, *Fissidens cristatus* and *Homalothecium lutescens*. 0–850 m (Caenlochan Glen). GB 54+29*, IR 2.

Dioecious; capsules unknown in the British Isles and very rare in Europe.

Widespread in cool regions of the Northern Hemisphere, especially in continental interiors, the boreal zone and Low Arctic, becoming rarer northwards to 78°N in Greenland. It extends southwards in mountains to Morocco, tropical and southern Africa, Himalaya, southern China (Yunnan), Mexico and Guatemala.

Inexplicably absent from the Carboniferous Limestone regions of Ireland (e.g. the Burren). In western N. America it can occur in *Picea glauca* forest, where it is fertile.

H. J. B. BIRKS

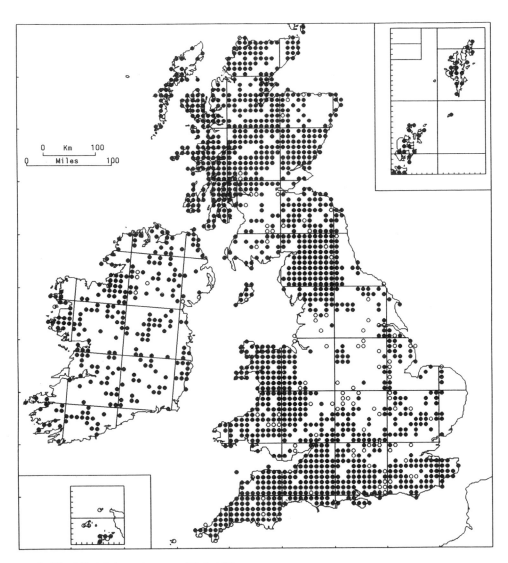

173/1. **Rhytidiadelphus triquetrus** (Hedw.) Warnst.

Forming coarse wefts among higher plants in sheltered situations, avoiding wet places and dense shade. Habitats include grassy banks, dunes, mountain slopes, ungrazed ledges, and open places in woodland. Generally most abundant on base-rich or mildly basic clay soils in woodland and in sheltered limestone and chalk grassland, it is also frequent on neutral to acid soils in the north and west. In Scotland it occurs among *Vaccinium myrtillus* in pine-woods, with *Hylocomium splendens* and *Ptilium crista-castrensis*, and it is dominant in some snow-beds. 0–980 m (Meall Garbh). GB 1417+111*, IR 264+8*.

Dioecious; sporophytes rare, winter.

Circumboreal. Europe including Faeroes and Iceland. N. Africa, Asia, N. America.

Decreasing in C. and S.E. England. Atmospheric pollution may be largely responsible (Jones, 1991), although cessation of coppicing may have played a part.

N. G. HODGETTS

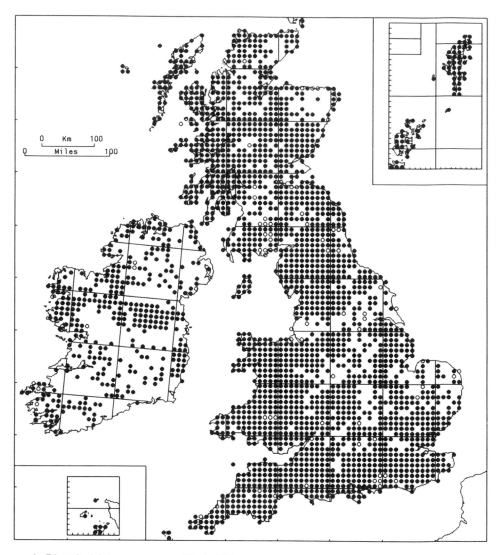

173/2. **Rhytidiadelphus squarrosus** (Hedw.) Warnst.

It has a wide ecological tolerance, occurring on all but the most acid soils in a variety of grassy habitats, including sheep pastures, roadside verges, woodland rides, lawns, dunes, streamsides, ditches and marshes. It also occurs on heaths. In lawns and low-lying pastures, it may be associated with *Lophocolea bidentata*, *Brachythecium rutabulum*, *Calliergon cuspidatum* and *Eurhynchium praelongum*. In mountain grassland and *Nardus* snow-beds, associates include *Hylocomium splendens*, *Pleurozium schreberi* and *Rhytidiadelphus loreus*. It is favoured by heavy grazing and mowing, and prefers damp conditions. 0–1225 m (Aonach Beag). GB 2069+71*, IR 362+9*.

Dioecious; sporophytes rare, winter.

Widespread throughout Europe. Macaronesia, Turkey, N. and E. Asia, N. America. Introduced in New Zealand.

The ubiquity of the species is remarkable, considering the rarity of sporophytes and lack of gemmae. Presumably it propagates vegetatively by stem fragments.

N. G. HODGETTS

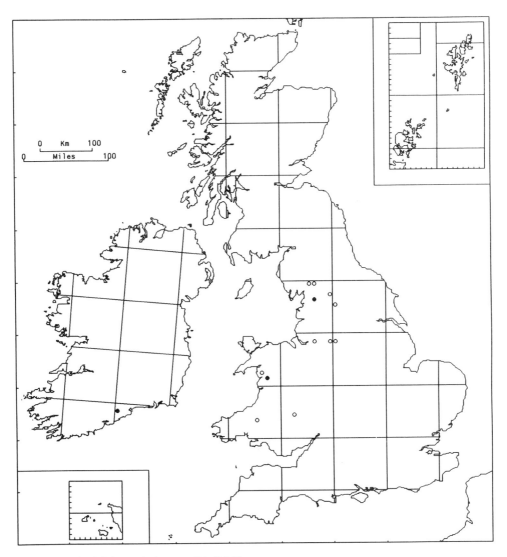

173/3. **Rhytidiadelphus subpinnatus** (Lindb.) Kop.

On damp, lightly shaded, grassy streamsides and banks in open woodland, with associates such as *Polytrichum formosum*, *Rhytidiadelphus loreus*, *R. squarrosus* and *Thuidium tamariscinum*. Lowland. GB 2+10*, IR 1.

Dioecious; sporophytes unknown in Britain.

Circumboreal: N. and C. Europe, N. and C. Asia, N. America.

This species has been misunderstood in Britain in the past; only records confirmed by M. O. Hill are shown on the map. According to Koponen (1971) *R. subpinnatus* has a more boreal distribution in Europe than *R. squarrosus*, being montane in C. Europe and rare in the lowlands south of Finland. Its occurrence in lowland Britain is therefore surprising. Although it has no doubt been overlooked in recent years, its apparent decline is likely to be real; many old records are from areas where *Rhytidiadelphus* spp. have been reduced by air pollution.

N. G. HODGETTS

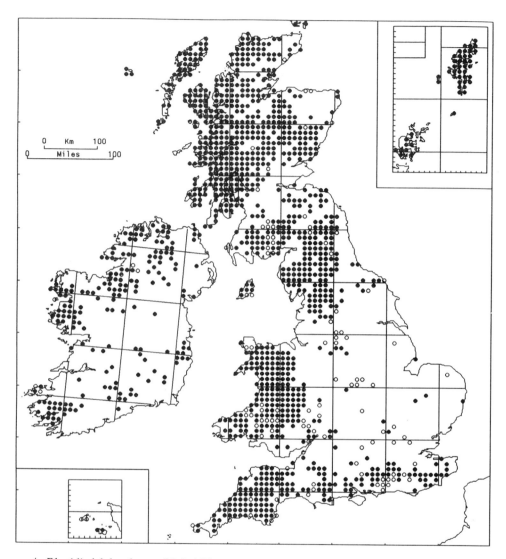

173/4. **Rhytidiadelphus loreus** (Hedw.) Warnst.

A calcifuge species of humid habitats in woodland, forming coarse wefts on soil, boulders, fallen logs and banks. Equally characteristic of montane habitats such as block-scree, cliff-ledges, damp heather moor, species-rich *Racomitrium* heath, montane willow scrub, alpine grassland, and snow-bed vegetation. In western woods, typical associates include *Dicranum majus, D. scoparium, Hylocomium splendens, Plagiothecium undulatum, Pleurozium schreberi, Sphagnum capillifolium* and *Thuidium tamariscinum*. In more open and upland habitats, the same associates occur, together with species such as *Barbilophozia floerkei, Ptilidium ciliare, Racomitrium lanuginosum* and *Rhytidiadelphus squarrosus*. 0–1225 m (Aonach Beag). GB 1050+106*, IR 183+6*.

Dioecious; sporophytes frequent in sheltered western woods but generally rare, winter to spring.

W. and C. Europe north to N. Norway, scarcely extending east of the Scandinavian mountains; montane in S. Europe. Macaronesia, western and eastern N. America.

It is sensitive to atmospheric pollution and is decreasing in some areas; it has almost disappeared from the S. Pennines. Recent records from E. England are from very sheltered, locally humid wooded thickets.

N. G. HODGETTS

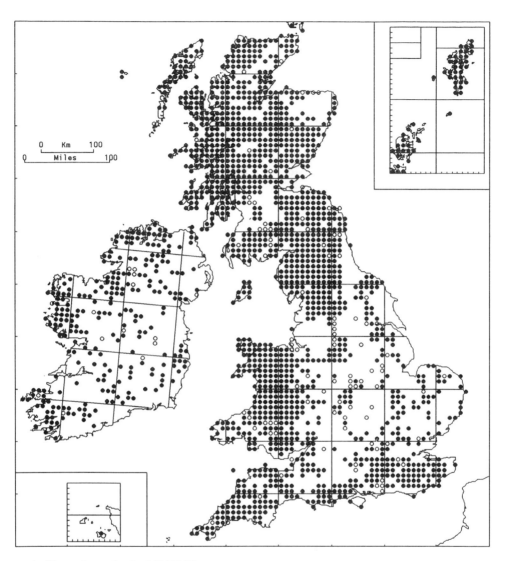

174/1. **Pleurozium schreberi** (Brid.) Mitt.

A calcifuge, often growing in abundance with other weft-forming pleurocarps, for example with *Hypnum jutlandicum* under *Calluna vulgaris* on lowland heaths and upland moors, and with *Hylocomium splendens* and *Rhytidiadelphus loreus* under conifers and in upland deciduous woodland. Other habitats include montane scree, sand-dunes, and acid grassland in the north and west. 0–1210 m (Aonach Beag). GB 1505 + 101*, IR 252 + 14*.

Dioecious; perichaetia develop in most populations but perigonia and sporophytes are generally rare in the British Isles, except in the Scottish Highlands (Longton & Greene, 1979).

Circumboreal, ranging from the Arctic south in mountains to S. Europe, Himalaya, southern China (Yunnan) and northern S. America.

A characteristic component of ground vegetation in coniferous forests throughout the boreal zone.

R. E. LONGTON

397

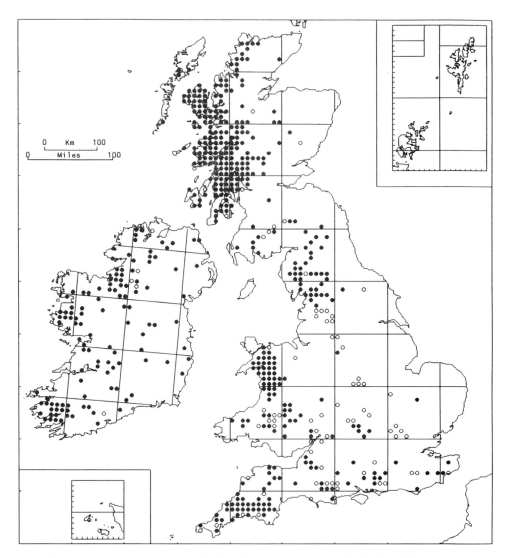

175/1. **Hylocomium brevirostre** (Brid.) Br. Eur. (*Loeskeobryum brevirostre* (Brid.) Fleisch.)

A mild calcicole, this species is almost always found on calcareous or basic rock or soil, or on other substrata in base-rich situations. It is locally frequent on shaded boulders and soil-banks in mixed deciduous woodlands (mainly hazel, ash, elm, oak and birch) associated with *Galium odoratum*, *Ctenidium molluscum*, *Eurhynchium striatum*, *Isothecium myurum* and *Thuidium* spp. and on bases and lower trunks of hazel and ash trees, particularly in extremely humid habitats such as shaded gullies and waterfalls. It also occurs on damp ungrazed ledges on cliffs. More rarely it grows on logs and wall-tops, on shaded banks in wooded lanes, in scrub on chalk, in flushes in moorland, and in damp fen woodlands. 0–650 m (Seana Bhraigh). GB 379+73*, IR 117+6*.

Dioecious; capsules rare, winter.

W. and C. Europe north to Faeroes and S. Scandinavia (but not Finland), east to the Balkans and Crimea. Azores, N. Africa, Turkey, Japan, Russian Far East (Kamchatka), eastern N. America, Guatemala.

Although generally found in permanent habitats, this species has considerable colonizing ability, having recently appeared in Wicken Fen (Lock, 1990) and in at least 30 localities in the Dutch polders (Bremer & Ott, 1990).

H. J. B. Birks

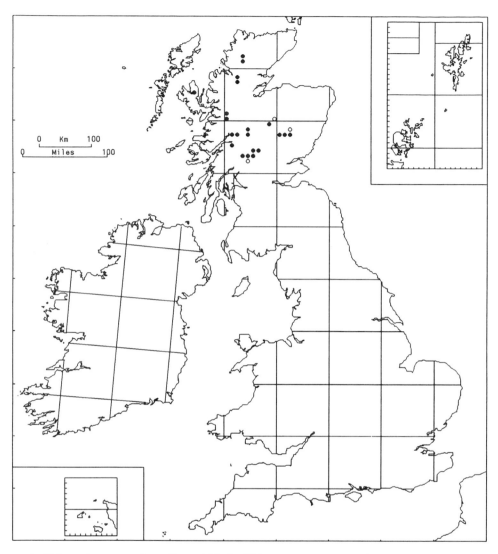

175/2. **Hylocomium pyrenaicum** (Spruce) Lindb. (*Hylocomiastrum pyrenaicum* (Spruce) Fleisch.)

It occurs as scattered stems amongst short turf at the base of cliffs, occasionally on cliff-ledges, and, more rarely, on rock-faces, in block-litters or at the margins of stony flushes. It favours, but is not restricted to, highly basic substrata (e.g. mica-schist, basalt) of north to east aspect. Consistent associates are few but include *Aulacomnium turgidum*, *Hylocomium splendens* and *Rhytidiadelphus loreus*. 670 m (Skye) to 1180 m (Ben Lawers). GB 21+3*.

Dioecious; capsules unknown in Britain.

Circumpolar, occurring widely from the boreal zone north to the Low Arctic, and in subalpine habitats on mountains further south. Iceland, Europe, Turkey, Caucasus, N. and C. Asia, Japan, Aleutian Islands, N. America, Greenland.

A curiously rare plant that is invariably present in small quantity and can easily be overlooked. It is likely to be commoner than the map indicates. In the boreal zone it occurs commonly in forests.

H. J. B. BIRKS

399

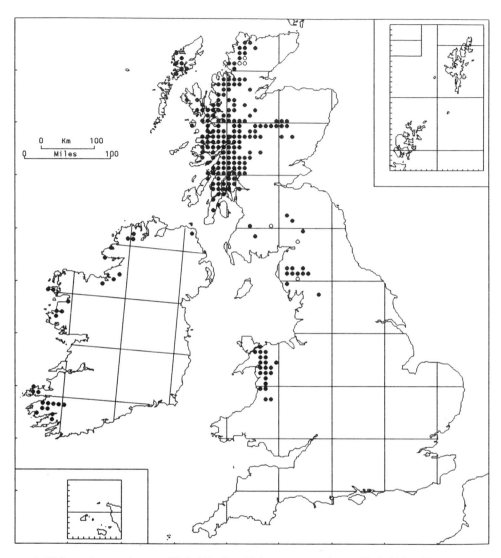

175/3. **Hylocomium umbratum** (Hedw.) Br. Eur. (*Hylocomiastrum umbratum* (Hedw.) Fleisch.)

A species of base-deficient or slightly basic substrata including bare rock, mineral soil, raw humus and shallow peat. It occurs frequently in natural or semi-natural deciduous woods, either in moss mats on dry shaded boulders with *Hylocomium splendens*, *Rhytidiadelphus loreus*, *Thuidium delicatulum* and *T. tamariscinum* or on the ground in humid rocky glens with *Anastrepta orcadensis* and *Plagiothecium undulatum*. Outside woodland it is mainly found under *Calluna vulgaris* and other dwarf shrubs on steep N.- to E.-facing block-strewn slopes in corries and below cliffs, associated with *Hymenophyllum wilsonii*, *Dicranodontium uncinatum* and large leafy liverworts. It also occurs on shaded mossy ledges in wooded ravines, between boulders in block-litters, and on cliff-ledges, especially on N.- to E.-facing slopes. In its more easterly Scottish localities it occurs mainly in situations where there is prolonged snow cover. 0–1170 m (Ben Lawers). GB 209+8*, IR 28.

Dioecious; capsules very rare, autumn.

A boreal-montane species, extending north in Europe to the Arctic and widespread in mountains further south. N. Africa, Caucasus, Urals, Siberia, Korea, Japan, Russian Far East (Kamchatka), western N. America (British Columbia), eastern N. America.

H. J. B. Birks

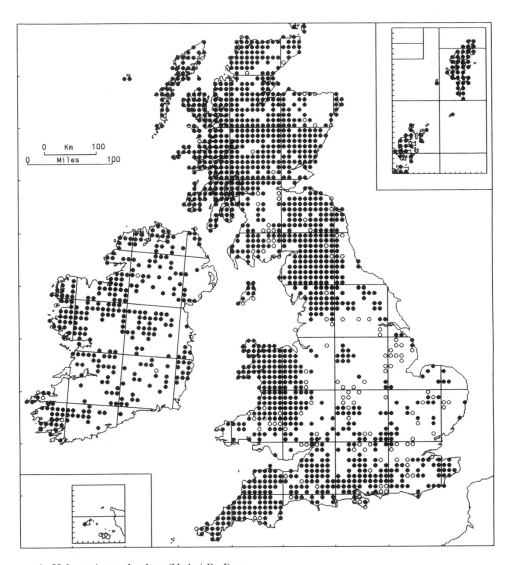

175/4. Hylocomium splendens (Hedw.) Br. Eur.

A mildly calcifuge species occurring on acid substrata or if on chalk and limestone then usually on deeper soils subject to leaching and some local acidification. In the north and west it is common under heather in dry or slightly damp moorland and heath associated with *Dicranum scoparium*, *Hypnum jutlandicum* and *Rhytidialphus loreus*, and in a range of upland grasslands, often with *Pseudoscleropodium purum* and *Rhytidiadelphus squarrosus*. In woodlands it is most frequent under a canopy of birch, pine, oak or birch-hazel. It also occurs on dry, often degraded, blanket-bog; between boulders on block-strewn slopes; in maritime heaths; in duneland and in chalk and limestone grassland; in grassy turf by roads and ditches; on damp shaded ledges on acid or mildly basic cliffs; and in shaded hedgebanks. 0–1180 m (Ben Lawers). GB 1387+138*, IR 314+9*.

Dioecious; capsules very rare, spring.

Widespread in cool and cold regions of the Northern Hemisphere north to 83°N on Ellesmere Island, south to N. Africa, Himalaya and, in N. America, to Oregon and N. Carolina.

In much of E. and S.E. England, it now behaves as a mild calcicole, having been eliminated from its more acid sites by atmospheric pollution (Farmer *et al.*, 1992).

H. J. B. BIRKS

BIBLIOGRAPHY

Titles of periodicals are abbreviated according to the *World List of Scientific Periodicals* (Brown & Stratton, 1963–65).

Adam, P., 1976. The occurrence of bryophytes on British saltmarshes. *J. Bryol.* **9**, 265–274.

Adams, K. J., 1974. The flora: bryophytes. *In:* S. T. Jermyn, *Flora of Essex*, pp. 227–271. Colchester.

———, 1984. *Zygodon forsteri* (With.) Mitt. in Epping Forest. *Bull. Br. bryol. Soc.* **43**, 26–27.

——— & Preston, C. D., 1992. Evidence for the effects of atmospheric pollution on bryophytes from national and local recording. *In:* P. T. Harding (ed.), *Biological Recording of Changes in British Wildlife*, pp. 31–43. London.

Anderson, L. E., Crum, H. A. & Buck, W. R., 1990. List of the mosses of North America north of Mexico. *Bryologist* **93**, 448–499.

Ando, H., 1973. Studies on the genus *Hypnum* Hedw. (II). *J. Sci. Hiroshima Univ., Ser. B, Div. 2*, **14**, 165–207.

———, 1986. Studies on the genus *Hypnum* Hedw. (IV). *Hikobia* **9**, 467–484.

———, 1987. Studies on the genus *Hypnum* Hedw. (V). *Hikobia* **10**, 43–54.

——— & Townsend, C. C., 1980. *Hypnum uncinulatum* Jur. reinstated as an Irish species. *J. Bryol.* **11**, 185–189.

Appleyard, J., 1970. A bryophyte flora of North Somerset. *Trans. Br. bryol. Soc.* **6**, 1–40.

———, 1986. Leaf gemmae in *Orthotrichum tenellum*. *J. Bryol.* **14**, 179–180.

Arts, T., 1986. The occurrence of moniliform tubers in *Pohlia melanodon* (Brid.) J. Shaw, the differences between juvenile plants of related species and their distribution in Belgium and the Grand-Duchy of Luxembourg. *Bull. Soc. r. Bot. Belg.* **119**, 114–120.

——— & Nordhorn-Richter, G., 1986. *Epipterygium tozeri* in Europe, its distribution and vegetative propagation. *J. Bryol.* **14**, 91–97.

———, ——— & Smith, A. J. E., 1987. *Pohlia muyldermansii* Wilcz. & Dem. var. *pseudomuyldermansii* var. nov., a new name for *P. muyldermansii* Wilcz. & Dem. sensu Lewis & Smith. *J. Bryol.* **14**, 635–647.

Ball, D. F., Radford, G. L. & Williams, W. M., 1983. *A land characteristic data bank for Great Britain*. Occasional Paper No. 13, Institute of Terrestrial Ecology, Bangor.

Barkman, J. J., 1955. *Brachythecium erythrorhizum* Br. et Schimp. new to Great Britain. *Trans. Br. bryol. Soc.* **2**, 568–570.

Bedford, T. H. B., 1938. Sex distribution in colonies of *Climacium dendroides* W. & M. and its relation to fruit bearing. *NWest. Nat.* **13**, 213–221.

———, 1940. The fruiting of *Breutelia arcuata* Schp. *Naturalist, Hull* **773**, 113–115.

Berkeley, M. J., 1863. *Handbook of British Mosses*. London.

Birks, H. J. B., 1987. Recent methodological developments in quantitative descriptive biogeography. *Annls zool. fenn.* **24**, 165–178

—— & Birks, H. H., 1974. Studies on the bryophyte flora and vegetation of the Isle of Skye. I. Flora. *J. Bryol.* **8**, 19–64,197–254.

—— & Dransfield, J., 1970. A note on the habitat of *Scorpidium turgescens* (T. Jens.) Loeske in Scotland. *Trans. Br. bryol. Soc.* **6**, 129–132.

Blockeel, T. L., 1987. The status of *Orthotrichum urnigerum* in Britain and a comparison of *O. urnigerum* and *O. anomalum*. *J. Bryol.* **14**, 649–651.

Boesen, D. F., Lewinsky, J. & Rasmussen, L., 1975. A check-list of the bryophytes of the Faroes. *Lindbergia* **3**, 69–78.

Bopp, M., Quader, H., Thoni, C., Sawidis, T. & Schnepf, E., 1991. Filament disruption in *Funaria* protonemata: formation and disintegration of tmema cells. *J. Pl. Physiol.* **137**, 273–284.

Braithwaite, R., 1870. *Anomodon attenuatus. Sci. Gossip* **6**, 83.

——, 1887–1905. *The British Moss-Flora*. 3 vols. London.

Brassard, G. R., 1980. The moss genus *Timmia*. 2. Sect. *Timmiaurea*. *Lindbergia* **6**, 129–136.

Bremer, P. & Ott, E. C. J., 1990. The establishment and distribution of bryophytes in the woods of the IJsselmeerpolders, The Netherlands. *Lindbergia* **16**, 3–18.

Brown, P. & Stratton, G. B. (eds), 1963–65. *World List of Scientific Periodicals*, edn 4. 3 vols. London.

Burrell, W. H., 1940. A field study of *Orthodontium gracile* (Wilson) Schwaegrichen and its variety *heterocarpum* Watson. *Naturalist, Hull* **785**, 295–302.

Cain, S. A., 1944. *Foundations of plant geography*. New York.

Clifford, M. H., 1937. A Mesolithic flora in the Isle of Wight. *Proc. Isle Wight nat. Hist. archaeol. Soc.* **2**, 582–594.

Coker, P. D., 1968a. Distribution maps of bryophytes in Britain. *Mielichhoferia elongata* (Hornsch.) Hornsch. *Trans. Br. bryol. Soc.* **5**, 598.

——, 1968b. *Mielichhoferia mielichhoferi* (Hook.) Wijk & Marg., new to the British Isles. *Trans. Br. bryol. Soc.* **5**, 448–451.

——, 1971. *Mielichhoferia elongata* (Hornsch.) Hornsch. and *Saelania glaucescens* (Hedw.) Broth. in Scotland. *Trans. Br. bryol. Soc.* **6**, 317–322.

Corley, M. F. V., 1990. *Brachythecium trachypodium* (Brid.) B., S. & G. in Scotland. *J. Bryol.* **16**, 173–177.

—— & Crundwell, A. C., 1991. Additions and amendments to the mosses of Europe and the Azores. *J. Bryol.* **16**, 337–356.

——, ——, Düll, R., Hill, M. O. & Smith, A. J. E., 1981. Mosses of Europe and the Azores; an annotated list of species, with synonyms from the recent literature. *J. Bryol.* **11**, 609–689.

—— & Hill, M. O., 1981. *Distribution of Bryophytes in the British Isles: A Census Catalogue of their Occurrence in Vice-counties*. Cardiff.

—— & Rothero, G. P., 1992. *Hygrohypnum styriacum* (Limpr.) Broth. in Scotland, new to the British Isles. *J. Bryol.* **17**, 107–110.

Crum, H. A. & Anderson, L. E., 1981. *Mosses of Eastern North America*. 2 vols. New York.

Crundwell, A. C., 1951. Report of the Annual Meeting, 1950. *Trans. Br. bryol. Soc.* **1**, 519–520.

————, 1953. *Pseudoleskea catenulata* var. *acuminata* in Britain and in America. *Trans. Br. bryol. Soc.* **2**, 278–282.

————, 1957. Some neglected British moss records. *Trans. Br. bryol. Soc.* **3**, 174–179.

————, 1959a. A revision of the British material of *Brachythecium glaciale* and *B. starkei. Trans. Br. bryol. Soc.* **3**, 565–567.

————, 1959b. *Plagiothecium laetum* in Britain. *Trans. Br. bryol. Soc.* **3**, 563–564.

————, 1962. *Bryum sauteri* and *B. klinggraeffii* in Britain. *Trans. Br. bryol. Soc.* **4**, 334.

————, 1978. *Rhizomnium magnifolium* (Horik.) Kop. in the British Isles. *J. Bryol.* **10**, 1–4.

————, 1981. Reproduction in *Myurium hochstetteri. J. Bryol.* **11**, 715–717.

————, 1982. *Pohlia scotica*, a new species of moss from the western Highlands of Scotland. *J. Bryol.* **12**, 7–10.

———— & Nyholm, E., 1962. A study of *Campylium hispidulum* and related species. *Trans. Br. bryol. Soc* **4**, 194–200.

———— & ————, 1964a. The European species of the *Bryum erythrocarpum* complex. *Trans. Br. bryol. Soc.* **4**, 597–637.

———— & ————, 1964b. *Amblystegium saxatile* Schimp. in Cornwall, new to the British Isles. *Trans. Br. bryol. Soc.* **4**, 638–641.

———— & ————, 1974. *Funaria muhlenbergii* and related European species. *Lindbergia* **2**, 222–229.

Dahl, E., 1956. *Rondane: Mountain Vegetation of South Norway and its Relation to the Environment*. Oslo.

Dickson, J. H., 1973. *Bryophytes of the Pleistocene*. Cambridge.

————, 1981. Mosses from a Roman well at Abingdon. *J. Bryol.* **11**, 559–560.

Dixon, H. N., 1924. *The Student's Handbook of British Mosses*, edn 3. Eastbourne.

Doyle, G. J., 1986. *Hypnum jutlandicum* Holmen & Warncke growing as a pendulous epiphyte on *Pinus contorta* in a plantation in the west of Ireland. *Lindbergia* **12**, 73–75.

Ducker, B. F. T. & Warburg, E. F., 1961. *Physcomitrium eurystomum* Sendtn. in Britain. *Trans. Br. bryol. Soc.* **4**, 95–97.

Duckett, J. G., 1973. Distribution maps of bryophytes in Britain. *Discelium nudum* (Dicks.) Brid. *J. Bryol.* **7**, 449.

———— & Ligrone, R., 1992. A survey of diaspore liberation mechanisms and germination patterns in mosses. *J. Bryol.* **17**, 335–354.

Duell, R., 1985. Distribution of the European and Macaronesian mosses (Bryophytina), Part II. *Bryol. Beitr.* **5**, 110–232.

————, 1992. Distribution of the European and Macaronesian mosses (Bryophytina), annotations and progress. *Bryol. Beitr.* **8/9**, 1–223.

Duell-Hermanns, I., 1981. Spezielle Untersuchungen zur modernen Taxonomie von *Thuidium abietinum* und der Varietät *hystricosum. J. Bryol.* **11**, 467–488.

Duncan, U.K., 1966. A bryophyte flora of Angus. *Trans. Br. bryol. Soc.* **5**, 1–82.

During, H. J., 1986. Longevity of spores of *Funaria hygrometrica* in chalk grassland soil. *Lindbergia* **12**, 132–134.

Edwards, S. R., 1978. Protonemal gemmae in *Schistostega pennata* (Hedw.) Web. et Mohr. *J. Bryol.* **10**, 69–72.

Farmer, A. M., Bates, J. W. & Bell, J. N. B., 1992. Ecophysiological effects of acid rain on bryophytes and lichens. *In:* J. W. Bates & A. M. Farmer (eds), *Bryophytes and Lichens in a Changing Environment*, pp. 284–313. Oxford.

Farrand, W. R., 1961. Frozen mammoths and modern geology. *Science, N.Y.* **133**, 729–735.

Field, J. H., 1986. Shootlet propagula on cultured material of *Philonotis fontana* from Sutton Park. *Bull. Br. bryol. Soc.* **47**, 20.

———, 1988. *Philonotis calcarea* – continuous production of shootlet propagula in the greenhouse. *Bull. Br. bryol. Soc.* **52**, 28.

Fremstad, E., 1978. *Campylium protensum* (Brid.) Kindb. in Norway. *Lindbergia* **4**, 333–336.

Frisvoll, A. A. & Blom, H. H., 1992. Trua moser i Norge med Svalbard; raud liste. *NINA Utredning* **42**, 1–55.

Fritsch, R., 1991. Index to bryophyte chromosome counts. *Bryophytorum Bibliotheca* **40**, 1–352.

Furness, S. B. & Gilbert, O. L., 1980. The status of *Thamnobryum angustifolium* (Holt) Crundw. *J. Bryol.* **11**, 139–144.

——— & Grime, J. P., 1982. Growth rate and temperature responses in bryophytes. 1. An investigation of *Brachythecium rutabulum*. *J. Ecol.* **70**, 513–523.

——— & Hall, R. H., 1981. An explanation of the intermittent occurrence of *Physcomitrium sphaericum* (Hedw.) Brid. *J. Bryol.* **11**, 733–742.

Gardiner, J. C., 1981. A bryophyte flora of Surrey. *J. Bryol.* **11**, 747–841.

Gilbert, O. L., 1968. Bryophytes as indicators of air pollution in the Tyne valley. *New Phytol.* **67**, 15–30.

———, 1969. The effect of SO_2 on lichens and bryophytes around Newcastle upon Tyne. *In:* Anon. (ed.), *Air Pollution* (Proceedings of the first European congress on the influence of air pollution on plants and animals, Wageningen, 1968), pp. 223–235. Wageningen.

———, 1970. A biological scale for the estimation of sulphur dioxide pollution. *New Phytol.* **69**, 629–634.

Goode, J. A., Stead, A. D. & Duckett, J. G., 1993. Studies of protonemal morphogenesis in mosses. II. *Orthotrichum obtusifolium* Brid. *J. Bryol.* **17**, 409–419.

Greene, S. W., 1958. *Leptodon smithii* (Hedw.) Mohr in Ireland. *Trans. Br. bryol. Soc.* **3**, 392–398.

Hedenäs, L., 1987. North European mosses with axillary rhizoids, a taxonomic study. *J. Bryol.* **14**, 429–439.

———, 1989a. The genera *Scorpidium* and *Hamatocaulis*, gen. nov., in northern Europe. *Lindbergia* **15**, 8–36.

———, 1989b. The genus *Sanionia* (Musci) in northwestern Europe, a taxonomic revision. *Annls bot. fenn.* **26**, 399–419.

———, 1990. The genus *Pseudocalliergon* in northern Europe. *Lindbergia* **16**, 80–99.

———, 1993. A generic revision of the *Warnstorfia-Calliergon* group. *J. Bryol.* **17**, 447–479.

———, Herben, T., Rydin, H. & Söderström, L., 1989. Ecology of the invading moss species *Orthodontium lineare* in Sweden: substrate preference and interactions with other species. *J. Bryol.* **15**, 565–581.

Hemerik, L., 1989. On the distinction between *Plagiothecium succulentum* (Wils.) Lindb. and *P. nemorale* (Mitt.) Jaeg.: a statistical analysis. *Lindbergia* **15**, 2–7.

Hill, M. O., 1988. A bryophyte flora of North Wales. *J. Bryol.* **15**, 377–491.

———, 1991. Patterns of species distribution in Britain elucidated by canonical correspondence analysis. *J. Biogeogr.* **18**, 247–255.

———, 1993. *Eurhynchium pulchellum* (Hedw.) Jenn. in Britain and Ireland. *J. Bryol.* **17**, 683–684.

———, Preston, C. D. & Smith, A. J. E., 1991. *Atlas of the Bryophytes of Britain and Ireland*, **1**. *Liverworts (Hepaticae and Anthocerotae)*. Colchester.

———, ——— & ———, 1992. *Atlas of the Bryophytes of Britain and Ireland*, **2**. *Mosses (except Diplolepideae)*. Colchester.

Hodgetts, N. G. & Blockeel, T. L., 1992. *Thamnobryum cataractarum*, a new species from Yorkshire, with observations on *T. angustifolium* and *T. fernandesii*. *J. Bryol.* **17**, 251–262.

Ireland, R. R., 1992. Studies of the genus *Plagiothecium* in Australasia. *Bryologist* **95**, 221–224.

Jones, E. W., 1953. A bryophyte flora of Berkshire and Oxfordshire. II. Musci. *Trans. Br. bryol. Soc.* **2**, 220–277.

———, 1986. Bryophytes in Chawley Brick Pit, Oxford, 1948–1985. *J. Bryol.* **14**, 347–358.

———, 1991. The changing bryophyte flora of Oxfordshire. *J. Bryol.* **16**, 513–549.

Jongman, R. H. G., Ter Braak, C. J. F. & van Tongeren, O. F. R., 1987. *Data analysis in community and landscape ecology*. Wageningen.

Kelly, M. G. & Huntley, B., 1987. *Amblystegium riparium* in brewery effluent channels. *J. Bryol.* **14**, 792.

King, A. L. K. & Scannell, M. J. P., 1960. Notes on the vegetation of a mineral flush in Co. Mayo. *Ir. Nat. J.* **13**, 137–140.

Koponen, A., 1990. Entomophily in the Splachnaceae. *Bot. J. Linn. Soc.* **104**, 115–127.

Koponen, T., 1971. *Rhytidiadelphus japonicus* and *R. subpinnatus*. *Hikobia*, **6**, 18–35.

Lewinsky, J., 1974. The genera *Leskeella* and *Pseudoleskeella* in Greenland. *Bryologist* **77**, 601–611.

Lewis, K. & Smith, A. J. E., 1978. Studies on some bulbiliferous species of *Pohlia* section *Pohliella*. II. Taxonomy. *J. Bryol.* **10**, 9–27.

Lock, J. M., 1990. Calcifuge bryophytes at Wicken Fen. *J. Bryol.* **16**, 89–96.

Lockhart, N. D., 1987. The occurrence of *Homalothecium nitens* (Hedw.) Robins. in Ireland. *J. Bryol.* **14**, 511–517.

Long, D. G., 1982a. Notes on the distribution and ecology of *Rhizomnium magnifolium* (Horikawa) Koponen in Scotland. *Bull. Br. bryol. Soc.* **39**, 39–41.

———, 1982b. *Campylium halleri* (Hedw.) Lindb. fruiting in Britain. *J. Bryol.* **12**, 115–116.

———, 1992. *Sanionia orthothecioides* (Lindb.) Loeske in Scotland, new to the British Isles. *J. Bryol.* **17**, 111–117.

———, 1993. *Sanionia orthothecioides* on the Scottish mainland. *J. Bryol.* **17**, 513–514.

Longton, R. E., 1981. Inter–population variation in morphology and physiology in the cosmopolitan moss *Bryum argenteum* Hedw. *J. Bryol.* **11**, 501–520.

——— & Greene, S. W., 1979. Experimental studies of growth and reproduction in the moss *Pleurozium schreberi* (Brid.) Mitt. *J. Bryol.* **10**, 321–338.

Malta, N., 1926. Die Gattung *Zygodon* Hook. et Tayl. Eine monographische Studie. *Latvijas Universitates Botaniske Darza Darbi* **1**, 1–185.

Margadant, W. D. & Meijer, W., 1950. Preliminary remarks on *Orthodontium* in Europe. *Trans. Br. bryol. Soc.* **1**, 266–274.

Marino, P. C., 1991a. Competition between mosses (Splachnaceae) in patchy habitats. *J. Ecol.* **79**, 1031–1046.

————, 1991b. Dispersal and coexistence of mosses (Splachnaceae) in patchy habitats. *J. Ecol.* **79**, 1047–1060.

McAdam, S. V. & Smith, A. J. E., 1981. *Brachythecium appleyardiae* sp. nov. in southwest England. *J. Bryol.* **11**, 591–598.

McVean, D. N. & Ratcliffe, D. A., 1962. *Plant Communities of the Scottish Highlands*. London.

Newton, M. E., 1971. A cytological distinction between male and female *Mnium undulatum* Hedw. *Trans. Br. bryol. Soc.* **6**, 230–243.

Nyholm, E., 1958. *Illustrated Moss Flora of Fennoscandia*, II. *Musci*, Fasc. 3. Lund.

————, 1989. *Illustrated Flora of Nordic Mosses*, Fasc. 2. *Pottiaceae–Splachnaceae–Schistostegaceae*. Lund.

———— & Crundwell, A. C., 1958. *Bryum salinum* Hagen ex Limpr. in Britain and in America. *Trans. Br. bryol. Soc.* **3**, 375–377.

Ochyra, R., Koponen, T. & Norris, D. H., 1991. Bryophyte flora of the Huon Peninsula, Papua New Guinea. XLVI. Amblystegiaceae (Musci). *Acta bot. fenn.* **143**, 91–106.

Orbán, S. & Pócs, T., 1976. *Rhodobryum ontariense* (Kindb.) Kindb. in Central Europe. *Acta bot. hung.* **23**, 437–448.

Paton, J. A, 1961. A bryophyte flora of South Hants. *Trans. Br. bryol. Soc.* **4**, 1–83.

————, 1968. *Eriopus apiculatus* (Hook. f. & Wils.) Mitt. established on Tresco. *Trans. Br. bryol. Soc.* **5**, 460–462.

————, 1969. A bryophyte flora of Cornwall. *Trans. Br. bryol. Soc.* **5**, 669–756.

Perry, A. R. & Dransfield, J., 1967. *Orthotrichum gymnostomum* in Scotland. *Trans. Br. bryol. Soc.* **5**, 218–221.

———— & Fitzgerald, R. D., 1963. *Hypnum vaucheri* Lesq. in Perthshire – new to the British Isles. *Trans. Br. bryol. Soc.* **4**, 418–421.

Petit, E., 1976. Les propagules dans le genre *Philonotis* (Musci). *Bull. Jard. bot. nat. Belg.* **46**, 221–226.

Pierrot, R. B., 1983. L'année bryologique dans le Centre-Ouest 1982. *Bull. Soc. bot. Centre-Ouest* N.S. **14**, 155–157.

Pigott, C. D., 1956. The vegetation of Upper Teesdale in the North Pennines. *J. Ecol.* **44**, 545–586.

Pócs, T., 1960. Die Verbreitung von *Leptodon smithii* (Dicks.) Mohr und die Verhältnisse seines Vorkommens. *Annls hist.-nat. Mus. natn. hung.* **52**, 168–176.

Praeger, R. L., 1932. The flora of the turloughs: a preliminary note. *Proc. R. Ir. Acad.* **41**B, 37–45.

Prentice, I. C., Cramer, W., Harrison, S. P., Leemans, R., Monserud, R. A. & Solomon, A. M., 1992. A global biome model based on plant physiology and dominance, soil properties and climate. *J. Biogeogr.* **19**, 117–134.

Proctor, M. C. F., 1959. A note on *Acrocladium trifarium* (W. & M.) Richards & Wallace in Ireland. *Trans. Br. bryol. Soc.* **3**, 571–574.

————, 1961. The habitat of *Zygodon forsteri* (Brid.) Mitt. in the New Forest, Hants. *Trans. Br. bryol. Soc.* **4**, 107–110.

————, 1962. The epiphytic bryophyte communities of the Dartmoor oak woods. *Rep. Trans. Devon. Ass. Advmt Sci.* **94**, 531–554.

————, 1967. The distribution of British liverworts: a statistical analysis. *J. Ecol.* **55**, 119–135.

———, Spooner, G. M. & Spooner, M. F., 1980. Changes in Wistman's Wood, Dartmoor: photographic and other evidence. *Rep. Trans. Devon. Ass. Advmt Sci.* **112**, 43–79.

Raeymaekers, G., 1983. *Philonotis rigida* Brid. in Europe. *Lindbergia* **9**, 29–33.

Ratcliffe, D. A., 1968. An ecological account of Atlantic bryophytes in the British Isles. *New Phytol.* **55**, 365–439.

——— & Oswald, P. H. (eds), 1988. *The Flow Country: the peatlands of Caithness and Sutherland*. Peterborough.

Rooy, J. van, 1991. The genus *Amphidium* in southern Africa. *Lindbergia* **17**, 59–63.

Rose, F., 1951. A bryophyte flora of Kent. III. Musci. *Trans. Br. bryol. Soc.* **1**, 427–464.

Rosman-Hartog, N. & Touw, A., 1987. On the taxonomic status of *Ulota bruchii* Hornsch. ex Brid., *U. crispa* (Hedw.) Brid. and *U. crispula* Bruch ex Brid. *Lindbergia* **13**, 159–164.

Schumacker, R., 1985. Working group for mapping bryophytes in Europe: objectives, and potential for British participation. *In:* R. E. Longton & A. R. Perry (eds), *British Bryological Society Diamond Jubilee*, pp. 31–42. Cardiff.

———, Lecointe, A., Touffet, J., de Zuttere, P., Leclercq, L. & Fabri, R., 1981. *Hyocomium armoricum* (Brid.) Wijk & Marg. en Belgique et dans le nord-ouest de la France (Ardenne, Bretagne, Normandie). *Cryptogamie, Bryol. Lichén.* **2**, 277–321.

——— & de Zuttere, P., 1982. *Sematophyllum demissum* (Wils.) Mitt. (Musci), espèce nouvelle pour la bryoflore belge. Étude critique de sa répartition en Europe. *Bull. Soc. r. Bot. Belg.* **115**, 14–22.

Selkirk, P. M., 1981. Protonemal gemmae on *Amblystegium serpens* (Hedw.) B., S. & G. from Macquarie Island. *J. Bryol.* **11**, 719–721.

Shaw, A. J., 1981a. A taxonomic revision of the propaguliferous species of *Pohlia* (Musci) in North America. *J. Hattori bot. Lab.* **50**, 1–81.

———, 1981b. Ecological diversification among nine species of *Pohlia* (Musci) in western North America. *Can. J. Bot.* **59**, 2359–2378.

——— & Crum, H., 1984. Peristome homology in *Mielichhoferia* and a taxonomic account of North American species. *J. Hattori bot. Lab.* **57**, 363–381.

Shaw, G. A., 1962. The distribution of *Zygodon gracilis* Wils. in West Yorkshire. *Trans. Br. bryol. Soc.* **4**, 206–208.

Side, A. G. & Whitehouse, H. L. K., 1987. Colourless tubers in *Discelium nudum* Brid. *J. Bryol.* **14**, 741–743.

Smith, A. J. E., 1973. On the differences between *Bryum creberrimum* Tayl. and *B. pallescens* Schleich. ex Schwaegr. *J. Bryol.* **7**, 333–337.

———, 1974a. Distribution maps of bryophytes in Britain and Ireland. *Bryum creberrimum* Tayl. *J. Bryol.* **8**, 126.

———, 1974b. *Philonotis marchica* (Hedw.) Brid. in Britain. *J. Bryol.* **8**, 5–8.

———, 1978. *The Moss Flora of Britain and Ireland*. Cambridge.

——— & Whitehouse, H. L. K., 1978. An account of the British species of the *Bryum bicolor* complex including *B. dunense* sp. nov. *J. Bryol.* **10**, 29–47.

Spence, D. H. N., 1967. Factors controlling the distribution of freshwater macrophytes with particular reference to the lochs of Scotland. *J. Ecol.* **55**, 147–170.

Steere, W. C., 1978. The mosses of Arctic Alaska. *Bryophytorum Bibliotheca* **14**, 1-508.

Stern, R. C., 1991. *Tortula freibergii* and *Eriopus apiculatus* in East Sussex. *J. Bryol.* **16**, 488–489.

Stevenson, [C.] R., 1986. Bryophytes from an archaeological site in Suffolk. *J. Bryol.* **14**, 182–184.

Syed, H., 1973. A taxonomic study of *Bryum capillare* Hedw. and related species. *J. Bryol.* **7**, 265–326.

Tallis, J. H., 1961. The distributions of *Thuidium recognitum* Lindb., *T. philiberti* Limpr. and *T. delicatulum* Mitt. in Britain. *Trans. Br. bryol. Soc.* **4**, 102–106.

Ter Braak, C. J. F., 1988. *CANOCO – a FORTRAN program for canonical community ordination by [partial] [detrended] [canonical] correspondence analysis, principal components analysis and redundancy analysis* (version 2.1). Wageningen.

Touw, A. & Rubers, W. V., 1989. *De Nederlandse Bladmossen.* Utrecht.

Townsend, C. C., 1982. *Pictus scoticus*, a new genus and species of pleurocarpous moss from Scotland. *J. Bryol.* **12**, 1–6.

Tuomikoski, R. & Koponen, T., 1979. On the generic taxonomy of *Calliergon* and *Drepanocladus* (Musci, Amblystegiaceae). *Annls bot. fenn.* **16**, 213–227.

Ukraintseva, V. V., 1986. On the composition of the forage of the large herbivorous mammals of the mammoth epoch. *Quärtarpaläontologie* **6**, 231–238.

Vitt, D. H., 1973. A revision of the genus *Orthotrichum* in North America, north of Mexico. *Bryophytorum Bibliotheca* **1**, 1–208.

Wallace, E. C., 1971. An *Eriopus* in Sussex. *Trans. Br. bryol. Soc.* **6**, 327–328.

———, 1972. Two mosses from Scotland, new to the British Isles. *J. Bryol.* **7**, 157–159.

Warburg, E. F., 1958. *Meesia tristicha* Bruch & Schimp. in the British Isles. *Trans. Br. bryol. Soc.* **3**, 378–381.

———, 1965. *Pohlia pulchella* in Britain. *Trans. Br. bryol. Soc.* **4**, 760–762.

——— & Perry, A. R., 1963. *Platygyrium repens* in Britain. *Trans. Br. bryol. Soc.* **4**, 422–425.

Watson, E. V., 1968. *Pohlia lutescens* (Limpr.) Lindb. f. in Britain and Ireland. *Trans. Br. bryol. Soc.* **5**, 443–447.

Whitehouse, H. L. K., 1963. *Bryum riparium* Hagen in the British Isles. *Trans. Br. bryol. Soc.* **4**, 389–403.

———, 1964. Bryophyta. *In:* F. H. Perring, P. D. Sell, S. M. Walters & H. L. K. Whitehouse (eds), *A Flora of Cambridgeshire*, pp. 281–328. Cambridge.

———, 1973. The occurrence of tubers in *Pohlia pulchella* (Hedw.) Lindb. and *Pohlia lutescens* (Limpr.) Lindb. fil. *J. Bryol.* **7**, 533–540.

———, 1983. Bryophytes. *In:* G. Crompton & H. L. K. Whitehouse, *A Checklist of the Flora of Cambridgeshire*, pp. 65–79. Cambridge.

———, 1987. Protonema-gemmae in European mosses. *Symp. biol. hung.* **35**, 227–231.

———, 1992. Tubers in *Bryum dixonii* Card. ex Nicholson. *J. Bryol.* **17**, 376–377.

Wilczek, R. & Demaret, F., 1976. Les espèces belges du "complexe *Bryum bicolor*" (Musci). *Bull. Jard. bot. nat. Belg.* **46**, 511–541.

Wilson, A., 1938. *The Flora of Westmorland.* Arbroath.

Wilson, P. & Norris, D. H., 1989. *Pseudoleskeella* in North America and Europe. *Bryologist* **92**, 387–396.

Yarranton, G. A., 1962. Bryophyte communities of the exposures of Breidden Hill (North Wales). *Rev. bryol. lichén.* **31**, 168–186.

Yeo, M. J. M. & Blackstock, T. H., 1988. *Amblystegium saxatile* Schimp. in North Wales. *J. Bryol.* **15**, 497–498.

LIST OF LOCALITIES CITED
IN THE TEXT

Localities mentioned in the text are given below with their grid references. Where possible, the 10-km square or squares are given but for larger areas (e.g. Breckland, Lake District) the 100-km squares are given. The accompanying map gives the numerical equivalents of the 100-km square alphabetical codes used below.

A
A'Bhuidheanach, NN49
Airlie Castle, NO25
Alderney, WA50,60
Allt Mhainisteir, NN58
Aonach Beag, NN17
Aonach Mor, NN17,27
Ardnamurchan, NM46,47,56,57,66
Ashdown Forest, TQ42,43,52,53

B
Beinn a'Bhuird, NO09
Beinn a'Chaorainn, NJ00
Beinn Dearg, NH28
Beinn Dorain, NN33
Beinn Fhada, NH01
Beinn Heasgarnich, NN43
Beinn Udlaidh, NN23
Ben Alder, NN47
Ben Alder Forest, NN57
Ben Hope, NC44,45
Ben Lawers, NN64
Ben Lui, NN22
Ben Macdui, NN99
Ben More, NM53
Ben Nevis, NN17
Ben Vorlich, NN21
Berwyn Mountains, SJ02,03,04,13,14
Bettyhill, NC76
Bidean nam Bian, NN15
Black Tor Copse, SX58
Bodmin Moor, SX
Braemar, NO19
Braeriach, NN99,NH90
Breadalbane, NN
Breckland, TF,TL
Breidden Hill, SJ21
Bressay, HU43,44,53,54
Burnham Beeches, SU98
Burnhope Head, NY94
Burren, M,R

Buscot Park, SU29

C
Caenlochan Glen, NO17
Cairn of Claise, NO17
Cairnacay, NJ23
Cairngorm Mountains, NH,NJ,NN,NO
Carn Eige, NH12
Carnedd Llewelyn, SH66
Carneddau, SO05
Cassington, SP41
Chawley, SP40
Clapham, SD76
Cliveden, SU98
Clocaenog Forest, SH95,SJ05
Clova, NO27,36,37
Coire an Lochain, NH90
Coire an t-Sneachda, NH90
Coire Cheap, NN47
Colt Park Wood, SE09
Colwyn Bay, SH87
Comberton, TL35
Corrie Kander, NO18
Craven Pennines, SD
Creag a'Ghaill, NM43
Creag an Dail Bheag, NO19
Creag an Duine, NN28
Creag an Lochain, NN54
Creag Meagaidh, NN48
Creag na Caillich, NN53
Crewe, SJ75
Crickhowell, SO21
Crooke Gill, SD87

D
Dale Head Farm, SE69
Dalwhinnie, NN68
Danna, Island of, NR67,77
Dart Mountain, H69
Den of Airlie, NO25
Dersingham Fen, TF62

Dugwm Rock, SO08

E
Edale, SK18
Edzell, NO66
Elcho, NO12
Epping Forest, TQ38,39,48,49,TL40

F
Fairlight, TQ81
Feith Buidhe, NH90
Foel Gasyth, SJ06
Fothringham, NO44

G
Glen Clova, NO27,37,36
Glen Coe, NN15
Glen Doll, NO27
Glen Duror, NN05
Glen Feshie, NN89
Glen Markie, NN59
Glen Nevis, NN16,17
Gragareth (=Greygarth Fell), SD67,68
Guernsey, WV

H
Halnaby Carr, NZ20

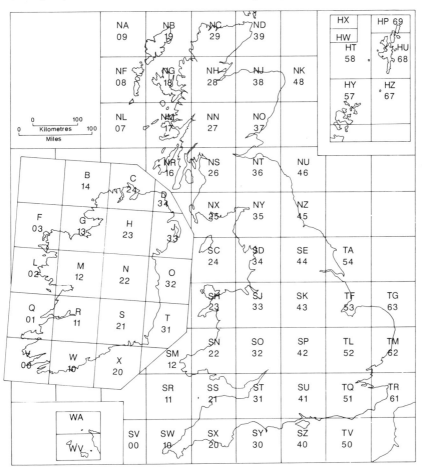

British and Irish National Grids

The map shows the numerical equivalents of the 100 km-square alphabetical codes used in the list of localities cited in the text. The Channel Islands are plotted on the UTM Grid.

Hampstead Heath, TQ28
Harthope Moor, NY83
Heyshott Down, SU81
High Force, NY82
Hood Bridge (=Riverford Bridge), SX76

I
Inchnadamph, NC22
Ingleborough, SD77
Inverlael Forest, NH18,28
Islay, NR

J
Jura, NR,NM

K
Kew, TQ17
Kielder Forest, NY68,69,78,79
Knock Fell, NY73
Knutsford Moor, SJ77

L
Ladhar Bheinn, NG80
Lake District, NY,SD
Lammermuir Hills, NT
Land's End, SW32
Linn of Corriemulzie, NO18
Lizard Peninsula, SW61,62,71,72
Llangattock, SO21
Llangollen, SJ24
Llyn Anafon, SH66
Llyn Brenig, SH95
Loch Baile a'Ghobhainn, NM84
Loch Beoraid, NM88
Loch Brandy, NO37
Loch Ericht, NN46,56,57,58,68
Loch Kander, NO18
Loch Maree, NG87,97,96,NH06
Lochnagar, NO28
London, TQ
Lough Currane, V56

M
Malham Tarn, SD86
Malham Tarn Moss, SD86
Meall Breac, NN96
Meall Garbh, NN64
Meall na Samhna, NN43
Meall nan Tarmachan, NN53
Meikle Kilrannoch, NO27
Mickle Fell, NY82
Mickledore, NY20
Moel Hebog, SH54
Monmouth, SO51
Mull, Island of, NM
Mynydd Hiraethog, SH95

N
New Forest, SU,SZ

O
Oxford, SP50

P
Peak District, SK
Pennines, SD,SE,SK,NY,NZ
Pen-y-Ghent, SD87
Point of Stoer, NC03
Portland, Isle of, SY66,67,77

R
Rothiemurchus Forest, NH80,90
River Teme, source of, SO18

S
St Kilda, NA00,10,NF09,19
Scafell Pike, NY20
Seana Bhraigh, NH28
Sgurr na Lapaich, NH13
Skipwith Common, SE63
Skye, Island of, NG
Smoo Cave, NC46
Snowdon, SH65
Southern Uplands, NS,NT,NX
Sruffaunnamuingabatia, N92
Storr, The, NG45

T
Tailbridge Hill, NY80
Tarnbrook Fell, SD55,65
Terrington North Carr, SE67
Touch Muir, NS78,79
Town Yetholm, NT82
Tresco, SV81,91
Tring, SP91
Twisleton Glen, SD77
Twll Du, SH65

U
Uisgnaval Mor, NB10
Upper Teesdale, NY

W
Walthamstow, TQ38
Weald, The, TQ
Wells, ST54
Wharfedale, SD,SE
Whernside, SD78
Whitney, SO24
Whittlesey Mere, TL29
Wicken Fen, TL56,57
Wistman's Wood, SX67

Y
Y Garn, SH65
Yorkshire Dales, SD,SE,NY,NZ

LIST OF TREES AND SHRUBS
CITED IN THE TEXT

This list includes all the trees and shrubs cited in the text by their English names.

alder – *Alnus glutinosa*
apple – *Malus domestica*
ash – *Fraxinus excelsior*
aspen – *Populus tremula*

beech – *Fagus sylvatica*
birch – *Betula* spp.
blackthorn – *Prunus spinosa*
buckthorn – *Rhamnus cathartica*
buckthorn, alder – *Frangula alnus*
buddleia – *Buddleja davidii*

chestnut, sweet – *Castanea sativa*

elder – *Sambucus nigra*
elm – *Ulmus* spp.
elm, wych – *Ulmus glabra*

gorse – *Ulex* spp.

hawthorn – *Crataegus monogyna*
hazel – *Corylus avellana*
heather – *Calluna vulgaris*
holly – *Ilex aquifolium*

ivy – *Hedera helix*

juniper – *Juniperus communis*

lime – *Tilia* spp.

maple, field – *Acer campestre*

oak – *Quercus* spp.

pear – *Pyrus communis*
pine – *Pinus sylvestris*
poplar – *Populus* spp.
privet – *Ligustrum* spp.

rhododendron – *Rhododendron ponticum*
rowan – *Sorbus aucuparia*

sallow – *Salix* spp.
spindle – *Euonymus europaeus*
sycamore – *Acer pseudoplatanus*

tamarisk – *Tamarix gallica*

willow – *Salix* spp.

yew – *Taxus baccata*

LIST OF RECORDERS
CONTRIBUTING
TO THE MAPPING SCHEME

The distribution maps published in these volumes include records made by many bryologists from the 17th century onwards. The older records have been extracted from literature sources or herbarium specimens. Most of the post-1950 records have been submitted to the Mapping Scheme by the recorders who made them. Other recent records have come from bryologists who have taken responsibility for collating the records for particular areas, have been extracted from recent literature, or are the result of surveys from which BRC has received data. The following list includes the recorders who have contributed, either personally or indirectly, at least fifty post-1950 records to one of the Atlas volumes, and those surveys from which we have received a similar number of records.

A
Adams, K. J.
Ambrose, F.
Anderson, Dr Margaret C.
Appleyard, Mrs J.
Argent, G. C. G.
Averis, A. B. G.

B
Ballard, D. W.
Bates, J. W.
Bell, F.
Benoit, P. M.
Bentley, F.
Birks, Dr Hilary H.
Birks, H. J. B.
Birse, E. L.
Blackstock, T. H.
Blockeel, T. L.
Bloom, G.
Booth, K. N.
Bourne, P. J.
Bowen, H. J. M.
Branson, F. E.
Bryce, D. M.
Bull, A.
Bullard, P. F.
Bunce, R. G. H.
Burton, Dr M. Agneta S.

C
Chamberlain, D. F.
Chandler, J. H.
Charman, D. J.
Clarke, G. C. S.

Cocking, Mrs K. M.
Corley, M. F. V.
Crundwell, A. C.
Curry, P.

D
Dalby, Miss M.
Dale, Miss K.
Davey, S.
Davies, R. H.
Dickson, J. H.
Dierssen, K.
Dransfield, J.
Driver, P.
Du Feu, Miss E. H.
Ducker, B. F. T.
Duckett, J. G.
Duncan, Dr Ursula K.

E
Eddy, A.
Ellis, D. E.

F
Field, J. A.
Finch, R. A.
Fisk, R. J.
Fitzgerald, Mrs J. W.
Fitzgerald, R. D.
Foster, W. D.
Fowler, B. R.

G
Game, J.
Gardiner, J. C.

Garlick, G. W.
Goriup, P. D.
Graham, G. G.

H
Hackney, P.
Halliday, G.
Harrington, A. J.
Haworth, C. C.
Henley, Miss P. A.
Hill, M. O.
Hodgetts, N. G.
Holmes, N. T. H.
Horrill, A. D.
Howard, J. A.

J
Jackson, P. E.
Jones, E. W.
Joyce, Mrs I. A.

K
Kay, Q. O. N.
Kelly, D. L.
Kenneth, A. G.
King, Mrs A. L. K.
Kungu, Mrs E. M.

L
Laflin, T.
Leslie, A. C
Little, E. R. B.
Lobley, Miss E. M.
Long, D. G.
Longton, R. E.

Lusby, P. S.

M
McAllister, H. A.
McFarlane, M. G.
McVean, D. N.
Martin, P.
Milne-Redhead, H.
Milnes-Smith, Mrs M. D.
Mountford, J. O.
Muirhead, Miss C. W.
Munro-Smith, D.
Murphy, Miss R. J.
Murray, Dr Barbara M.

N
NCC Scottish Freshwater
    Loch Survey
NCC Welsh Field Unit
Newton, Miss A. E.
Newton, Dr Martha E.
Northern Ireland Lake Survey

O
O'Shea, B. J.
Orange, A.
Ormand, E.
Outen, A. R.

P
Parker, J.
Parker, R. E.
Paton, Mrs J. A.
Perry, A. R.

Peterken, J. H. G.
Pettifer, A. J.
Pitkin, P. H.
Pool, M.
Poore, M. E. D.
Porley, R. D.
Port, P. J.
Preston, C. D.
Proctor, M. C. F.
Pucknell, R. D.
Pyner, T.

R
Ranwell, D. S.
Ratcliffe, D. A.
Read, P. D.
Revell, R. D.
Richards, P. W. M.
Richter, R.
Roberts, F. J.
Robertson, Miss J.
Robertson, J. S.
Rose, F.
Rothero, G. P.

S
Sargent, Dr Caroline M.
Seaward, M. R. D.
Side, Mrs A. G.
Skinner, R. N.
Smith, A. C.
Smith, A. J. E.
Smith, D. G.
Smith, Mrs J. E.

Smith, T.
Snow, Mrs L.
Southey, Ms J. F.
Stace, C. A.
Stern, R. C.
Stevenson, C. R.
Stirling, A. McG.
Swann, E. L.
Synnott, D. M.

T
Thompson, B. H.
Tidswell, R. J.
Townsend, C. C.

U
Urquhart, U. H.

W
Wade, A. E.
Walker, R.
Wallace, E. C.
Wanstall, P. J.
Warburg, E. F.
Warren, W. E.
Watson, E. V.
Whitehouse, H. L. K.
Wigginton, M. J.
Willmot, A.
Wilson, G. P.
Woods, R. G.

Y
Yeo, M. J. M.

INDEX TO SPECIES IN VOLUME 3